T0393972

Land Subsidence Mitigation

Aquifer Recharge Using Treated Wastewater Injection

Land Subsidence Mitigation
Aquifer Recharge Using Treated Wastewater Injection

Frank R. Spellman

CRC Press
Taylor & Francis Group
Boca Raton London New York

CRC Press is an imprint of the
Taylor & Francis Group, an **informa** business

CRC Press
Taylor & Francis Group
6000 Broken Sound Parkway NW, Suite 300
Boca Raton, FL 33487-2742

Library of Congress Cataloging-in-Publication Data

Names: Spellman, Frank R., author.
Title: Land subsidence mitigation : aquifer recharge using treated wastewater injection / Frank R. Spellman.
Description: Boca Raton : Taylor & Francis, a CRC title, part of the Taylor & Francis imprint, a member of the Taylor & Francis Group, the academic division of T&F Informa, plc, [2017] | Includes bibliographical references.
Identifiers: LCCN 2017009372 | ISBN 9781138050761 (hardback : acid-free paper) | ISBN 9781315110943 (ebook)
Subjects: LCSH: Water reuse. | Subsidences (Earth movements)--Prevention. | Groundwater recharge.
Classification: LCC TD429 .S68 2017 | DDC 628.1/14--dc23
LC record available at http://lccn.loc.gov/2017009372

Visit the Taylor & Francis Web site at
http://www.taylorandfrancis.com

and the CRC Press Web site at
http://www.crcpress.com

Printed and bound in the United States of America by
Edwards Brothers Malloy on sustainably sourced paper

Dedication

For Kathern Welsh

Contents

Preface

Land is vanishing at various locations in the southern Chesapeake Bay region. The impact of this vanishing act is particularly noticeable in Norfolk, Virginia. In this area, known as the Tidewater or Hampton Roads region, sea-level rise currently ranges from 0.04331 inch (1.1 mm) to 0.189 inch (4.8 mm) per year. Looking at the high end of this range, 0.189 inch, one might think, "What's the big deal? Why should we be worried about a 0.189-inch rise in sea level?" Such questions typically surface when you talk with residents of the area—that is, until a coastal storm or hurricane hits the area, when it is not uncommon for Norfolk's downtown and Ocean View areas to be impassable with as much as 8 feet of water in the streets.

Let's look at the high end number of sea-level rise again, 0.189 inch per year. If we multiply this number by 10 years, we obtain a total calculated rise of approximately 1.89 inches (4.8 mm). Some might think that we can live with 2 inches of sea-level rise every 10 years. Maybe. Let's multiply by 100 years, which gives us a sea-level rise of almost 19 inches (480 mm). Now we are talking about a serious impact to the area—a 19-inch rise in sea level is not to be shrugged off. This is especially the case if one factors in coastal storm events when waves and tides bring the sea level to damaging levels.

The projected sea-level rise for the southern Chesapeake Bay region and in particular Norfolk is a serious problem, even though there are those who would agree that this could be a problem for those who are around 100 years from now but don't consider it to be a big deal today (but try driving though Norfolk during a storm event at high tide—not recommended). The problem is complicated, though. Because water levels are measured relative to the land, relative sea-level rise in the southern Chesapeake Bay region has two components: global water level and land subsidence. Melting glaciers and the thermal expansion of seawater, both attributed to global climate change, contribute to *eustatic*, or worldwide, sea-level rise. *Isostatic*, or local, factors contribute to relative sea-level rise through sinking of the land or *subsidence*. In the southern Chesapeake Bay area, subsidence of land has occurred because of excessive withdrawal of groundwater and the geologic features of the area. This combination has contributed to a relative increase in sea level in the southern Chesapeake Bay area of 1 foot in the last century, and an even greater increase in sea level is projected over the next 100 years.

The take-home message is that land subsidence due to excessive groundwater withdrawal is a problem; however, as with all of my environmental presentations, I take the position that every problem has a solution. This book is about solving the land subsidence problem caused by excessive groundwater withdrawal. The focus of the book is on the Norfolk, Virginia, area, but the discussion can be applied to any area of the world where land subsidence due to groundwater withdrawal is an issue. Particular attention is paid to the Norfolk area because of the progressive, far-reaching actions of the Hampton Roads Sanitation District (HRSD). The forward-thinking, governor-appointed HRSD Commission has taken a bold step forward with their Sustainable Water Initiative for Tomorrow (SWIFT). The SWIFT program has many facets but the focus here is on its innovative pilot study and hoped-for positive results of injecting treated wastewater (treated to drinking water level) into the Potomac Aquifer in an attempt to ease and slow land subsidence in the southern Chesapeake Bay region due to excessive groundwater withdrawal. This book chronicles the efforts of the U.S. Geological Survey (USGS); HRSD's consulting engineering firm, CH2M; and HRSD internal operations personnel working to make SWIFT a reality. *Land Subsidence Mitigation: Aquifer Recharge Using Treated Wastewater Injection* provides a fair and balanced discussion about the concerns centering on land subsidence due to excessive groundwater withdrawal. It provides comprehensive coverage of all aspects of the issue, including a discussion of Class V injection wells and their specific intended purpose. This book is intended as a reference book and road map for administrators, legal professionals, research engineers, graduate students, wastewater or sanitary engineers, non-engineering professionals, general readers, and anyone else concerned with land subsidence and clean water injection into underground aquifers.

A final word to readers: This book is written in the conversational, engaging, and reader-friendly style that is the author's trademark. Simply, I never apologize for attempting to communicate.

Acknowledgments

This book would not have been possible without the help provided by Ted Henifin, General Manager of HRSD, in addition to the following other HRSD personnel: Jennifer See, Safety Manager; Jennifer Cascio, Executive Secretary; John Dano, Chief of Planning and Design; Germano Salazar-Benites, Recycling Project Manager; and many Nansemond employees. Also of tremendous help were the staff of engineers and environmental professionals at the consulting engineering firm CH2M and, finally, my co-illustrator, artist Kathern Welsh.

Author

Frank R. Spellman, PhD, is a retired adjunct assistant professor of environmental health at Old Dominion University, Norfolk, Virginia, and the author of more than 110 books covering topics ranging from concentrated animal feeding operations (CAFOs) to all areas of environmental science and occupational health. Many of his texts are readily available online, and several have been adopted for classroom use at major universities throughout the United States, Canada, Europe, and Russia; two have been translated into Spanish for South American markets. Dr. Spellman has been cited in more than 850 publications. He serves as a professional expert witness for three law groups and as an incident/accident investigator for the U.S. Department of Justice and a northern Virginia law firm. In addition, he consults on homeland security vulnerability assessments for critical infrastructures, including water/wastewater facilities, and conducts pre-Occupational Safety and Health Administration and Environmental Protection Agency audits throughout the country. Dr. Spellman receives frequent requests to co-author with well-recognized experts in several scientific fields; for example, he is a contributing author to the prestigious text *The Engineering Handbook*, 2nd ed. Dr. Spellman lectures on wastewater treatment, water treatment, and homeland security, as well as on safety topics, throughout the country and teaches water/wastewater operator short courses at Virginia Tech in Blacksburg. In 2011, he traced and documented the ancient water distribution system at Machu Picchu, Peru, and surveyed several drinking water resources in Amazonia, Ecuador. He has also studied and surveyed two separate potable water supplies in the Galapagos Islands, in addition to studying Darwin's finches while there. Dr. Spellman earned a BA in public administration, a BS in business management, an MBA, and both an MS and a PhD in environmental engineering.

Conversions

This book uses English units. To determine metric values, use the conversion factors listed below:

To Convert	Multiply by	To Get
Length		
Inch	25.4	Millimeter
Foot	0.3048	Meter
Mile	1.609	Kilometer
Area		
Square foot	0.09290	Square meter
Square mile	2.590	Square kilometer
Acre	0.4047	Hectare
Volume		
Acre-foot	1233	Cubic meter
Cubic foot	0.02832	Cubic meter
Gallon	3.785	Liter
Mass		
Ounce	28.35	Gram
Pound	0.4356	Kilogram
Ton (short)	0.9072	Megagram
Temperature		
°Farenheit	(°F – 32)/1.8	°Celsius

Vertical Datum

In this book, *sea level* refers to the National Geodetic Vertical Datum of 1929 (formerly known as the Sea Level Datum of 1929), a geodetic datum derived from a general adjustment of the first-order level nets of both the United States and Canada. *Mean sea level* is not used with reference to any particular vertical datum; where used, the phrase means the average surface of the ocean as determined by calibration of measurements at tidal stations.

Acronyms and Abbreviations

°C	Degrees Celsius
AB	Army Base
AFT	Alternate filtration technology
ALCR	Air–liquid conversion ratio
AOP	Advanced oxidation process
ASTM	American Society for Testing and Materials
AWT	Advanced water treatment
AWTP	Advanced water treatment plant
BAC	Biological activated carbon
BH	Boat Harbor
BNR	Biological nitrogen removal
BOD	Biochemical oxygen demand
BOD_5	5-day biochemical oxygen demand
BV	Bed volume
CA	Cellulose acetate
$CaCO_3$	Calcium carbonate
CAS	Conventional activated sludge
CBOD	Carbonaceous biochemical oxygen demand
CEC	Contaminant of emerging concern
CFU	Coliform forming unit
Cl_2	Chlorine
COD	Chemical oxygen demand
CPES	CH2M's Parametric Cost Estimating System
CWA	Clean Water Act
DBP	Disinfection byproduct
DO	Dissolved oxygen
DOC	Dissolved organic carbon
DPR	Direct potable reuse
EDC	Endocrine-disrupting compound
FBG	Feet below grade
FCV	Flow control valve
FLOC	Flocculation
GAC	Granular activated carbon
GMF	Granular media filtration
gpm/sf	Gallons per minute per square foot
H_2O_2	Hydrogen peroxide
HFF	Hollow fine fiber
hp	Horsepower
HPC	Heterotrophic plate count
HRSD	Hampton Roads Sanitation District
IMS	Integrated membrane system
IPR	Indirect potable reuse

JR	James River
kWh	Kilowatt hour
LRV	Log reduction value
MBR	Membrane bioreactor
MCC	Motor control center
MCL	Maximum contaminant level
MCF	Membrane cartridge filtration
MF	Microfiltration
mg/L	Milligrams per liter
MGD	Million gallons per day
μg/L	Micrograms per liter
mJ/cm^2	Millijoules per square centimeter
mL	Milliliter
N	Nitrogen
N/A	Not applicable
ND	No data
NDMA	N-nitrosodimethylamine
NF	Nanofiltration
ng/L	Nanograms per liter
NH$_3$	Ammonia
NH$_3$-N	Ammonia nitrogen
NOM	Natural organic matter (humic and fulvic acids)
NO$_x$-N	Nitrate/nitrite-nitrogen
NP	Nansemond Plant
NPDES	National Pollutant Discharge Elimination System
NTU	Nephelometric turbidity unit
O&M	Operations and maintenance
P	Phosphorus
PAS	Potomac Aquifer System
PCV	Pressure control valve
PPCPs	Pharmaceuticals and personal care products
PSI	Pounds per square inch
RO	Reverse osmosis
SAT	Soil aquifer treatment
SDI	Silt density index
SDWA	Safe Drinking Water Act
SED	Sedimentation
SWTR	Surface Water Treatment Rule
TCEP	Tris(2-carboxyethyl)phosphine
TCEQ	Texas Commission on Environmental Quality
TDS	Total dissolved solids
THMs	Trihalomethanes
TKN	Total Kjeldahl nitrogen
TMDL	Total maximum daily load
TN	Total nitrogen
TOC	Total organic carbon

TOX	Total organic halides
TP	Total phosphorus
TSS	Total suspended solids
TTHMs	Total trihalomethanes
UF	Ultrafiltration
UOSA	Upper Occoquan Service Authority
USDA	U.S. Department of Agriculture
USEPA	U.S. Environmental Protection Agency
UV	Ultraviolet
UVAOP	Ultraviolet advanced oxidation process
VFD	Variable frequency drive
VIP	Virginia Initiative Plant
WB	Williamsburg
WWTP	Wastewater treatment plant
YR	York River

Coastal Environs

We often hear that Hampton Roads is among the U.S. areas most vulnerable to sea level rise. I double-checked the numbers the other day to make sure that claim still rings true. I'm sorry to say that it does.

—**Dave Mayfield,** *The Virginia-Pilot*, **January 1, 2017**

1 The Upside-Down Sombrero

METRIC UNIT PYRAMID

1 kilometer (km) = 0.621 mile
1 meter (m) = 0.00062137119 mile
4000 cubic kilometers (km^3) = 960 cubic miles
Energy (kilojoules) of average hurricane = 7.5×10^{11}
Energy (kilojoules) of exploding hydrogen bomb = 8.4×10^{16}
Energy (kilojoules) of impact of the meteor that wiped out dinosaurs = 7.5×10^{23}

INTRODUCTION

Date: Today
Time: Right now
Place: First Landing State Park, Virginia Beach, Virginia, at the very spot of land settled by the English in 1607
Surroundings: On the edge of Chesapeake Bay in a maritime forest filled with lagoons, bald cypress swamps, and rare flora and fauna, all of which have withstood strong winds, periodic flooding, and salt spray
Protagonists: Mr. Rabbit and Mr. Grasshopper

During a Mr. Grasshopper and Mr. Rabbit conversation, Mr. Rabbit said to Mr. Grasshopper, "The ancient mystery has been revealed. Mr. Poag and Mr. Powar's research has answered the question, solved the mystery." Mr. Grasshopper's curiosity now acutely piqued, he asked, "What mystery are you talking about?" Mr. Rabbit replied, "Well, you see, Mr. Grasshopper, this particular mystery involved the Chesapeake Bay area, right here. Like, what caused the salty groundwater, the rivers changing their course, the land sinking, and the unusual erosion patterns? Today, the mystery has been solved. And it was something really big." Scratching his abdomen with a hind leg, Mr. Grasshopper replied, "That's good news, Mr. Rabbit, but I still don't understand what the mystery is and how it was solved."

There was a moment of silence and reflection and an awareness of their surroundings (when you are a rabbit or a grasshopper, it is wise to be on guard at all times). Finally, Mr. Rabbit spoke up: "I will reveal all that to you soon, Mr. Grasshopper, soon … very soon." Mr. Grasshopper replied, "Well, if it was really something that big, I'm surprised I missed it. I'm not sure where I was, Mr. Rabbit." Mr. Rabbit replied, "Right on, Grasshopper, right on. The fact is that at this exact moment you and I are standing right on the edge of it all—right on one edge of an upside-down sombrero. Furthermore, be glad we missed it, Mr. Grasshopper, because if we hadn't missed it we wouldn't be here now. I'll give you a hint, though. It happened here in this exact area about, well, about 33 to 35 million years ago."

ABOUT 33 TO 35 MILLION YEARS AGO

This book is a geologically based presentation of facts as we know them or as we think we know them. To begin this discussion, we need to begin at the beginning. Thus, Table 1.1 is provided in order to give the reader an understanding of the beginning as it relates to this book. Among the various epochs presented in the table, we are primarily concerned with the Eocene (from the Greek *eos* for "dawn" and *kainos* for "recent"), which sets the stage for material presented throughout the text.

CHESAPEAKE BAY BOLIDE: A WHOLE BUNCH OF SHOCK AND VERY LITTLE AWE

At a catastrophic moment in Earth's history, it was as if Mother Nature had reached out into the universe; grabbed hold of a fiery, flaming, scorching mass; and wound up to throw Earth a massive curve ball, a white-hot curve ball at least two times brighter than Earth's moon. It was at least 2 miles wide. From the northwest horizon it crossed paths with Earth at more than 76,000 miles per hour, more than 1260 miles per minute, and roughly 21 miles per second. It moved too quickly to be heard and its white-hot light would have blinded had it not killed before optic nerves could signal brain matter (Tennant and Hall, 2001). This was an age of prehistoric sharks, whales, tiny camels, modern ungulates, bats, sea cows, eagles, pelicans, quails, and vultures. Were all of these life forms present when Mother Nature's fireball impacted Earth? Probably not, but we will never really know for sure who was there and who was not there. What we do know is that those who were present died in a wink.* Some might call the Chesapeake Bay bolide impact the ultimate shock and awe event; however, because the event was instantaneously deadly to all those within impact affected areas it would be better stated to say that the event was a whole bunch of shock (to minerals, especially quartz) with very little awe. When death is instant, who is around, at least in the immediate area, to be awed?

Fast-forward to the impact result. For millennia, humans had no knowledge that this event had occurred or of the literal impact it would have on southeastern Virginia, the formation of Chesapeake Bay, and the present-day local problems with groundwater quality and land subsidence. Recently, though, a handful of modern-day, Sherlock Holmes-type detectives, with Dr. C.W. Poag in the lead, figured out what happened. Through intuition, bore-holing operations, and applying scientific protocols and a lot of common sense they determined that at the end of the Eocene Mother Nature's fiery curveball impacted the coastline of what is now known as Virginia's Chesapeake Bay region. Today, many refer to this fireball as the

---

* For those who love dinosaurs (and don't we all?), not to worry. The mighty *Tyrannosaurus rex*, nasty *Utahraptor*, and affable, chicken-sized *Velociraptor* were nowhere to be seen; they had all perished at least 30 million years earlier during the Cretaceous–Tertiary mass extinction.

TABLE 1.1
Geologic Time Scale

Erathem or Era	System, Subsystem or Period, Subperiod	Series or Epoch
Cenozoic (Age of Recent Life) 65 MYA to present	**Quaternary** 1.8 MYA to the present	**Holocene** ~11,477 (±85) years ago to the present Greek *holos* (entire) and *ceno* (new)
		Pleistocene (Great Ice Age) 1.8 MYA to ~11,477 (±85) years ago Greek *pleistos* (most) and *ceno* (new)
	Tertiary 65.5 to 1.8 MYA	**Pliocene** 5.3 to 1.8 MYA Greek *pleion* (more) and *ceno* (new)
		Miocene 23.0 to 5.3 MYA Greek *meion* (less) and *ceno* (new)
		Oligocene 33.9 to 23.0 MYA Greek *oligos* (little, few) and *ceno* (new)
		Eocene 55.8 to 33.9 MYA Greek *eos* (dawn) and *ceno* (new)
		Paleocene 65.5 to 58.8 MYA Greek *palaois* (old) and *ceno* (new)
Mesozoic (Age of Medieval Life) 251.0 to 65.5 MYA	**Cretaceous** (Age of Dinosaurs) 145.5 to 65.5 MYA	Late or Upper Early or Lower
	Jurassic 199.6 to 145.5 MYA	Late or Upper Middle Early or Lower
	Triassic 251.0 in 199.6 MYA	Late or Upper Middle Early or Lower
Paleozoic (Age of Ancient Life) 542.0 to 251.0 MYA	**Permian** 299.0 to 251.0 MYA	Lopingian Guadalupian Cisuralian
	Pennsylvanian (Coal Age) 318.1 to 299.0 MYA	Late or Upper Middle Early or Lower
	Mississippian 359.2 to 318.1 MYA	Late or Upper Middle Early or Lower

(continued)

TABLE 1.1 (continued)
Geologic Time Scale

Erathem or Era	System, Subsystem or Period, Subperiod	Series or Epoch
Paleozoic (cont.) (Age of Ancient Life) 542.0 to 251.0 MYA	**Devonian** 416.0 to 359.2 MYA	Late or Upper Middle Early or Lower
	Silurian 443.7 to 416.0 MYA	Pridoli Ludlow Wenlock Llandovery
	Ordovician 488.3 to 443.7 MYA	Late or Upper Middle Early or Lower
	Cambrian 542.0 to 488.3 MYA	Late or Upper Middle Early or Lower
Precambrian ~4 billion years ago to 542.0 MYA		

Note: MYA = million years ago.

meteor that hit Chesapeake Bay, but scientists in the know understand that the term "meteor" is best replaced with the term "bolide," which is defined below, along with other relevant meteorite terms (PWNET, 2016; USGS, 1998).

METEORITE TERMS

- *Asteroid*—A rocky body orbiting the sun that is usually great than 100 m in diameter; most asteroids orbit between Mars and Jupiter.
- *Bolide*—In her bestselling book, *Dark Matter and the Dinosaurs*, Randall (2015) defined a bolide as "an object from space that disintegrates in the atmosphere" (p. 127). Although there is no consensus on the definition of a bolide, the term generally refers to an extraterrestrial body in the 1- to 10-km size range that impacts the Earth at velocities of literally faster than a speeding bullet (Mach 75 = 20 to 70 mg/sec), explodes upon impact, and creates a large crater (USGS, 1998). The bottom line is that it is a generic term used to imply that we do not know the precise nature of the impacting body (e.g., whether it is a rocky or metallic asteroid or any icy comet) (see Figure 1.1). In this book, a bolide is defined as an extraterrestrial body of some size and of some composition that impacts Earth and creates a crater; for our purposes, the descriptor "impact" is key.

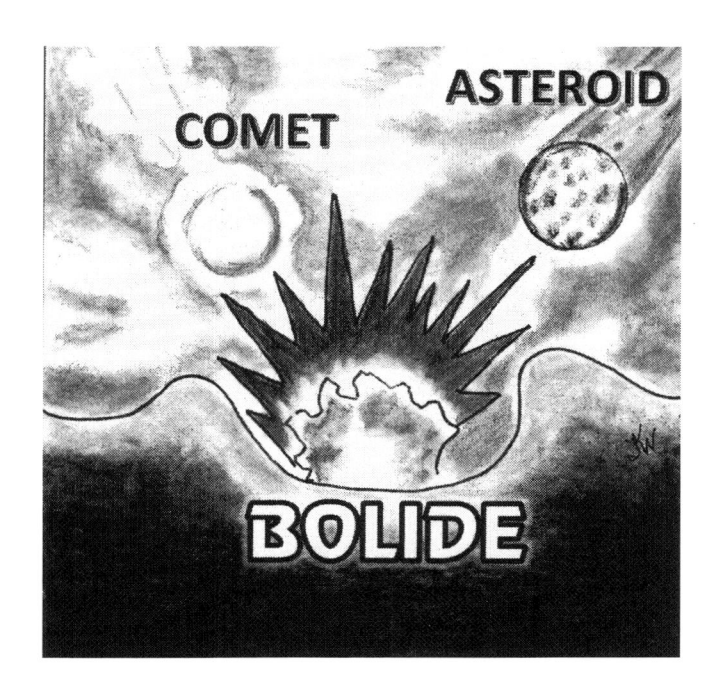

FIGURE 1.1 Bolide impact. (Illustration by F.R. Spellman and Kathern Welsh, adapted from USGS, *The Chesapeake Bay Bolide: A New View of Coastal Plain Evolution*, USGS Fact Sheet 049-98, U.S. Geological Survey, Woods Hole, MA, 2016.)

- *Cape Charles, Virginia*—Location of a huge peaking impact crater whose center is located near this Eastern Shore town (see Figure 1.2).
- *Central peak*—A small mountain that forms at the center of a crater in reaction to the force of the impact.
- *Chesapeake Bay bolide*—One of the largest known impact structures found in North America. This event has been dated at about 35 to 33 million years ago during the Eocene Epoch of the Cenozoic Era. (*Note:* For the sake of simplicity and continuity, this text assumes that the bolide impact occurred 35 million years ago.) The nature of the impact substantially affected the geology of the Atlantic continental crust and is suspected to affect the nature and quality of groundwater in southeastern Virginia. The diameter has been estimated to be about 85 km (~55.8 miles), and the crater may be about 1.3 km (~0.81 miles) deep.
- *Crater*—The result of a bolide body impacting the surface of another planetary body; the resulting explosion leaves a round hole.
- *Ejecta*—Debris that shoots out of the impact site when a crater forms.
- *Eocene Epoch*—Time period that occurred 58 to 33.8 million years ago and is marked by the emergence of mammals as the dominant land animals. The fossil record reveals many mammals quite unlike anything we have

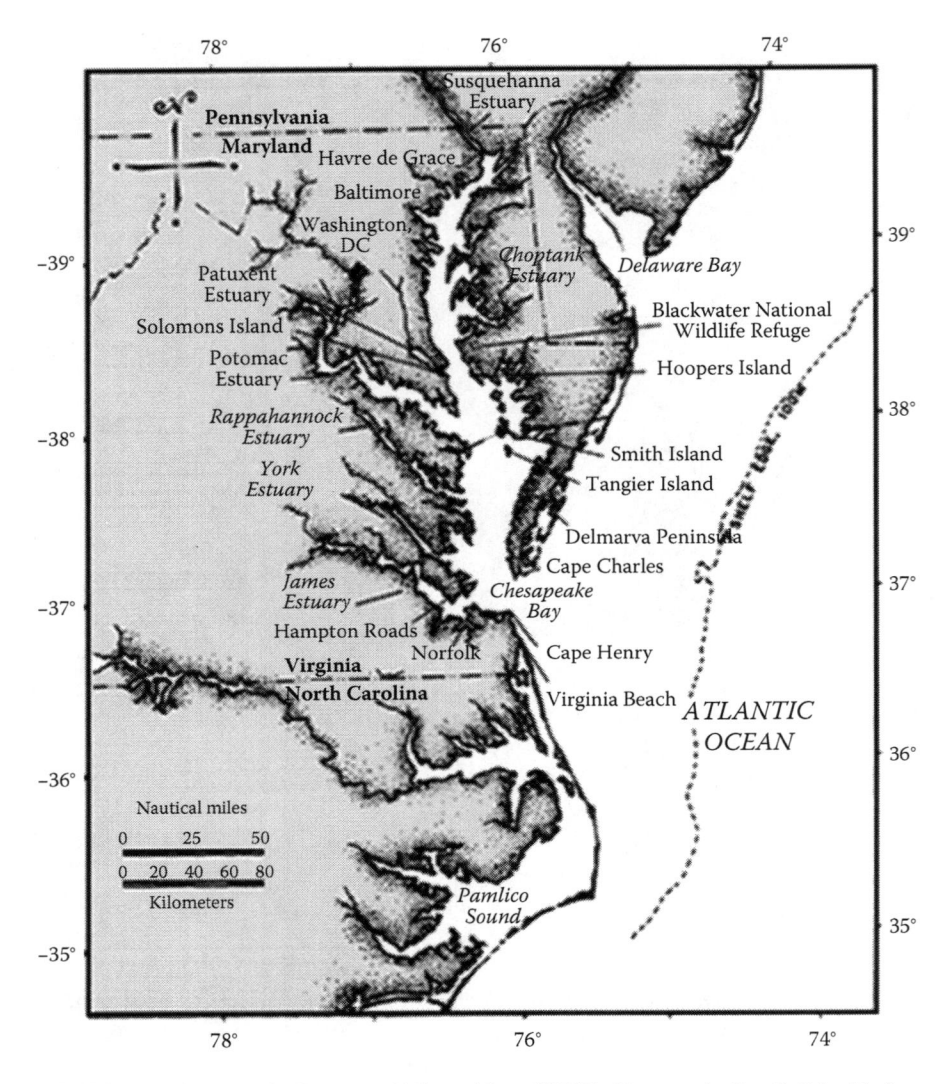

FIGURE 1.2 Chesapeake Bay area. (Adapted from USGS, *Chesapeake Bay Bolide: Modern Consequences of an Ancient Cataclysm*, U.S. Geological Survey, Washington, DC, 1998.)

today; however, there were increasing numbers of forest plants, freshwater fish, and insects that were much like those today. In fact, the term Eocene means "dawn of the recent."

- *Fault fracture/ground fissure*—Fractures caused by tectonic movement or impact events (see Figure 1.3).
- *Floor*—The bottom part of an impact crater; it can be flat or rounded and is often lower than the surrounding surface of the planet or moon.
- *Impact breccia*—Rubble sediment that contains a mix of debris resulting from an impact event.

FIGURE 1.3 Fault fracture/land fissure due to land subsidence. (Illustration by F.R. Spellman and Kathern Welsh, adapted from USGS, *Chesapeake Bay Bolide: Modern Consequences of an Ancient Cataclysm*, U.S. Geological Survey, Washington, DC, 1998.)

- *Iridium*—Very hard and brittle metal (atomic number 77) that is often associated with meteorite impacts.
- *Mass*—A measure of the inertia of an object (i.e., how heavy it is); mass is not the same as weight, which measures the gravitational force on an object.
- *Meteor*—Visible as a bright steak of light in the sky caused by a meteoroid or small icy particle entering Earth's atmosphere; also known as a *shooting star*. Meteor showers occur when the Earth passes through debris left behind by an orbiting comet.
- *Meteorite*—Small rocky remains of meteoroids that survive a fiery journey through Earth's atmosphere and land on Earth.
- *Micrometeorite*—Very small meteorite with a diameter less than 1 millimeter.
- *Ray, bright*—Lines of debris projecting from the edges of craters.
- *Rim*—The highest point along the edge of a crater hole.
- *Rubble bed*—Jumbled sediments and aged dated fossils that are associated with the Exmore beds of the Chesapeake impact structure.
- *Shocked materials*—Minerals, especially quartz, that show the result of tremendous forces, such as those found in impact events that alter and distort the normal optical qualities of a quartz crystal.
- *Tektites*—Glass beads that are 1 millimeter to 1 centimeter in size (see Figure 1.4) derived from sediments melted by a bolide impact.

FIGURE 1.4   Tektites. (Photograph by F.R. Spellman.)

- *Tsunami*—A very large ocean wave usually associated with underwater earthquakes or volcanic eruptions. Tsunamis may also be associated with large meteorite impacts in the oceans.
- *Wall*—The sides of the bowl of a crater.
- *Wave material*—Material left along the trajectory of a meteor after the head of the meteor has passed.

THE EARTH-CHANGING EVENT

A bolide blasting into the Earth and into the shallow sea that covered Virginia from Cape Henry to Richmond has to be classified as an Earth-changing event. It certainly changed things from the southwest to upper Virginia. The Chesapeake Bay bolide exploded with more force than the combined nuclear arsenal of today's world powers (Tennant and Hall, 2001). The impact cracked the Earth as deep as 7 miles. The bolide blasted into creation a crater 85 miles wide, resulting in a flash of evaporating ocean water (millions of gallons evaporated instantly; millions more were

---

DID YOU KNOW?

In this text, the work of Dr. C. Wylie Poag is referenced often due to his flagship work on investigating and finding the Chesapeake Bay bolide crater. Dr. Poag's work as a senior research scientist with the U.S. Geological Survey is highlighted here by sharing a short version of his biography. Dr. Poag's research emphasizes the integration of subsurface geophysical, geological, and paleontological data to reconstruct the stratigraphic framework and depositional history of the Atlantic and Gulf Coast margins of the United States. His 30-year geological career includes experience as a petroleum explorationist, a university professor, and project coordinator of the National Science Foundation's Deep Sea Drilling Project. Dr. Poag has published more than 200 abstracts, articles, and books. A highlight of Dr. Poag's research has been the identification of the largest impact craters in the United States buried beneath the lower part of Chesapeake Bay and its surrounding peninsulas (USGS, 1997).

hurled 60 miles into the atmosphere) and a volume of ejected bedrock that may have risen a towering 30 miles high (Mayell, 2001). Most of the debris fell back into the crater. Some shocked quartz and tiny glass beads—tektites (see Figure 1.4)—were scattered as far as New Jersey. The bolide acted like a giant drill and, unimpeded by ocean, ripped through almost a mile of sediment and sand, penetrating and mincing the 600-million-year-old granite bedrock. Huge boulders, chunks, and grains of solid earth were propelled upward. Keep in mind that when the bolide was drilling its way into Earth the front end slowed down a bit but the back end was still flying at supersonic speeds. The bedrock at the front end and at the back end squeezed together like a huge sponge, and rebounded. The bedrock splintered, and massive faults split open. In the chaos of searing heat and ferocity, the bolide vaporized, ultimately leaving a crater 55+ miles wide with a network of fractures and fissures spreading more than 40 miles beyond its rim.

As the rocks blasted out into the air, they were ignited by friction and subsequently sparked firestorms for hundreds of miles. Then the heavy hand of gravity reclaimed the boulders and water from the sky and returned them to the gaping crater. We all know that when we toss a small stone into a river pool concentric circles ripple outward as the stone drops through pool bottom. For a brief instant, we are struck by the obvious: the stone sinks to the bottom, following the laws of gravity. Eventually, the ripples die away, leaving as little mark as the usual human lifespan creates in the waters of the world. Imagine, however, the bolide and its ejected materials rolling out, creating swells, and forming concentric circles headed across the ocean to Greenland and Europe, in addition to the East Coast of the United States. These swells raced at extraordinary speeds across the ocean floor and rose with the land and exploded. Even the Blue Ridge Mountains felt the impact from 1000-foot-high tsunamis (Tennant and Hall, 2001). Adhering to the time-worn axiom applied within the realm of the bounds of Earth and based on scientific fact that what goes up will eventually come back, the water came back with a vengeance. Remember, there are very few substances, if any, that are more destructive than water on the move. All that tsunami water that ran up the Blue Ridge and other places and loosened house-sized boulders and other objects rushed back, carrying tons of rock, soil, trees, and wildlife that filled the gigantic empty crater that was as deep as the Grand Canyon; what escaped the crater was carried on to the sea.

When it finally became quiet and peaceful, with the surrounding area steaming and barren, there was no one around to remember that the impact had happened at all. Unlike the stone thrown in a pond that creates ripples that leave no evidence that they were ever formed, the Chesapeake Bay bolide had a much greater impact; however, it was somewhat similar to a common river pool momentarily disturbed by a pebble thrown into its surface in that the bolide-struck Atlantic coastline became still. Below the surface, though, in the murky depths, scars and evidence of its occurrence were left behind, but it took time to discover what was hidden so well.

It was mentioned earlier that the bolide struck the Atlantic coastline, not Chesapeake Bay. This was the case, of course, because at the time of impact there was no Chesapeake Bay or Eastern shore; it was certainly not a coastline formation as we know it today. The Atlantic Ocean at that time hugged the area near what is now known as Richmond, Virginia, roughly in the path of Interstate-95 today. During the late Eocene and at the time of the bolide impact, the area was covered by the ocean; the sea level was much higher than it is today due to the climate being much warmer. The Chesapeake Bay itself did not form until after the Wisconsin Glaciation ice sheet melted 18,000 years ago. As described in a story in the *Richmond Times-Dispatch* (Anon., 2005):

> The waves nearly overlapped the Blue Ridge Mountains before washing back into the horrible gash, then covered the superheated water beneath a thick blanket of debris, rock and sediment. Over time, as this new geologic formation settled, it set the stage for Virginia's baffling coastal groundwater system, with its pockets of salty groundwater. USGS geologist Wylie Poag, another co-discoverer of the bay's ancient depression, has called it "probably the most dramatic geological event that ever took place on the Atlantic margin of North America."

Over the last 35 million years, erosion has deposited sediments on top of the crater, and shifts in the path of the Susquehanna River have formed the peninsula of the Eastern Shore. Today, the impact crater is buried under 1500 feet of gravel, sand, silt, and clay, with the center of the crater underneath the Delmarva Peninsula (Gohn et al., 2009). Most people know what an impact crater looks like. Earth's moon gives us some idea of what an impact crater looks like, as does Meteor Crater in Arizona, some 38 miles east of Flagstaff, the archetypical example of what cratering experts call a *simple crater*. It is a shallow, bowl-shaped excavation, 1 km in diameter, with an upraised subcircular rim, and it is extraordinarily well preserved. Craters wider than 10 km are classified as *complex craters*, because they exhibit additional features. Complex craters differ from simple bowl-shaped craters in that the objects that created them hit hard and fast enough to melt the rock and splash it all in the center like a skyscraper, where it hardened. Like simple craters, the outer margin of complex craters is marked by a raised rim. Inside the rim is a broad, flat, circular plain, called the *annular trough*. Large slump blocks fall away from the crater's outer wall and slide out over the floor of the annular trough toward the crater center. The inner edge of the annular trough is marked by a central mountainous peak, a ring of peaks (a peak ring), or both. Inside the peak ring is the deepest part of the crater, called the *inner basin*. The Chesapeake Bay crater has all the characteristics of a peak-ring crater and is said to look like an upside-down sombrero with an upturned outer rim, a trough, and then a high peak in the center (see Figures 1.5, 1.6, and 1.7).

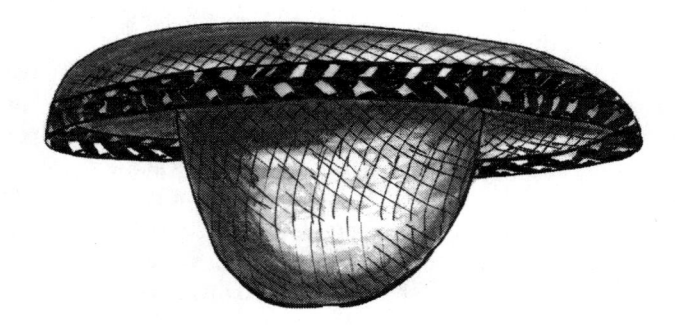

FIGURE 1.5   An upside-down sombrero. (Illustration by F.R. Spellman and Kathern Welsh.)

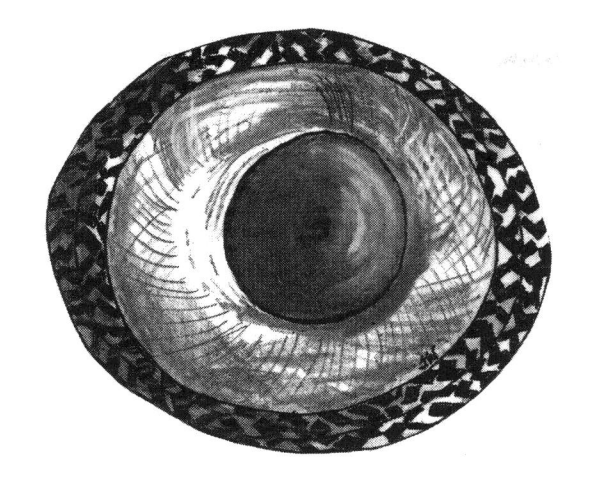

FIGURE 1.6 Looking up into an upside-down sombrero. (Illustration by F.R. Spellman and Kathern Welsh.)

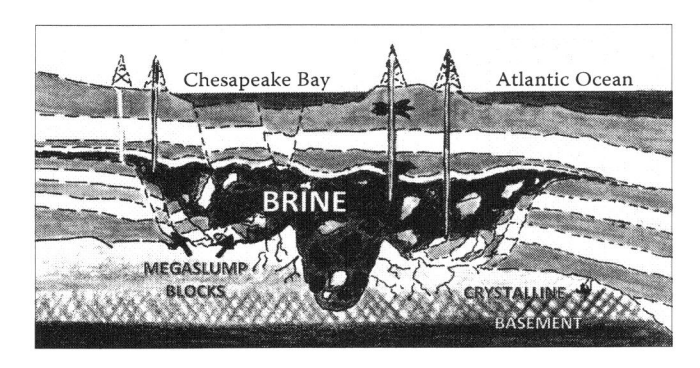

FIGURE 1.7 Chesapeake Bay bolide impact crater, resembling an upside-down sombrero. (Illustration by F.R. Spellman and Kathern Welsh, adapted from USGS, *Ancient Cataclysm*, U.S. Geological Survey, Washington, DC, 1997).

EFFECTS OF THE CHESAPEAKE BAY BOLIDE IMPACT

One might think that, after the bolide hit, water blasted out of the area would be sucked back in along with whatever debris was nearby, including sediments laid down over 35 million years, and that all of this would be enough to fill the enormous gorge left behind. Although water and debris were sucked back into the chasm, they were not enough to eliminate the low spot in the floor of what is now Chesapeake Bay. Keep in mind that we did not even know until relatively recently that the crater even existed in the floor of Chesapeake Bay and the surrounding area. The discovery of the giant crater by Dr. Poag and others has completely revised our understanding of Atlantic Coastal Plain evolution. What many people in the Hampton Roads (Tidewater) Region do not know is that Dr. Poag's studies revealed several consequences of the ancient cataclysm that still affect citizens in the bay area today. These consequences include the location of Chesapeake Bay itself, river diversion, ground instability, disruption of coastal aquifers, and land subsidence. This text focuses primarily on disruption of local aquifers and land subsidence related to bolide impact and relative sea level rise and the ongoing rebound efforts brought about by Hampton Roads Sanitation District (HRSD) and U.S. Geological Survey (USGS). But, it is also important to point out that the effects of the bolide impact are multiple, and they are discussed briefly below.

LOCATION OF CHESAPEAKE BAY

Over 35 million years ago, Chesapeake Bay did not exist. In fact, as late as 18,000 years ago, the bay region was dry land; the last great ice sheet was at its maximum over North America, and sea level was about 200 m lower than at present. This sea level exposed the area that now is the bay bottom and continental shelf. Because sea level was so low, the major East Coast rivers had to cut narrow valleys across the region all the way to the shelf edge. About 10,000 years ago, however, the ice sheets began to melt rapidly, causing the sea level to rise and flood the shelf and coastal river valleys. The flooded valleys became major modern estuaries, such as Delaware Bay and Chesapeake Bay. The impact crater created a long-lasting topographic depression, which helped determined the eventual location of Chesapeake Bay (USGS, 2016).

RIVER DIVERSION

The rivers of the Chesapeake region converged at a location directly over the buried crater. Some might think that the convergence of these rivers is merely a coincidence, but it is not. Notice that in Figure 1.8 the important river channels in the area change course significantly just after they cross the rim of the buried bolide crater. These channels are actually successive buried ice-age channels of the ancient Susquehanna River (formed from 450,000 years ago to 20,000 years ago). The river diversion and seismic evidence that post-impact units sag and thicken over the crater indicate that the ground surface over the crater remained lower than the areas outside the crater for 35 million years. Why? Why does the Rappahannock River flow southeastward to the Atlantic but the York River and James River make sharp turns to the northeast near the outer rim of the crater? The courses of the York River and James River in the lower bay region are the result of the ongoing influence of *differential subsidence* over the bolide crater.

Two factors have caused subsidence in the region. First, subsidence is the result of loading during the past 35 million years since the impact. Second, subsidence is also due to compaction of the *breccia*—rock composed of broken fragments of minerals or gravel cemented together by a fine-grained matrix. The crater breccia is 1.2 km thick and was deposited as a water-saturated, sandy, rubble-bearing, non-gelled slurry. The sediment layers surrounding it were already partly consolidated, so the mushy breccia compacted much more rapidly under its subsequent sediment load than the surrounding strata. One might ask what all this has to do with the local

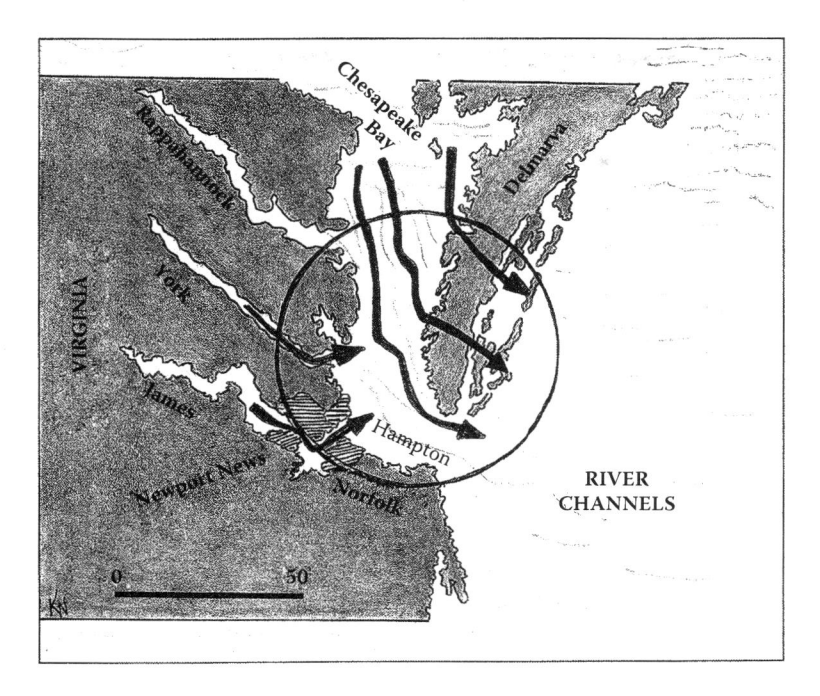

FIGURE 1.8 River channels in the Chesapeake Bay region. (Illustration by F.R. Spellman and Kathern Welsh, adapted from USGS, *Ancient Cataclysm*, U.S. Geological Survey, Washington, DC, 1997.)

bay rivers. The combination of these two factors produced a subsidence differential, causing the land surface over the breccia to remain lower than the land surface outside the crater. Therefore, the river valleys covered over the crater and were located in those particular places when rising sea levels flooded them. In short, the impact crater created a long-lasting topographic depression, which helped predetermine the eventual location of Chesapeake Bay (USGS, 1997). Finally, keep in mind that one of the main focuses of this book is land subsidence in the Chesapeake Bay area, which will be discussed in detail later.

GROUND INSTABILITY DUE TO FAULTING

Seismic profiles across the crater show many faults that cut the sedimentary beds above the breccia and extend upward toward the bay floor (see Figure 1.9). The current resolution of the seismic profiles allows us to trace the faults to within 10 m of the bay floor. These faults represent another result of the differential compaction and subsidence of the breccia. As the breccia continues to subside under the load of post-impact deposits, it subsides unevenly due to its viable content of sand and huge clasts. This eventually causes the overlying beds to bend and break and to slide apart along the fault planes. These faults are zones of crustal weakness and have the potential for continued slow movement or sudden large offsets if reactivated by earthquakes. It is

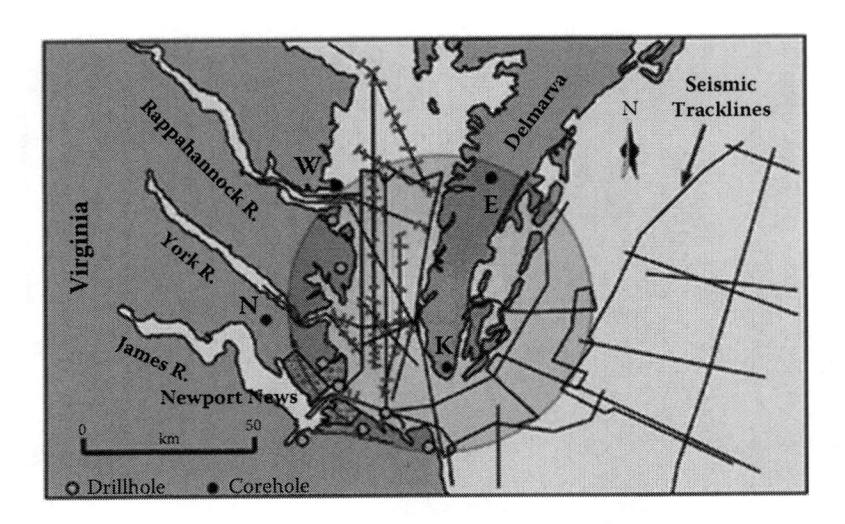

FIGURE 1.9 Location of faults (short dashes) where they cross seismic profiles. Large circle shows the extent of the buried crater. The brick pattern shows the three main cities of the lower Chesapeake Bay. Capital letters mark the locations of Newport News, Windmill Point, Exmore, and Kiptopeke coreholes. (From USGS, *The Chesapeake Bay Bolide: A New View of Coastal Plain Evolution*, USGS Fact Sheet 049-98, U.S. Geological Survey, Woods Hole, MA, 2016.)

important to know in detail the location, orientation, and amount of offset of these compaction faults because of the potential for the faulting to separate adjacent sides of the confining unit over the saltwater reservoir. If this occurred, salty water could flow upward and contaminate the freshwater supply. Using the seismic profiles currently available, Dr. Poag identified and is mapping more than 100 faults or fault clusters around and over the crater, which reach or are near the bay floor (USGS, 2016).

DISRUPTION OF COASTAL AQUIFERS

The hydrogeological framework thought to be typical of southeastern Virginia, in cross-section, consists of groundwater aquifers alternating with confining beds. The aquifers are mainly sands, with water-filled pore spaces between the sand grains. The pore spaces are connected, which allows the water to flow slowly though the aquifers. The confining beds are mainly clay beds, which have only very fine pores, and these are poorly interconnected, which greatly retards or prevents the flow of water. Before we knew about the Chesapeake Bay crater, this framework of alternating aquifers and confining units was applied to models of groundwater flow and water-quality assessments in the lower Chesapeake Bay region.

Based on core samples, researchers have determined that in the crater area itself the orderly stack of aquifers seen outside the crater does not exist; instead, they were truncated and excavated by the bolide impact. In place of those aquifers, there is now a single huge reservoir with a volume of 4000 km^3. That's enough breccia to cover all of Virginia and Maryland with a layer 30 m thick. But the most startling part is that this huge new reservoir does not contain freshwater like the aquifers it replaced;

DID YOU KNOW?

The parameters for salinity are

- Freshwater—Less than 1000 ppm
- Slightly saline water—From 1000 ppm to 3000 ppm
- Moderately saline water—From 3000 ppm to 10,000 ppm
- Highly saline water—From 10,000 ppm to 35,000 ppm

Ocean water contains about 35,000 ppm of salt, but Chesapeake Bay crater water contains about 1.5 times more salt than that.

rather, the pore spaces are filled with briny water that is 1.5 times saltier than normal seawater. This water is too salty to drink or to use in industry (USGS, 2016). It is interesting to note that for decades geohydrologists and others in the Hampton Roads region scratched their collective heads and wondered why, in locations away from the crater, water wells yielded good quality freshwater suitable for potable purposes, but whenever wells were drilled within the crater ring or close to it salty water was all that could be found.

The presence of this hypersaline aquifer has some practical implications for groundwater management in the lower bay region. For example, we need to know how deeply buried the breccia is in order to avoid drilling into it inadvertently and contaminating the overlying freshwater aquifers. Its presence also limits the availability of freshwater. On the Delmarva Peninsula, over the deepest part of the crater, only the aquifers above the breccia are available for freshwater. The crater investigation suggests that it is necessary to be especially conservative of groundwater use in the area (USGS, 2016).

LAND SUBSIDENCE

Land subsidence and the potential for land rebound due to the injection of treated wastewater are, as the title of the book suggests, the focus of this text. Later much more will be presented about this topic, particularly with regard to how it is related to the Chesapeake Bay bolide impact crater. For now, it is important to point out that there is growing evidence that accelerated land subsidence is reflected in the geology and topography of the modern land surfaces around the crater. As noted earlier, the breccia is 1.2 km thick and was deposited as a water-saturated, sandy, rubble-bearing slurry (like concrete before it hardens). The sediment layers surrounding the crater, on the other hand, were already partly consolidated, so the mushy breccia compacted much more rapidly under its subsequent sediment load than the surrounding strata. The compaction differences produced a subsidence differential (i.e., the difference in subsidence between two points on the crater), causing the land surface over the breccia (due to breccia compaction) to remain lower than the land surface over sediments outside the crater. During Dr. Poag's investigation, he and his team observed that the boundary between older surface rocks and younger surface rocks coincides with the position and orientation of the crater rim on all three peninsulas that cross the rim. The older beds have sagged over the subsiding breccia, and the younger rocks have been deposited in the resulting topographic depression. The topography also reflects the differential subsidence. The Suffolk Scarp and the Ames Ridge are elevated landforms (10 to 15 m high) located at, and oriented parallel to, the crater rim. Crater-related ground subsidence also may play a role in the high rate of relative sea-level rise documented for the Chesapeake Bay region. One of the locations of highest relative sea-level rise is at Hampton Roads, the lower part of the James River, located over the crater rim.

THE BOTTOM LINE

This chapter has set the first row of foundation blocks for the material to follow. Specifically, this chapter described the late Eocene period in the Virginia Coastal Plain area when the formerly quiescent geological regime was dramatically transformed after a bolide struck in the vicinity of the Delmarva Peninsula. This consequential event produced the following principal consequences (USGS, 1998):

- The bolide carved a roughly circular crater twice the size of the state of Rhode Island (~6400 km^2) and nearly as deep as the Grand Canyon (1.3 km deep).
- The excavation truncated all existing groundwater aquifers in the impact area by gouging ~4300 km^3 of rock from the upper lithosphere, including Proterozoic and Paleozoic crystalline basement rocks and Middle Jurassic to upper Eocene sedimentary rocks.
- A structural and topographic low formed over the crater.
- The impact crater may have predetermined the present-day location of Chesapeake Bay.
- A porous breccia lens, 600 to 1200 m thick, replaced local aquifers, resulting in groundwater ~1.5 times saltier than normal seawater.
- Long-term differential compaction and subsidence of the breccia lens spawned extensive fault systems in the area, which are potential hazards for local population centers in the Chesapeake Bay area.

REFERENCES AND RECOMMENDED READING

Anon. (2005). Drill explores blast: research seeks insight into explosion that carved huge crater under the Chesapeake. *Richmond Times Dispatch*, September 8.

Gohn, G.S., Koeberl, C., Miller, K., and Reimold, W.U., Eds. (2009). *Chesapeake Bay Impact Structure: Results from the Eyreville Core Holes*, GSA Special Papers 458. Boulder, CO: Geological Society of America.

Mayell, H. (2001). Chesapeake Bay crater offers clues to ancient cataclysm. *National Geographic*, November 13 (http://news.nationalgeographic.com/news/2001/11/1113_chesapeakcrater.html).

PWNET. (2016). *The Impact Crater.* Manassas, VA: Prince William Network, U.S. Geological Survey (https://meteor.pwnet.org/impact_event/impact_crater.htm).

Randall, L. (2015). *Dark Matter and the Dinosaurs.* New York: Harper Collins.

Tennant, D. and Hall, M. (2001). The Chesapeake Bay meteor: a mystery, meteors and one man's quest for the truth. *The Virginian-Pilot*, June 24.

USGS. (1997). *Location of Chesapeake Bay.* Washington, DC: U.S. Geological Survey (http://woodshole.er.usgs.gov/epubs/bolide/location_of_bay.html).

USGS. (1998). *Chesapeake Bay Bolide: Modern Consequences of an Ancient Cataclysm.* Washington, DC: U.S. Geological Survey (https://woodshole.er.usgs.gov/epubs/bolide/introduction.html).

USGS. (2016). *The Chesapeake Bay Bolide: A New View of Coastal Plain Evolution*, USGS Fact Sheet 049-98. Woods Hole, MA: U.S. Geological Survey (https://pubs.usgs.gov/fs/fs49-98/).

2 Soil Basics

We stand on soil, not on earth.

Illich et al. (1991)

Weekend gardeners tend to think of soil as the first few inches below the Earth's surface—the thin layer that needs to be weeded and that provides a firm foundation for plants. But the soil actually extends from the surface down to the Earth's hard rocky crust. It is a zone of transition, and, as in many of nature's transition zones, the soil is the site of important chemical and physical processes. In addition, because plants need soil to grow, it is arguably the most valuable of all the mineral resources on Earth.

Beazley (1992)

A chapter on soil basics? But this book is about land subsidence. Yes, it is, for sure. But what is land, exactly? Land can be an estate, an area, a country, grounds, a nation, a parcel, a region, terra firma … and it can be soil. And soil, in turn, can be clay, dust, ground, grime, loam, soot, terra firma, and, finally, land. The point is that, when we are talking about land subsidence, we are essentially talking about soil. Subsidence is a function of what happens to soil; therefore, we need to have a clear understanding of soil. When we talk of land decline, land decrease, or land reduction, we are talking about land subsidence, which is the same as soil subsidence.

INTRODUCTION*

We take soil for granted. It's always been there, with the implied corollary that it will always be there. But, where does soil come from? Of course, soil was formed and, in a never-ending process, is still being formed; however, soil formation is a slow process—one at work over the course of millennia as mountains are worn away to dust through bare rock succession. Any activity, human or natural, that exposes rock to air begins the process. Through the agents of physical and chemical weathering, through extremes of heat and cold, through storms and earthquake and entropy, bare rock is gradually broken, reduced, and worn away. As its exterior structures are exposed and weakened, plant life appears to speed the process along.

Lichens cover the bare rock first, growing on the surface of the rock, etching it with mild acids and collecting a thin film of soil that is trapped against the rock and clings. This changes the conditions of growth so much that the lichens can no longer survive, and they are replaced by mosses. The mosses establish themselves in the soil trapped and enriched by the lichens and collect even more soil. They hold moisture to the surface of the rock, setting up another change in environmental

---

* Information in the following sections is adapted from Spellman, F.R., *The Science of Environmental Pollution*, 3rd ed., CRC Press, Boca Raton, FL, 2017.

conditions. Well-established mosses hold enough soil to allow herbaceous plant seeds to invade the rock. Grasses and small flowering plants move in, sending out fine root systems that hold more soil and moisture and work their way into minute fissures in the surface of the rock. More and more organisms join the increasingly complex community. Weedy shrubs are the next invaders; they have heavier root systems that find their way into every crevice. Each stage of succession affects the decay of the surface of the rock and adds its own organic material to the mix. Over the course of time, mountains are worn away, eaten away to soil, as time, plants, weather, and extremes of weather work on them. The parent material, the rock, becomes smaller and weaker as the years, decades, centuries, and millennia go by, thus producing the rich, varied, and valuable mineral resource we call soil.

SOIL: WHAT IS IT?

Perhaps no term causes more confusion in communication among typical people, soil scientists, soil engineers, and earth scientists than the word soil. itself. In simple terms, soil can be defined as the topmost layer of decomposed rock and organic matter which usually contains air, moisture, and nutrients and therefore can support life. Most people would have little difficulty in understanding and accepting this simple definition, then why the confusion over the exact meaning of the word soil? Quite simply, confusion reigns because soil is not simple—it is quite complex. In addition, the term has different meanings to different groups (like pollution, the exact definition of soil is a personal judgment call). Let's take a look at how some of these different groups view soil.

Typical people seldom give soil a thought, because it usually doesn't directly impact their lives. They seldom think about soil as soil, but they might think of soil in terms of dirt. First of all, soil is not dirt. Dirt is misplaced soil—soil where we don't want it, such as on our hands, clothes, automobiles, or floors. Dirt is something we try to clean up and to keep out of our living environments. Second, soil is too special to be called dirt, because soil is mysterious and, whether we realize it or not, essential to our existence. Because we think of it as common, we relegate soil to an ignoble position. As our usual course of action, we degrade it, abuse it, throw it away, contaminate it, and ignore it. We treat it like dirt; only feces hold a more lowly status. Soil deserves better. Why? Because soil is not dirt; moreover, it is not filth, or grime, or squalor. Soil is composed of clay, air, water, sand, loam, and organic detritus of former life forms. If water is Earth's blood and air is Earth's breath, then soil is its flesh and bone and marrow. People with much exposure to soil would know that soil is a natural, three-dimensional body at the Earth's surface capable of supporting plants. It has properties resulting from the integrated effects of climate and living matter acting on earthy parent material that is conditioned by topographic relief and by the passage of time.

Soil scientists (or pedologists) are people interested in soils as a medium for plant growth. Their focus is on the upper meter or so beneath the land surface—the *weathering zone*, which contains the organic-rich material that supports plant growth and is directly above the unconsolidated *parent material*. Soil scientists have developed a classification system for soils based on the physical, chemical, and biological properties that can be observed and measured in the soil.

Soils engineers are typically soils specialists who look at soil as a medium that can be excavated using tools. Soils engineers are not concerned with the plant-growing potential of a particular soil but rather are concerned with the ability of a particular soil to support a load. They attempt to determine (through examination and testing) the particle size, particle-size distribution, and plasticity of the soil.

Earth scientists (or geologists) have a view that typically falls between pedologists and soils engineers; they are interested in soils and the weathering processes as past indicators of climatic conditions and in relation to the geologic formation of useful materials ranging from clay deposits to metallic ores.

To gain a new understanding of soil, go out to a plowed farm field, pick up a handful of soil, and look at it very closely. What are you holding in your hand? Read the two descriptions that follow to gain a better understanding of what soil actually is and why it is critically important to us all (Spellman and Whiting, 2006):

1. A handful of soil is alive, a delicate living organism that is as lively as an army of migrating caribou and as fascinating as a flock of egrets. Literally teeming with life of incomparable forms, soil deserves to be classified as an independent ecosystem or, more correctly stated, as many ecosystems.
2. When we reach down and pick up a handful of soil, it should remind us (and maybe startle some of us) that without its thin living soil layer Earth would be a planet as lifeless as our own moon.

KEY TERMS DEFINED[*]

Each branch of science, including soil science, has its own language. To work even at the edge of soil science, soil pollution, and soil pollution remediation, it is necessary to be familiar with the vocabulary used. This text has a glossary at the end, but some terms need to be defined here to aid in our understanding of the material as we proceed:

Ablation till—A superglacial, coarse-grained sediment or till that accumulates as the subadjacent ice melts and drains away and is finally deposited on the exhumed subglacial surface.

Absorption—Movement of ions and water into the plant roots as a result of either metabolic processes by the root (active absorption) or diffusion along a gradient (passive absorption).

Acid rain—Atmospheric precipitation with pH values less than about 5.6, the acidity being due to inorganic acids such as nitric and sulfuric acids that are formed when oxides of nitrogen and sulfur are emitted into the atmosphere.

Acid soil—A soil with a pH value of <7.0 or neutral. Soils may be naturally acidic from their rocky origin or due to leaching, or they may become acidic from decaying leaves or from soil additives such as aluminum sulfate (alum). Acid soils can be neutralized by the addition of lime products.

---

[*] This section was compiled and adapted from several sources, including USDA, *Soil Taxonomy*, 2nd ed., U.S. Department of Agriculture, Washington, DC, 1999; SCSA, *Resource Conservation Glossary*, ED044296, Soil Conservation Society of America, Ankeny, IA, 1982; SSSA, *Soil Science Glossary*, Soil Science Society of America, Madison, WI, 2008.

Actinomycetes—A group of organisms intermediate between the bacteria and the true fungi that usually produce a characteristic branched mycelium; includes many (but not all) organisms belonging to the order of Actinomycetales.

Adhesion—Molecular attraction that holds the surfaces of two substances (e.g., water and sand particles) in contact.

Adsorption—The attraction of ions or compounds to the surface of a solid.

Aeration, soil—The process by which air in the soil is replaced by air from the atmosphere. In a well-aerated soil, the soil air is similar in composition to the atmosphere above the soil. Poorly aerated soils usually contain more carbon dioxide and correspondingly less oxygen than the atmosphere above the soil.

Aerobic—Organisms growing only in the presence of molecular oxygen.

Aggregates, soil—Soil structural units of various shapes, composed of mineral and organic material, formed by natural processes, and having a range of stabilities.

Agronomy—A specialization of agriculture concerned with the theory and practice of field crop production and soil management; the scientific management of land.

Air capacity—Percentage of soil volume occupied by air spaces or pores.

Air porosity—The proportion of the bulk volume of soil that is filled with air at any given time or under a given condition, such as a specified moisture potential; usually the large pores.

Alkali—A substance capable of liberating hydroxide ions in water, having a pH of more than 7.0, and possessing caustic properties; it can neutralize hydrogen ions, with which it reacts to form a salt and water, and is an important agent in rock weathering.

Alluvium—A general term for unconsolidated, granular sediments deposited by rivers.

Amendment, soil—Any substance other than fertilizers (such as compost, sulfur, gypsum, lime, and sawdust) used to alter the chemical or physical properties of a soil, generally to make it more productive.

Ammonification—The production of ammonia and ammonium–nitrogen through the decomposition of organic nitrogen compounds in soil organic matter.

Anaerobic—Without molecular oxygen.

Anion—An atom that has gained one or more negatively charged electrons and is thus itself negatively charged.

Aspect (of slopes)—The direction that a slope faces with respect to the sun.

Assimilation—The taking up of plant nutrients and their transformation into actual plant tissues.

Atterburg limits—Water contents of fine-grained soils at different states of consistency.

Autotrophs—Plants and microorganisms capable of synthesizing organic compounds from inorganic materials by photosynthesis or oxidation reactions.

Available water—The portion of water in a soil that can be readily absorbed by plant roots; the amount of water released between the field capacity and the permanent wilting point.

Bedrock—The solid rock underlying soils and the regolith in depths ranging from zero (where exposed by erosion) to several hundred feet.

Biological function—The role played by a chemical compound or a system of chemical compounds in living organisms.

Biomass—The total weight of living biological organisms within a specified unit (area, community, population).

Biome—A major ecological community extending over large areas.

Blowout—A sandy depression caused by wind eroding the face of a vegetated dune.

Breccia—A rock composed of coarse angular fragments that are cemented together.

Calcareous soil—Containing sufficient calcium carbonate (often with magnesium carbonate) to effervesce visibly when treated with hydrochloric acid.

Caliche—A layer near the surface, more or less cemented by secondary carbonates of calcium or magnesium precipitated from the soil solution. It may occur as a soft, thin soil horizon; as a hard, thick bed just beneath the solum; or as a surface layer exposed by erosion.

Capillary water—Water held within the capillary pores of soils; mostly available to plants.

Catena—The sequence of soils that occupy a slope transect, from the topographic divide to the bottom of the adjacent valley.

Cation—An atom that has lost one or more negatively charged electrons and is thus itself positively charged.

Chelate (from Greek *chele* for "claw")—A complex organic compound containing a central metallic ion surrounded by organic chemical groups.

Class, soil—A group of soils having a definite range in a particular property such as acidity, degree of slope, texture, structure, land-use capability, degree of erosion, or drainage.

Clay—A soil separate that consists of particles <0.0002 mm in equivalent diameter.

Cohesion—Holding together; a force holding a solid or liquid together due to attraction between like molecules; decreases with rise in temperature.

Colloidal—Matter of very fine particle size.

Convection—A process of heat transfer in a fluid involving the movement of substantial volumes of the fluid concerned. Convection is very important in the atmosphere and to a lesser extent in the oceans.

Denitrification—The biochemical reduction of nitrate or nitrite to gaseous nitrogen, either as molecular nitrogen or as an oxide of nitrogen.

Detritus—Debris from dead plants and animals.

Diffusion—The movement of atoms in a gaseous mixture or ions in a solution, primarily as a result of their own random motion.

Drainage—The removal of excess water, both surface and subsurface, from plants. All plants (except aquatics) will die if exposed to an excess of water.

Duff—The matted, partly decomposed organic surface layer of forest soils.

Erosion—Wearing away of the land surface by running water, wind, ice, or other geological agents, including such processes as gravitational creep.

Eutrophication—A process of lake aging whereby aquatic plants are abundant and waters are deficient in oxygen. The process is usually accelerated by enrichment of waters with surface runoff containing nitrogen and phosphorus.

Evapotranspiration—The combined loss of water from a given area, during a specified period of time, by evaporation from the soil surface and by transpiration from plants.

Exfoliation—Mechanical or physical weathering that involves the disintegration and removal of successive layers of rock mass.

Fertility, soil—The quality of a soil that enables it to provide essential chemical elements in quantities and proportions for the growth of specified plants.

Fixation—The transformation in soil of a plant nutrient from an available to an unavailable state.

Fluvial—Deposits of parent materials laid down by rivers or streams.

Friable—A soil consistency term pertaining to the ease of crumbling of soils.

Heaving—The partial lifting of plants, buildings, roadways, fence posts, etc., out of the ground as a result of freezing and thawing of the surface soil during the winter.

Heterotroph—An organism capable of deriving energy for life processes only from the decomposition of organic compounds and incapable of using inorganic compounds as sole sources of energy or for organic synthesis.

Horizon, soil—A layer of soil, approximately parallel to the soil surface, differing in properties and characteristics from adjacent layers below or above it.

Humus—More or less stable fraction of the soil organic matter (usually dark in color) remaining after the major portions of added plant and animal residues have decomposed.

Hydration—The incorporation of water into the chemical composition of a mineral, converting it from an anhydrous to a hydrous form; the term is also applied to a form of weathering in which hydration swelling creates tensile stress within a rock mass.

Hydraulic conductivity—The rate at which water is able to move through a soil.

Hydrolysis—The reaction between water and a compound (commonly a salt). The hydroxyl from the water combines with the anion from the compound undergoing hydrolysis to form a base; the hydrogen ion from the water combines with the cation from the compound to form an acid.

Hygroscopic coefficient—The amount of moisture in a dry soil when it is in equilibrium with some standard relative humidity near a saturated atmosphere (about 98%), expressed in terms of percentage on the basis of oven-dry soil.

Infiltration—The downward entry of water into the soil.

Ions—Atoms that have lost or gained one or more negatively charged electrons.

Land classification—The arrangement of land units into various categories based on the properties of the land and its suitability for some particular purpose.

Leaching—The removal of materials in solution from the soil by percolating waters.

Liebig's law—The growth and reproduction of an organism are determined by the nutrient substance (oxygen, carbon dioxide, calcium, etc.) that is available in minimum quantity with respect to organic needs, the limiting factor.

Loam—The textural class name for soil having moderate amounts of sand, silt, and clay.

Loess—An accumulation of wind-blown dust (silt) that may have undergone mild digenesis.

Marl—An earthy deposit consisting mainly of calcium carbonate, usually mixed with clay. Marl is used for liming acid soils. It is slower acting than most lime products used for this purpose.

Mineralization—The conversion of an element from an organic form to an inorganic state as a result of microbial decomposition.

Nitrogen fixation—The biological conversion of elemental nitrogen (N_2) to organic combinations or to forms readily utilized in biological processes.

Osmosis—The movement of a liquid across a membrane from a region of high concentration to a region of low concentration. Water and nutrients move into roots independently.

Oxidation—The loss of electrons by a substance.

Parent material—The unconsolidated and more or less chemically weathered mineral or organic matter from which the solum of soils is developed by pedogenic processes.

Ped—A unit of soil structure such as an aggregate, crumb, prism, block, or granule, formed by natural processes.

Pedogenic/pedological process—Any process associated with the formation and development of soil.

pH—The degree of acidity or alkalinity of the soil. Also referred to as *soil reaction*, this measurement is based on the pH scale where 7.0 is neutral, values from 0.0 to 7.0 are acid, and values from 7.0 to 14.0 are alkaline. The pH of soil is determined by a simple chemical test where a sensitive indicator solution is added directly to a soil sample in a test tube.

Photosynthesis—The process by which green leaves of plants, in the presence of sunlight, manufacture their own needed materials from carbon dioxide in the air and water and minerals taken from the soil.

Porosity, soil—The volume percentage of the total bulk not occupied by solid particles.

Profile, soil—A vertical section of the soil through all its horizons and extending into the parent material.

Reduction—The gain of electrons and therefore the loss of positive valence charge by a substance.

Regolith—The unconsolidated mantle of weathered rock and soil material on the Earth's surface; loose earth materials above solid rock.

Rock—The material that forms the essential part of the Earth's solid crust, including loose incoherent masses such as sand and gravel, as well as solid masses of granite and limestone.

Rock cycle—The global geological cycling of lithospheric and crustal rocks from their igneous origins through all or any stages of alteration, deformation, resorption, and reformation.

Runoff—The portion of the precipitation on an area that is discharged from the area through stream channels.

Salinization—The process of accumulation of salts in soil.

Sand—A soil particle between 0.05 and 2.0 mm in diameter; a soil textural class.

Silt—A soil separate consisting of particles between 0.05 and 0.002 mm in equivalent diameter; a soil textural class.

Slope—The degree of deviation of a surface from horizontal, measured in a numerical ratio, percent, or degrees.

Soil—An assemblage of loose and normally stratified granular minerogenic and biogenic debris at the land surface; it is the supporting medium for the growth of plants.

Soil air—The soil atmosphere; the gaseous phase of the soil, which is the volume not occupied by soil or liquid.

Soil horizon—A layer of soil, approximately parallel to the soil surface, with distinct characteristics produced by soil-forming processes. These characteristics form the basis for systematic classification of soils.

Soil profile—A vertical section of the soil from the surface through all its horizons, including C horizons.

Soil structure—The combination or arrangement of primary soil particles into secondary particles, units, or peds. These secondary units may be, but usually are not, arranged in the profile in such a manner as to give a distinctive characteristic pattern. The secondary units are characterized and classified on the basis of size, shape, and degree of distinctness into classes, types, and grades, respectively.

Soil texture—The relative proportions of the various soil separates in a soil.

Soluble—Dissolves easily in water.

Solum (pl. *sola*)—The upper and most weathered part of the soil profile; the A, E, and B horizons.

Subsoil—That part of the soil below the plow layer.

Till—Unstratified glacial drift deposited directly by the ice and consisting of clay, sand, gravel, and boulders intermingled in any proportion.

Tilth—The physical condition of soil as related to its ease of tillage, fitness as a seedbed, and its impedance to seedling emergence and root penetration.

Topsoil—The layer of soil moved in cultivation.

Weathering—All physical and chemical changes produced in rocks at or near the Earth's surface by atmospheric agents.

ALL ABOUT SOIL

Before we begin a journey that takes us through the territory that is soil and examine soil from micro to macro levels, we need to stop for a moment and discuss why, beyond the obvious reason, soil is so important to us, to our environment, to our very survival.

FUNCTIONS OF SOIL

We normally relate soil to our backyards, to farms, to forests, or to a regional watershed. We think of soil as the substance upon which plants grow. Soils play other roles, though. There are six main functions of soil that are important to us (Figure 2.1):

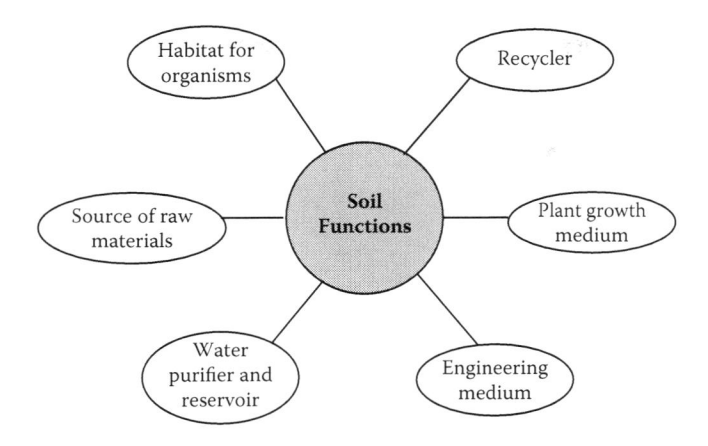

FIGURE 2.1   Functions of soil.

(1) soils serve as a medium for plant growth, (2) soils regulate our water supplies, (3) soils are recyclers of raw materials, (4) soils provide a habitat for organisms, (5) soils are used as an engineering medium, and (6) soils provide materials. Let's take a closer look at each of these functions of soils.

Soil as a Plant Growth Medium

We are all aware of the primary function of soil—to serve as a plant growth medium, a function that becomes more important with each passing day as Earth's population continues to grow. Although soil is a medium for plant growth, soil itself is actually alive as well. We depend on soil for life, and at the same time soil depends on life. Its very origin, its maintenance, and its true nature are intimately tied to living plants and animals. What does this mean? Let's take a look at how the elegant prose of renowned environmental writer Rachel Carson (1962) explained this paradox:

> The soil community ... consists of a web of interwoven lives, each in some way related to the others—the living creatures depending on the soil, but the soil in turn is a vital element of the earth only so long as this community within it flourishes.

The soil might say to us if it could, "Don't kill off the life within me and I will do the best I can to provide life that will help to sustain your life." What we have here is a tradeoff—one vitally important to both soil and ourselves. Remember that most of Earth's people are tillers of the soil; the soil is their source of livelihood, and those soil tillers provide food for us all.

As a plant growth medium, soil provides vital resources and performs essential functions for the plant. To grow in soil, plants must have water and nutrients. Soil provides both of these. To grow and to sustain its growth, a plant must have a root system. Soil provides pore spaces for roots. To grow and maintain growth, a plant's roots must have oxygen for respiration and carbon dioxide exchange and ultimate diffusion out of the soil. Soil provides the air and pore spaces (the ventilation system) for this. To continue to grow, a plant must have support. Soil provides this support.

If a plant seed is planted in a soil and is exposed to the proper amount of sunlight for growth to occur, the soil still must provide nutrients through a root system that has space to grow, a continuous stream of water (about 500 g of water are required to produce 1 g of dry plant material) for root nutrient transport and plant cooling, and a pathway for both oxygen and carbon dioxide transfer. Just as important, soil water provides the plant with the normal fullness or tension (turgor) it needs to stand—the structural support it needs to face the sun for photosynthesis to occur.

In addition to the functions stated above, soil is also an important moderator of temperature fluctuations. If you have ever dug in a garden on a hot summer day, you probably noticed that the soil was warmer (even hot) on the surface but much cooler just a few inches below the surface.

Soil as a Regulator of Water Supplies

When we walk on land, few of us probably realize that we are actually walking across a bridge. This bridge (in many areas) transports us across a veritable ocean of water below us, deep—or not so deep—under the surface of the Earth. Consider what happens to rain. Where does the rain water go? Some, falling directly over water bodies, becomes part of the water body again, but an enormous amount falls on land. Some of the water, obviously, runs off—always following the path of least resistance (Figure 2.2). In modern communities, stormwater runoff is a hot topic. Cities have taken giant steps to try to control runoff, to send it where it can be properly handled to prevent flooding. Let's take a closer look at precipitation and the sinks it pours into, then relate this usually natural operation to soil water. We begin with surface water, then move on to that ocean of water below the soil's surface: groundwater.

Surface water (water on the Earth's surface as opposed to subsurface water, or groundwater) is mostly a product of precipitation: rain, snow, sleet, or hail. Surface water is exposed or open to the atmosphere and results from overland flow, the movement of water on and just under the Earth's surface. This overland flow is the same thing as surface runoff, which is the amount of rainfall that passes over the Earth's surface. Specific sources of surface water include rivers, streams, lakes, impoundments, shallow wells, rain catchments, tundra ponds, or meskegs (peat bogs).

Most surface water is the result of surface runoff. The amount and flow rate of surface runoff are highly variable. This variability stems from two main factors: (1) human interference (influences), and (2) natural conditions. In some cases, surface water runs quickly off land. Generally, this is undesirable from a water resources standpoint because the water does not have enough time to infiltrate into the ground and recharge groundwater aquifers. Other problems associated with quick surface water runoff are erosion and flooding. Probably the only good thing that can be said about surface water that quickly runs off land is that it does not have enough time (normally) to become contaminated with high mineral content. Surface water running slowly off land can be expected to have all the opposite effects. Surface water travels over the land to what amounts to a predetermined destination. What factors influence how surface water moves? Surface water's journey over the face of the Earth typically begins at its drainage basin, sometimes referred to as its *drainage area*, *catchment*, or *watershed*. For a groundwater source, this is known as the *recharge area*, the area from which precipitation flows into an underground water source.

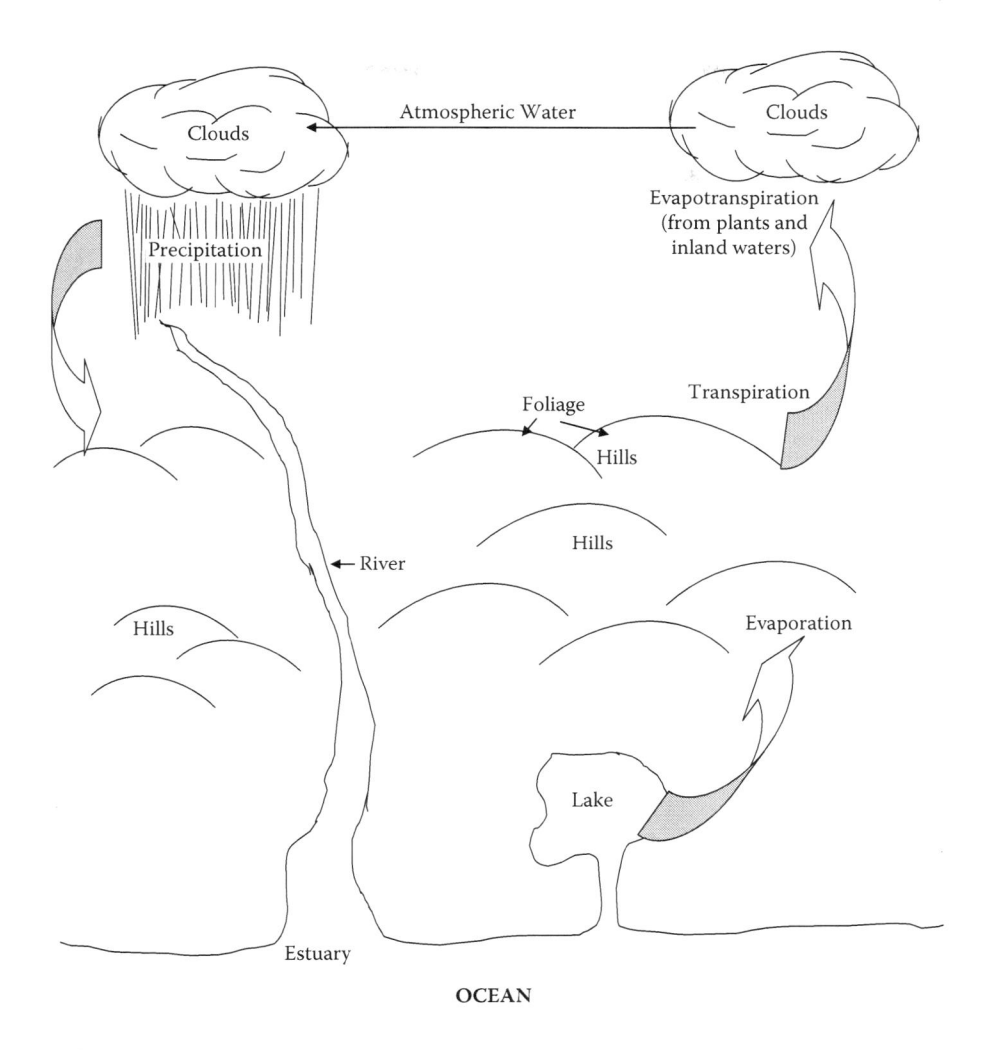

FIGURE 2.2 Water cycle.

The area of a surface water drainage basin is usually measured in square miles, acres, or sections. If a city takes water from a surface water source, essential information for the assessment of water quality includes how large the drainage basin is and what lies within it.

We know that water does not run uphill; instead, surface water runoff (like the flow of electricity) follows along the path of least resistance. Generally speaking, water within a drainage basin will naturally be shunted by the geological formation of the area toward one primary watercourse (a river, stream, creek, brook) unless some manmade distribution system diverts the flow. Various factors directly influence the flow of surface water over land:

- *Rainfall duration*—Length of the rainstorm affects the amount of runoff. Even a light, gentle rain will eventually saturate the soil if it lasts long enough. When the saturated soil can absorb no more water, rainfall builds up on the surface and begins to flow as runoff.
- *Rainfall intensity*—The harder and faster it rains, the more quickly soil becomes saturated. With hard rains, the surface inches of soil quickly become inundated; with short, hard storms, most of the rainfall may end up as surface runoff because the moisture is carried away before significant amounts of water are absorbed into the ground.
- *Soil moisture*—Obviously, if the soil is already laden with water from previous rains, the saturation point will be reached sooner than if the soil were dry. Frozen soil also inhibits water absorption; up to 100% of snow melt or rainfall on frozen soil will end up as runoff because frozen ground is impervious.
- *Soil composition*—The runoff amount is directly affected by soil composition. Hard rock surfaces will shed all rainfall, obviously, but so will soils with a heavy clay composition. Clay soils possess small void spaces that swell when wet. When these void spaces close, they form a barrier that does not allow additional absorption or infiltration. On the opposite end of the spectrum, coarse sand allows easy water flowthrough, even in a torrential downpour.
- *Vegetation cover*—Runoff is limited by groundcover. Roots of vegetation, pine needles and cones, leaves, and branches create a porous layer (of decaying natural organic substances) above the soil. This porous organic sheet (groundcover) readily allows water into the soil. Vegetation and organic waste also act as a cover to protect the soil from hard, driving rains. Hard rains can compact bare soils, close off void spaces, and increase runoff. Vegetation and groundcover work to maintain the infiltration and water-holding capacity of the soil. Note that vegetation and groundcover also reduce evaporation of soil moisture as well.
- *Ground slope*—Flat-land water flow is usually so slow that large amounts of rainfall can infiltrate the ground. Gravity works against infiltration on steeply sloping ground, where up to 80% of rainfall may become surface runoff.
- *Human influences*—Various human activities have a definite impact on surface water runoff. Most human activities tend to increase the rate of water flow. For example, canals and ditches are usually constructed to provide steady flow, and agricultural activities generally remove groundcover that would work to retard the runoff rate. At the opposite extreme, manmade dams are generally built to retard the flow of runoff.

Human habitations, with their paved streets, tarmac, paved parking lots, and buildings create surface runoff potential, because so many surfaces are impervious to infiltration. All these surfaces hasten the flow of water, and they also increase the possibility of flooding, often with devastating results. Because of urban increases in runoff, a whole new industry has developed: stormwater management. Paving over natural surface acreage has another serious side effect. Without enough area available for water to infiltrate the ground and percolate through the soil to eventually reach and replenish (recharge) groundwater sources, those sources may eventually fail, with devastating impact on the local water supply.

Now let's shift gears and take a look at groundwater. Water falling to the ground as precipitation normally follows three courses. Some runs off directly to rivers and streams, some infiltrates to ground reservoirs, and the rest evaporates or transpires through vegetation. The water in the ground is invisible and may be thought of as a temporary natural reservoir (ASTM, 1969; Spellman, 2015). Almost all groundwater is in constant motion toward rivers or other surface water bodies.

Groundwater is defined as water below the Earth's crust but above a depth of 2500 feet. Water located between the Earth's crust and the 2500-foot level is considered usable (potable) freshwater. In the United States, it is estimated that at least 50% of total available freshwater storage is in underground aquifers (Kemmer, 1979).

In this text, we are concerned with that amount of water retained in the soil to ensure plant life and growth. Recall that earlier we stated that 500 g of water are required to produce 1 g of dry plant material. Note that about 5 g of this water become an integral part of the plant. Unless rainfall is frequent, you don't have to be a rocket scientist to figure out that the ability of soil to hold water against the force of gravity is very important. Thus, one of the vital functions of soil is to regulate the water supply to plants. Also, we are concerned with the ability of soil to hold and retain water in aquifers, thus allowing wells to be dug to provide a potable water supply where one is needed and to maintain a groundwater level that will prevent or retard land subsidence.

Soil as a Recycler of Raw Materials

Imagine what it would be like to step out into the open air and be hit by a stench that not only would offend your olfactory sense but could almost reach out and grab you (like the situation we had in the cave earlier, but worse). Imagine looking out upon the cluttered fields in front of your domicile and seeing nothing but stack upon stack upon stack of the sources of the horrible, putrefied, foul, decaying, gagging, choking, retching stench. We are talking about plant and animal remains and waste, mountains of it, reaching toward the sky and surrounded by colonies of flies of all varieties. "Ugh," you say. Well, if it were not for the power of the soil to recycle waste products, then this scene or something very much like it would be possible. Of course, this scenario is impossible because under these conditions there would be no more life to die and stack up.

Soil is a recycler—probably the premier recycler on Earth. The simple fact is that if it were not for the incredible recycling ability of soil, plants and animals would have run out of nourishment long ago. Soil recycles in other ways; for example, consider the geochemical cycles (i.e., the chemical interactions between soil, water, air, and life on Earth) in which soil plays a major role.

Soil possesses the incomparable ability and capacity to assimilate great quantities of organic wastes, turn them into beneficial organic matter (humus), and then convert the nutrients in the wastes to forms that can be utilized by plants and animals. In turn, the soil returns carbon to the atmosphere as carbon dioxide, where it again will eventually become part of living organisms through photosynthesis. Soil performs several different recycling functions—most of them good, some of them not so good.

TABLE 2.1

A Representative Sample of Soil Organisms

Microorganisms (Protists)	Arthropod Animals
Bacteria	Springtails
Fungi	Mites
Actinomycetes	Millipedes and centipedes
Algae	Harvestman
Protozoa	Ants
	Diplopoda
Nonarthropod Animals	Diptera
Nematodes	Crustacea
Earthworms and potworms	
Vertebrates	
Mice, moles, voles	
Rabbits, gophers, squirrels	

Consider one recycling function of soil that may not be so good. Soils have the capacity to accumulate large amounts of carbon as soil organic matter, which can have a major impact on global changes such as the greenhouse effect. Moreover, it is important that wastes be applied in appropriate amounts and not contain toxic and environmentally harmful elements or compounds that could poison soils, wastes, and plants.

Soil as a Habitat for Soil Organisms

> Life not only formed the soil, but other living things of incredible abundance and diversity now exist within it; if this were not so the soil would be a dead and sterile thing.
>
> **Carson (1962)**

One thing is certain, most soils are not dead and sterile things. The fact is, a handful of soil is an ecosystem. It may contain up to billions of organisms belonging to thousands of species. Table 2.1 lists a few (very few) of these organisms. Obviously, communities of living organisms inhabit the soil. What is not so obvious is that they are as complex and intrinsically valuable as are those organisms that roam the land surfaces and waters of Earth.

Soil as an Engineering Medium

We usually think of soil as being firm and solid (solid ground, *terra firma*). As solid ground, soil is usually a good substrate upon which to build highways and structures; however, not all soils are firm and solid. Some are not as stable as others. Whereas construction of buildings and highways may be suitable in one location on one type of soil, it may be unsuitable in another location with different soil. To construct structurally sound and stable highways and buildings, construction on soils and with soil materials requires knowledge of the diversity of soil properties. Note that working with manufactured building materials that have been engineered to withstand certain stresses and forces is much different than working with natural soil materials, even though engineers have the same concerns about soils as they do with manmade building materials (concrete and steel). It is much more difficult to evaluate the ability of a soil to resist compression or to remain in place and to determine its bearing strength, shear strength, and stability than it is to make these same determinations for manufactured building materials.

Soil as a Source of Materials

In addition to providing valuable minerals for various purposes, soil is commonly used to provide materials for road building and dam construction.

CONCURRENT SOIL FUNCTIONS

Soils perform specific critical functions no matter where they are located (USDA, 2009), and they perform more than one function at the same time, as described below and shown in Figure 2.3:

- Soils act like sponges, soaking up rainwater and limiting runoff. Soils also impact groundwater recharge and flood-control potentials in urban areas.
- Soils act like faucets, storing and releasing water and air for plants and animals to use.
- Soils act like supermarkets, providing valuable nutrients and air and water to plants and animals. Soils also store carbon and prevent its loss into the atmosphere.
- Soils act like strainers or filters, filtering and purifying water and air that flow through them.
- Soils buffer, degrade, immobilize, detoxify, and trap pollutants, such as oil, pesticides, herbicides, and heavy metals, and keep them from entering groundwater supplies. Soils also store nutrients for future use by plants and animals above ground and by microbes within soils.

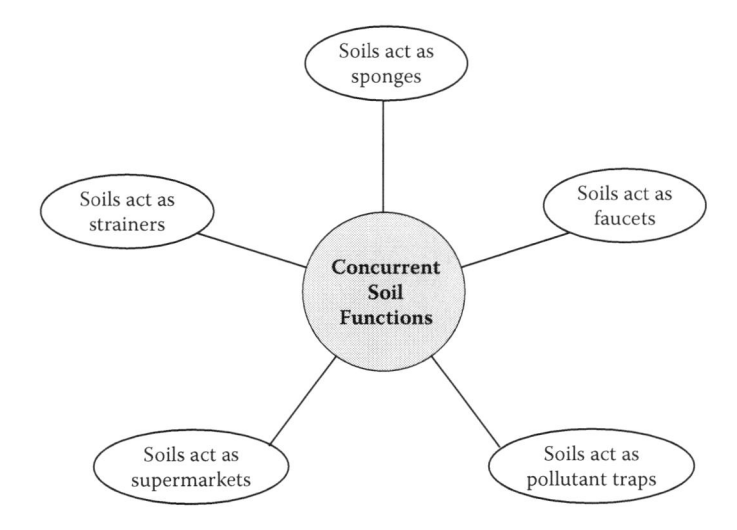

FIGURE 2.3 Concurrent soil functions.

SOIL BASICS

Any fundamental discussion about soil should begin with a definition of what soil is. The word *soil* is derived through Old French from the Latin *solum*, which means floor or ground. John Steinbeck referenced the scars, crusts, and crusting of soil. The Swiss writer Charles-Ferdinand Ramuz referred to soil as that soft stuff under the feet. A student of Hippocrates talked about soil as an immense quantity of forces. A more current and concise definition is made difficult by the great diversity of soils throughout the globe; however, here is a generalized definition from the Soil Science Society of America (SSSA, 2008):

> Soil is unconsolidated mineral matter on the surface of the earth that has been subjected to and influenced by genetic and environmental factors of: parent material, climate, macro- and microorganisms, and topography, all acting over a period of time and producing a product—soil—that differs from the material from which it is derived in many physical, chemical, and biological properties, and characteristics.

Engineers might define soil by saying that soil occupies the unconsolidated mantle of weathered rock making up the loose materials on the Earth's surface, commonly known as the *regolith* (see Figure 2.4). Soil can be described as a three-phase system,

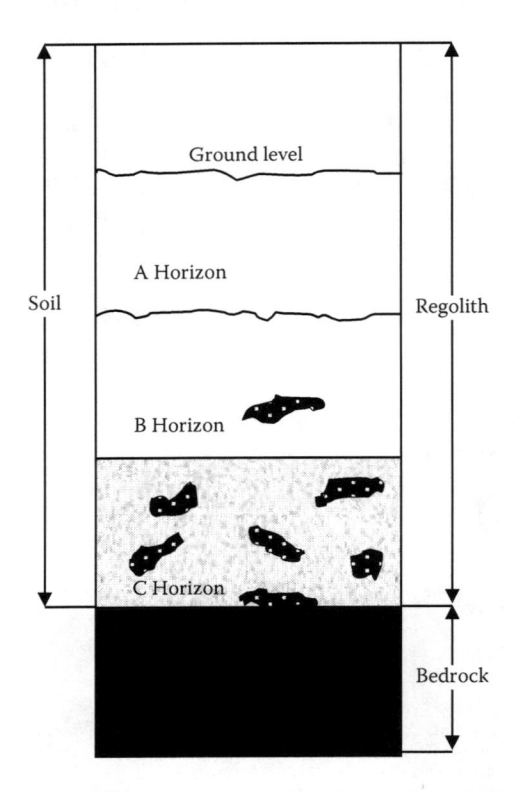

FIGURE 2.4   Relative positions of the regolith, its soil, and the underlying bedrock.

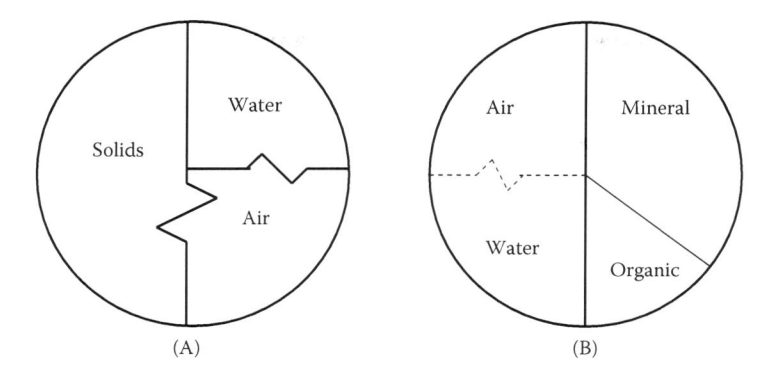

(A)   (B)

FIGURE 2.5   (A) Three phases of soil: solids, water, and air. Broken lines indicate that these phases are not constant but change with conditions. (B) Another view of soil (a loam surface soil) as being comprised of air, water, and mineral and organic solids.

composed of a solid, liquid, and gaseous phase (see Figure 2.5A). This phase relationship is important in dealing with soil pollution, because each of the three phases of soil is in equilibrium with the atmosphere and with rivers, lakes, and the oceans. Thus, the fate and transport of pollutants are influenced by each of these components.

Soil is also commonly described as a mixture of air, water, mineral matter, and organic matter (see Figure 2.5B); the relative proportions of these four components greatly influence the productivity of soils. The interface of these soil components, where the regolith meets the atmosphere, is what concerns us here.

Keep in mind that the four major ingredients that make up soil are not mixed or blended like cake batter. Instead, a major and critically important constituent of soil is the pore spaces, which are vital to air and water circulation, as they provide space for roots to grow and microscopic organisms to live. Without sufficient pore space, soil would be too compacted to be productive. Ideally, the pore space will be divided roughly equally between water and air, with about one-quarter of the soil volume consisting of air and one-quarter consisting of water. The relative proportions of air and water in a soil typically fluctuate significantly as water is added and lost. Compared to surface soils, subsoils tend to contain less total pore space, less organic matter, and a larger proportion of micropores, which tend to be filled with water.

Let's take a closer look at the four major components that make up soil. Soil air circulates through soil pores in the same way air circulates through a ventilation system. Only when the pores (the ventilation ducts) become blocked by water or other substances does the air fail to circulate. Although soil pores normally connect to interface with the atmosphere, soil air is not the same as atmospheric air. It differs in composition from place to place. Soil air also normally has a higher moisture content than the atmosphere. The content of carbon dioxide (CO_2) is usually higher and that of oxygen (O_2) lower than the accumulations of these gases in the atmosphere.

Earlier we stated that only when soil pores are occupied by water or other substances does air fail to circulate in the soil. For proper plant growth, this is of particular importance, because in soil pore spaces that are water dominated, air oxygen content is low and carbon dioxide levels are high, which restricts plant growth.

The presence of water in soil (often reflective of climatic factors) is essential for the survival and growth of plant and other soil organisms. Soil moisture is a major determinant of the productivity of terrestrial ecosystems and agricultural systems. Water moving through soil materials is a major force behind soil formation. Along with air, water, and dissolved nutrients, soil moisture is critical to the quality and quantity of local and regional water resources.

Mineral matter varies in size and is a major constituent of nonorganic soils. Mineral matter consists of large particles (rock fragments), including stones, gravel, and coarse sand. Many of the smaller mineral matter components are made of a single mineral. Minerals in the soil (for plant life) are the primary source of most of the chemical elements essential for plant growth.

Soil organic matter consists primarily of living organisms and the remains of plants, animals, and microorganisms that are continuously broken down (biodegraded) in the soil into new substances that are synthesized by other microorganisms. These other microorganisms continually use this organic matter and reduce it to carbon dioxide via respiration until it is depleted, making repeated additions of new plant and animal residues necessary to maintain soil organic matter (Brady and Weil, 2007).

Now that we have defined soil, let's take a closer look at a few of the basics pertaining to soil and some of the common terms used in any discussion related to soil basics. Soil is the layer of bonded particles of sand, silt, and clay that covers the land surface of the Earth. Most soils develop in multiple layers. The topmost layer, topsoil, is the layer of soil moved in cultivation and in which plants grow. This topmost layer is actually an ecosystem composed of both biotic and abiotic components—inorganic chemicals, air, water, and decaying organic material that provides vital nutrients for plant photosynthesis, as well as living organisms. Below the topmost layer is the subsoil, the part of the soil below the plow level, usually no more than a meter in thickness. Subsoil is much less productive, partly because it contains much less organic matter.

Below that is the parent material, the unconsolidated (and more or less chemically weathered) bedrock or other geologic material from which the soil is ultimately formed. The general rule of thumb is that it takes about 30 years to form 1 inch of topsoil from subsoil; it takes much longer than that for subsoil to be formed from parent material, with the length of time depending on the nature of the underlying matter (Franck and Brownstone, 1992).

PHYSICAL PROPERTIES OF SOIL

From the soil pollution technologist's point of view (regarding land conservation and methodologies for contaminated soil remediation through reuse and recycling), five major physical properties of soil are of interest: soil texture, slope, structure, organic matter, and soil color. Soil texture, or the relative proportions of the various soil separates in a soil (Figure 2.6), is a given and cannot be easily or practically changed significantly. It is determined by the size of the rock particles (sand, silt, and clay particles) or the soil separates within the soil. The largest soil particles are gravel, which consist of fragments larger than 2.0 mm in diameter.

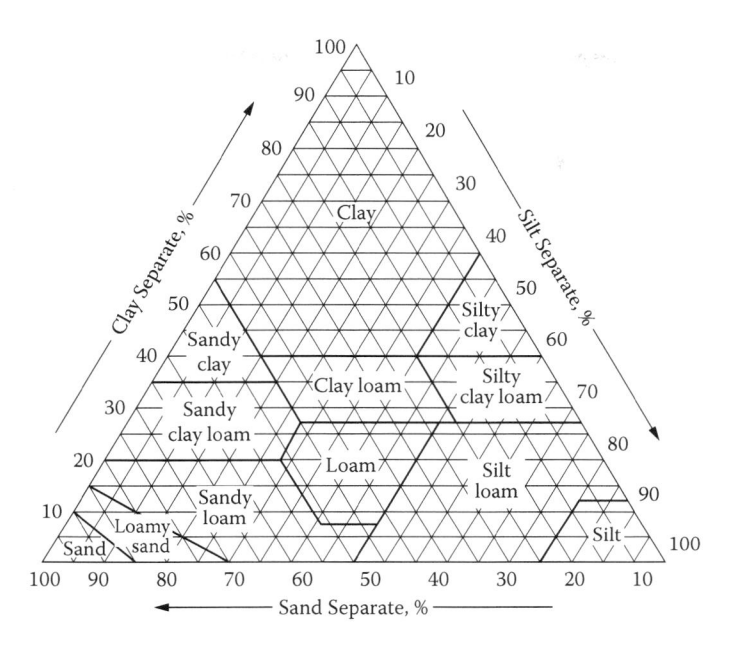

FIGURE 2.6 Textural triangle. (Adapted from USDA, *Urban Soil Primer*, U.S. Department of Agriculture, Washington, DC, 2009.)

Particles between 0.05 and 2.0 mm are classified as sand. Silt particles range from 0.002 to 0.05 mm in diameter, and the smallest particles, clay, are less than 0.002 mm in diameter. Clays are composed of the smallest particles, but these particles have stronger bonds than silt or sand; once broken apart, though, they erode more readily. Particle size has a direct impact on erosion. Rarely does a soil consist of only one single size of particle; most are a mixture of various sizes.

The slope (or steepness of the soil layer) is another given, important because the erosive power of runoff increases with the steepness of the slope. Slope also allows runoff to exert increased force on soil particles, which breaks them apart more readily and carries them farther away.

Soil structure (tilth) should not be confused with soil texture—they are different. In fact, in the field, the properties determined by soil texture may be considerably modified by soil structure. Soil structure refers to the combination or arrangement of primary soil particles into secondary particles (units or peds). Simply stated, soil structure refers to the way various soil particles clump together. Clusters of soil particles, called *aggregates*, can vary in size, shape, and arrangement; they combine naturally to form larger clumps called *peds*. Sand particles do not clump because sandy soils lack structure. Clay soils tend to stick together in large clumps. Good soil develops small friable (easily crumbled) clumps. Soil develops a unique, fairly stable structure in undisturbed landscapes, but agricultural practices break down the aggregates and peds, lessening erosion resistance.

The presence of decomposed or decomposing remains of plants and animals (organic matter) in soil helps not only fertility but also soil structure—especially the ability of water to store water. Live organisms such as protozoa, nematodes, earthworms, insects, fungi, and bacteria are typical inhabitants of soil. These organisms work to either control the population of organisms in the soil or to aid in the recycling of dead organic matter. All soil organisms, in one way or another, work to release nutrients from the organic matter, changing complex organic materials into products that can be used by plants.

Just about anyone who has looked at soil has probably noticed that soil color is often different from one location to another. Soil colors range from very bright to dull grays, to a wide range of reds, browns, blacks, whites, yellows, and even greens. Soil color is dependent primarily on the quantity of humus and the chemical form of iron oxides present. Soil scientists use a set of standardized color charts (the *Munsell Color System*) to describe soil colors. They consider three properties of color—hue, value, and chroma—in combination to come up with a large number of color chips to which soil scientists can compare the color of the soil being investigated.

SOIL SEPARATES

As pointed out in the previous section, soil particles have been divided into groups (*soil separates*) based on their size (sand, silt, and clay; see Figure 2.7) by the International Soil Science Society System, the U.S. Public Roads Administration, and the U.S. Department of Agriculture (USDA). In this text, we use the classification established by the USDA. The size ranges in these separates reflect major changes in how the particles behave and in the physical properties they impart to soils. Table 2.2 lists the separates, their diameters, and the number of particles in 1 g of soil (according to the USDA).

Sand ranges in diameter from 0.05 to 2 mm and is divided into five classes (see Table 2.2). Sand grains are more or less spherical (rounded) in shape, with variable angularity, depending on the extent to which they have been worn down by abrasive processes such as rolling around by flowing water during soil formation.

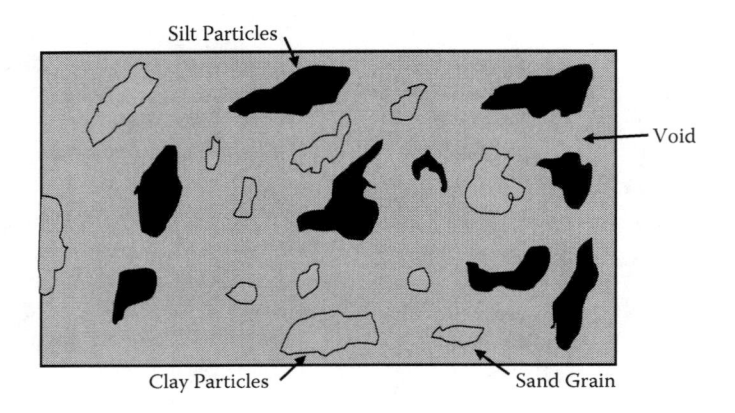

FIGURE 2.7　Enlarged view: cross-section of sandy soil.

TABLE 2.2

Characteristics of Soil Separates (USDA)

Soil Separate	Diameter (mm)	Number of Particles per Gram
Very coarse sand	2.00–1.00	90
Coarse sand	1.00–0.50	720
Medium sand	0.50–0.25	5700
Fine sand	0.25–0.10	46,000
Very fine sand	0.10–0.05	722,000
Silt	0.05–0.002	5,776,000
Clay	Below 0.002	90,260,853,000

Sand forms the framework of soil and gives it stability when in a mixture of finer particles. Sand particles are relatively large, which allows voids that form between each grain to also be relatively large. This promotes free drainage of water and the entry of air into the soil. Sand is usually composed of a high percentage of quartz; because quartz is most resistant to weathering, its breakdown is extremely slow. Many other minerals are found in sand, depending on the rocks from which the sand was derived. In the short term (on an annual basis), sand contributes little to plant nutrition in the soil; however, in the long term (thousands of years of soil formation), soils with a lot of weatherable minerals in their sand fraction develop a higher state of fertility.

Silt (essentially microsand), though spherically and mineralogically similar to sand, is smaller, too small to be seen with the naked eye (see Figure 2.7). It weathers faster and releases soluble nutrients for plant growth faster than sand. Too fine to be gritty, silt imparts a smooth feel (like flour) without stickiness. The pores between silt particles are much smaller than those in sand (sand and silt are just progressively finer and finer pieces of the original crystals in the parent rocks). In flowing water, silt is suspended until it drops out when flow is reduced. On the land surface, silt, if disturbed by strong winds, can be carried great distances and is deposited as loess.

The clay soil separate is (for the most part) much different from sand and silt (see Figure 2.6). Clay is composed of secondary minerals that were formed by the drastic alteration of the original forms or by the recrystallization of the products of their weathering. Because clay crystals are plate-like (sheeted) in shape they have a tremendous surface area-to-volume ratio, giving clay a great capacity to adsorb water and other substances on its surfaces. Clay actually acts as a storage reservoir for both water and nutrients. There are many kinds of clay, each with different internal arrangements of chemical elements which give them individual characteristics.

SOIL FORMATION

Everywhere on Earth's land surface is either rock formation or exposed soil. When rocks formed deep in the Earth are thrust upward and exposed to the Earth's atmosphere, the rocks adjust to the new environment, and soil formation begins. Soil is formed as a result of physical, chemical, and biological interactions in specific locations. Just as vegetation varies among biomes, so do the soil types that support

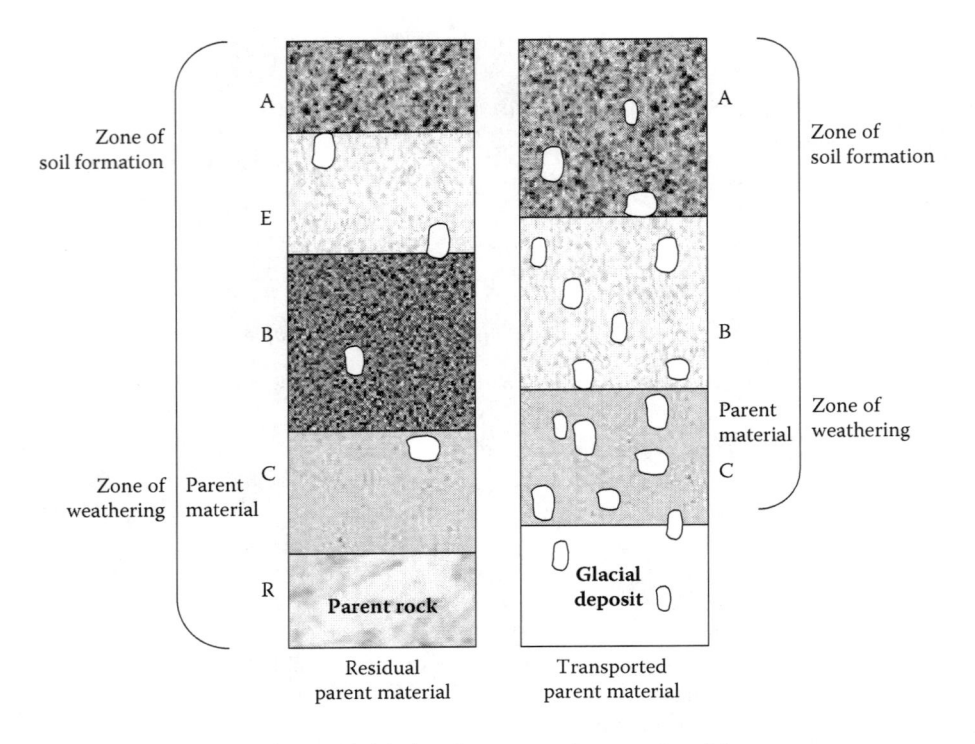

FIGURE 2.8   Soil profiles on residual and transported parent materials.

that vegetation. The vegetation of the tundra and that of the rainforest differ vastly from each other and from vegetation of the prairie and coniferous forest; soils differ in similar ways.

In the soil-forming process, two related, but fundamentally different, processes are occurring simultaneously. The first is the formation of soil parent materials by *weathering* of rocks, rock fragments, and sediments. This set of processes is carried out in the zone of weathering. The end point is producing parent material for the soil to develop in and is referred to as C horizon material (see Figure 2.8). It applies in the same way for glacial deposits as for rocks. The second set of processes is the formation of the soil profile by *soil-forming processes*, which gradually change the C horizon material into A, E, and B horizons. Figure 2.8 illustrates two soil profiles, one on hard granite and one on a glacial deposit.

WEATHERING

Soil development takes time and is the result of two major processes: weathering and morphogenesis, which can be described as bare rock succession. Weathering, the breaking down of bedrock and other sediments that have been deposited on the bedrock by wind, water, volcanic eruptions, or melting glaciers, happens physically, chemically, or by a combination of both. Weathering is the first step in the erosion process; again, it causes the breakdown of rocks, either to form new minerals that are stable on the surface of Earth or to break the rocks down to smaller particles. Simply, weathering (which projects itself on all surface material above the water table) is the general term used for all the ways in which a rock may be broken down. The factors that influence weathering include the following:

- *Rock type and structure*—Each mineral contained in rocks has a different susceptibility to weathering. A rock with bedding planes, joints, and fractures provides pathways for the entry of water, leading to more rapid weathering. Differential weathering (rocks erode at differing rates) can occur when rock combinations consist of some rocks that weather faster than more resistant rocks.
- *Slope*—On steep slopes, weathering products may be quickly washed away by rains. Wherever the force of gravity is greater than the force of friction holding particles upon a slope, these tend to slide downhill.
- *Climate*—Higher temperatures and high amounts of water generally cause chemical reactions to run faster. Rates of weathering are higher in warmer than in colder dry climates.
- *Animals*—Rodents, earthworms, and ants that burrow into soil bring material to the surface where it can be exposed to the agents of weathering.
- *Time*—Length of time depends on slope, animals, and climate.

Although weathering processes are separated, it is important to recognize that these processes work in tandem to break down rocks and minerals to smaller fragments. Geologists recognize two categories of weathering processes:

- *Physical (or mechanical) weathering*—Disintegration of rocks and minerals by a physical or mechanical process.
- *Chemical weathering*—Decomposition of rock by chemical changes or solution.

Physical Weathering

Physical weathering involves the disintegration of a rock by physical processes. These include freezing and thawing of water in rock crevices, disruption by plant roots or burrowing animals, and the changes in volume that result from chemical weathering with the rock. These and other physical weathering processes are discussed below:

- *Development of joints*—Joints are another way that rocks yield to stress. Joints are fractures or cracks in which the rocks on either side of the fracture have not undergone relative movement. Joints form as a result of expansion due to cooling or relief of pressure as overlying rocks are removed by erosion. Joints are free spaces in rock through which other agents of chemical or physical weathering can enter (unlike faults that show offset across the fracture). They play an important part in rock weathering as zones of weakness and water movement.

- *Crystal growth*—As water percolates through fractures and pore spaces it may contain ions that precipitate to form crystals. When crystals grow they can cause the necessary stresses needed for mechanical rupturing of rocks and minerals.
- *Heat*—It was once thought that daily heating and cooling of rocks was a major contributor to the weathering process. This view is no longer shared by most practicing geologists, but it should be pointed out that sudden heating of rocks from forest fires may cause expansion and eventual breakage of rock.
- *Biological activities*—Plant and animal activities are important contributors to rock weathering. Plants contribute to the weathering process by extending their root systems into fractures and growing, causing expansion of the fracture. The effects of plant growth are evident in many places, such as when they are planted near cement work (streets, brickwork, and sidewalks). Animal burrowing in rock cracks can break rock.
- *Frost wedging*—Frost wedging is often produced by the alternating freezing and thawing of water in rock pores and fissures. Expansion of water during freezing causes the rock to fracture. Frost wedging is more prevalent at high altitudes where there may be many freeze–thaw cycles. One classic and striking example of weathering of Earth's surface rocks by frost wedging is illustrated by the formation of hoodoos in Bryce Canyon National Park, Utah. Bryce Canyon receives a meager 18 inches of precipitation annually, but it is amazing what this little bit of water can do under the right circumstances (NPS, 2008). Approximately 200 freeze–thaw cycles occur annually in Bryce. During these periods, snow and ice melt in the afternoon and water seeps into the joints of the Bryce or Claron Formation. When the sun sets, temperatures plummet and the water refreezes, expanding up to 9% as it becomes ice. This frost wedging process exerts tremendous pressure or force on the adjacent rock and shatters the weak rock. The assault from frost wedging is a powerful force, but, at the same time, rainwater (the universal solvent), which is naturally acidic, slowly dissolves away the limestone, rounding off the edges of fractured rocks and washing away the debris. Small rivulets of water run down Bryce's rim, forming gullies. As gullies are cut deeper, narrow walls of rock known as fins being to emerge. Fins eventually develop holes known as windows. Windows grow larger until their roofs collapse, creating hoodoos. As old hoodoos age and collapse, new ones are born.

DID YOU KNOW?

Bryce Canyon National Park lies along the high eastern escarpment of the Paunsaugunt Plateau in the Colorado Plateau region of southern Utah. Its extraordinary geological character is expressed by thousands of rock chimneys (hoodoos) that occupy amphitheater-like alcoves in the Pink Cliffs, whose bedrock host is the Eocene-age Claron Formation (Davis and Pollock, 2003).

> **DID YOU KNOW?**
>
> Hoodoo (n.), pronounced "hü-dü," is perhaps derived from voodoo (*vodou*). A hoodoo is a natural column of rock in western North American, often in fantastic form (http://www.merriam-webster.com/dictionary/hoodoo).

Chemical Weathering

Chemical weathering involves the decomposition of rock by chemical changes or solution. Rocks that are formed under conditions found deep within the Earth are exposed to quite different conditions when uplifted onto the surface; for example, temperatures and pressures are lower on the surface, and copious amounts of free water and oxygen are available. The chief chemical weathering processes are oxidation, carbonation, hydration, and solution in water.

Persistent Hand of Water

Because of its unprecedented impact on shaping and reshaping Earth, at this point in the text it is important to point out that, given time, nothing, absolutely nothing, on Earth is safe from the heavy hand of water. The effects of water sculpting by virtue of movement and accompanying friction will be covered later in the text. For now, with regard to water exposure and chemical weathering, the main agent responsible for chemical weathering reactions is not water movement but water and weak acids formed in water. The acids formed in water are solutions that have abundant free H^+ ions. The most common weak acid that occurs in surface waters is carbonic acid. Carbonic acid (H_2CO_3) is produced when atmospheric carbon dioxide dissolves in water; it exists only in solution. Hydrogen ions are quite small and can easily enter crystal structures, releasing other ions into the water:

$$H_2O + CO_2 \rightarrow H_2CO_3 \rightarrow H^+ + HCO_3$$

Water + Carbon dioxide $\rightarrow$ Carbonic acid $\rightarrow$ Hydrogen ion + Bicarbonate ion

> **DID YOU KNOW?**
>
> Plants such as mosses and lichens penetrate rock and loosen particles. Bare rocks are also subjected to chemical weathering due to chemical attack and dissolution of rock. Accomplished primarily through oxidation via exposure to oxygen gas in the atmosphere, acidic precipitation (after having dissolved small amounts of carbon dioxide gas from the atmosphere), and acidic secretions of microorganisms (bacteria, fungi, and lichens), chemical weathering speeds up in warm climates and slows down in cold ones.

Types of Chemical Weathering Reactions

As mentioned, chemical weathering breaks rocks down by adding or removing chemical elements and changing them into other materials. Again, as stated, chemical weathering consists of chemical reactions, most of which involve water. Types of chemical weathering include the following:

- Hydrolysis, which is a water–rock reaction that occurs when an ion in the mineral is replaced by H^+ or OH^-
- Leaching, which occurs when ions are removed by dissolution into water
- Oxidation, which is a result of oxygen being plentiful near the Earth's surface and reacting with minerals to change the oxidation state of an ion
- Dehydration, which occurs when water or a hydroxide ion is removed from a mineral
- Complete dissolution

Bare Rock Succession

Physical and chemical weathering does not always (if ever) occur independently of each other. Instead, both types of weathering normally work in combination. A classic example of the effect—the power of their simultaneous actions—can be seen in an ecological process known as *bare rock succession*, described earlier.

FINAL STAGES OF SOIL FORMATION

The final stages of soil formation consist of the processes of morphogenesis, or the production of a distinctive soil profile with its constituent layers or horizons. The soil profile (the vertical section of the soil from the surface through all its horizons, including C horizons) gives the environmental scientist critical information. When properly interpreted, soil horizons can provide warnings regarding potential problems in using the land and can tell much about the environment and history of a region. Soil profiles allow us to describe, sample, and map soils.

Soil horizons are distinct layers, roughly parallel to the surface, which differ in color, texture, structure, and content of organic matter. The clarity with which horizons can be recognized depends on the relative balance of the migration, stratification, aggregation, and mixing processes that take place in the soil during morphogenesis. In podzol-type soils (formed mainly in cool, humid climates), striking horizonation is quite apparent; in vertisol-type soils (soils high in swelling clays), the horizons are less distinct. When horizons are studied, they are given a letter symbol to reflect the genesis of the horizon.

Certain processes work to create or destroy clear soil horizons. Processes that tend to create clear horizons by vertical redistribution of soil materials include the leaching of ions in soil solutions, movement of clay-sized particles, upward movement of water by capillary action, and surface deposition of dust and aerosols. Clear soil horizons are destroyed by mixing processes that occur because of organisms, cultivation practices, creep processes on slopes, frost heave, and swelling and shrinkage of clays—all part of the natural soil formation process.

SOIL CHARACTERIZATION

Classification schemes of natural objects seek to organize knowledge so that the properties and relationships of the objects may be most easily remembered and understood for some specific purpose. The ultimate purpose of soil classification is maximum satisfaction of human wants that depend on use of the soil. This requires grouping soils with similar properties so that lands can be efficiently managed for crop production. Furthermore, soils that are suitable or unsuitable for pipelines, roads, recreation, forestry, agriculture, wildlife, building sites, and so forth can be identified.

Foth (1978)

When people become ill, they may go to a doctor to seek a diagnosis of what is causing the illness and perhaps a prognosis regarding how their illness will progress. What do diagnosis and prognosis have to do with soil? Actually, quite a lot. The diagnostic techniques used by a physician to identify the causative factors leading to a particular illness are analogous to the soil practitioner using diagnostic techniques to identify a particular soil. Sound farfetched? It shouldn't, because it isn't. Soil scientists must be able to determine the types of soil they study or work with.

Determining the type of soil makes sense, but what does prognosis have to do with all this? Soil practitioners must be able to identify or classify a soil type, because this information allows them to correctly predict how a particular pollutant will react or respond when spilled in that type of soil. The fate of the pollutant is important in determining the possible damage inflicted on the environment—soil, groundwater, and air—because ultimately a spill could easily affect all three. Thus, the soil practitioner not only must use diagnostic tools in determining soil type but also must be familiar with the soil type to determine how a particular pollutant or contaminant will respond when spilled in that soil type.

Let's take a closer look at the genesis of soil classification. From the time humans first advanced from hunter–gatherer status to cultivators of crops, they noticed differences in productive soils and unproductive soils. The ancient Chinese, Egyptians, Romans, and Greeks all recognized and acknowledged the differences in soils as media for plant growth. These early soil classification practices were based primarily on texture, color, and wetness.

Soil classification as a scientific practice did not gain a foothold until the later 18th and early 19th centuries, when the science of geology was born. This is when such terms (with an obvious geological connotation) as *limestone soils* and *lake-laid soils*, as well as *clayey* and *sandy soils*, came into being. The Russian scientist V.V. Dokuchaev was the first to suggest, in the late 1800s, that soils were natural bodies, and he developed a generic classification of soils that was later expanded. The system was based on the theory that each soil has a definite form and structure (morphology) related to a particular combination of soil-forming factors. This system was used until 1960, when the USDA published its original *Soil Classification: A Comprehensive System*. This classification system places major emphasis on soil morphology and gives less emphasis to genesis or the soil forming factors as compared to previous systems. In 1975, this text was replaced by *Soil Taxonomy: A Basic System of Soil Classification for Making and Interpreting Soil Surveys*, now in its second edition (USDA, 1999). *Soil Taxonomy* classifies objects according to their natural relationships, and soils are classified based on measurable properties of soil profiles.

Note that no clear delineation or line of demarcation can be drawn between the properties of one soil and those of another. Instead, a gradation (sometimes quite subtle, like comparing one shade of white to another) occurs in soil properties as one moves from one soil to another. Brady and Weil (2007, p. 58) noted that, "The gradation in soil properties can be compared to the gradation in the wavelengths of light as you move from one color to another. The changing is gradual, and yet we identify a boundary that differentiates what we call 'green' from what we call 'blue.'"

To properly characterize the primary characteristics of a soil, a soil must be identified down to the smallest three-dimensional characteristic sample possible; however, to accurately perform a particular soil sample characterization, a sampling unit must be large enough so the nature of its horizons can be studied and the range of its properties identified. The *pedon* (rhymes with head-on) is this unit. The pedon is roughly polygonal in shape and designates the smallest characteristic unit that can still be called a soil. Because pedons occupy a very small space (from approximately 1 to 10 m^2), they cannot, obviously, be used as the basic unit for a workable field soil classification system. To solve this problem, a group of pedons, termed a *polypedon*, is of sufficient size to serve as a basic classification unit (or, as it is commonly referred to, a *soil individual*). In the United States, these groupings have been called a *soil series*.

There is a difference between *a* soil and *the* soil. This difference is important in the soil classification scheme. A soil is characterized by a sampling unit (pedon), which as a group (polypedons) forms a soil individual. The soil, on the other hand, is a collection of all of these natural ingredients and is distinguishable from other bodies such as water, air, solid rock, and other parts of the Earth's crust. By incorporating the difference between a soil and the soil, a classification system has been developed that is effective and widely used.

DIAGNOSTIC HORIZONS, TEMPERATURE, AND MOISTURE REGIMES

Soil taxonomy uses a strict definition of soil horizons called *diagnostic horizons*, which are used to define most of the orders. Two kinds of diagnostic horizons are recognized: surface and subsurface. The surface diagnostic horizons are *epipedons* (Greek *epi*, "over"; *pedon*, "soil"). The epipedons include the dark (organic rich) upper part of the soil and the upper eluvial horizons, and sometimes both. Soils beneath the epipedon horizons are *subsurface diagnostic horizons*. Each of these layers is used to characterize different soils in soil taxonomy.

In addition to using diagnostic horizons to strictly define soil horizons, soil moisture regime classes can also be used. A soil moisture regime refers to the presence of plant-available water or groundwater at a sufficiently high level. The control section of the soil (ranging from 10 to 30 cm for clay and from 30 to 90 cm for sandy soils) designates that section of the soil where water is present or absent during given periods in a year. The control section is divided into upper and lower sections. The upper portion is defined as the depth to which 2.5 m^3 of water will penetrate within 24 hours. The lower portion is the depth that 7.5 m^3 of water will penetrate.

TABLE 2.3
Soil Moisture Regimes

Moisture Regime	Percent of Global Area Occupied
Aridic	35.9
Xeric	3.5
Ustic	18.0
Udic	33.1
Perudic	1.0
Aquic	8.3

Source: Adapted from Eswaran, H., *Pedologie*, 43, 19–39, 1993.

Six soil moisture regimes have been identified (see Table 2.3):

Aridic—Characteristic of soils in arid regions
Xeric—Characteristic of having long periods of drought in the summer
Ustic—Soil moisture generally high enough to meet plant needs during growing season
Udic—Common soil in humid climatic regions
Perudic—An extremely wet moisture regime annually
Aquic—Soil saturated with water and free of gaseous oxygen

In soil taxonomy, several soil temperature regimes are also used to define classes of soils. These soil temperature regimes, shown in Table 2.4, are based on mean annual soil temperature, mean summer temperature, and the difference between

TABLE 2.4
Soil Temperature Regimes

Soil Temperature Regimes (Mean Annual Temperature)	Percent of Global Area Occupied
Pergelic (0°C)	10.9
Cryic (0–8°C)	13.5
Frigid (0–8°C)	1.2
Mesic (8–15°C)	12.5
Thermic (15–22°C)	11.4
Hyperthermic (>22°C)	18.5
Isofrigid (0–8°C)	0.1
Isomesic (8–15°C)	0.3
Isothermic (15–22°C)	2.4
Isohyperthermic (>22°C)	26.0
Water (NA)	1.2
Ice (NA)	1.4

Source: Adapted from Eswaran, H., *Pedologie*, 43, 19–39, 1993.

mean summer and winter temperatures. The diagnostic horizons and moisture/temperature regimes just discussed are the main criteria used to define the various categories in soil taxonomy.

SOIL TAXONOMY

The U.S. Soil Conservation Service's soil taxonomy (which is based on measurable properties of soil profiles) places soils in six categories (see Table 2.5):

Order—Soils not too dissimilar in their genesis. There are 12 soil orders in soil taxonomy. The names and major characteristics of each soil order are shown in Table 2.6.

TABLE 2.5
Subdivision of Soil Taxonomy Classification System (in Hierarchical Order)

Category	Number of Taxa
Order	12
Suborder	64
Great group	~300
Subgroup	~1200
Family	~7500
Series	~18,500 (in United States)

TABLE 2.6
Soil Orders

Order	Description
Alfisol	Mild forest soil with gray to brown surface horizon, medium to high base supply (amount of interchangeable cations that remain in soil), and a subsurface horizon of clay accumulation
Andisol	Formed on volcanic ash and cinders and lightly weathered
Aridsol	Dry soil with pedogenic (soil forming) horizon, low in organic matter
Entisol	Recent soil without pedogenic horizons
Gelisol	Soils of very cold climates; defined as containing permafrost
Histosol	Organic (peat or bog) soil
Inceptisol	Soil at the beginning of the weathering process with weakly differentiated horizons
Mollisol	Soft soil with a nearly black, organic-rich surface horizon and high base supply
Oxisol	Oxide-rich soil principally a mixture of kaolin, hydrated oxides, and quartz
Spodosol	Soil that has an accumulation of amorphous materials in the subsurface horizons
Ultisol	Soil with a horizon of silicate clay accumulation and low base supply
Vertisol	Soil with high-activity clays (cracking clay soil)

Source: Adapted from Soil Survey Staff, *Soil Classification: A Comprehensive System*, 7th approximation, U.S. Department of Agriculture, Washington, DC, 1960; Soil Survey Staff, *Keys to Soil Taxonomy*, 4th ed., Virginia Polytechnic Institute and State University, Blacksburg, VA, 1990.

Suborder—The 64 subdivisions of orders emphasize properties that suggest some common features of soil genesis.

Great group—Diagnostic horizons are the major bases for differentiating approximately 300 great groups.

Subgroup—Approximately 1200 subdivisions of the great groups.

Family—Approximately 7500 soils with subgroups having similar physical and chemical properties.

Series—A subdivision of the family, and the most specific unit of the classification system. More than 18,000 soil series are recognized in the United States.

Soil Orders

The 12 soil orders constitute the first category of classification (see Table 2.6).

Soil Suborders

Soil orders are further divided into 64 suborders, based primarily on the chemical and physical properties that reflect either the presence or absence of water logging or genetic differences caused by climate and vegetation. Aqualfs (from *aqua*, for "wet") are formed under wet conditions, and alfisols become saturated with water sometime during the year. The suborder names all have two syllables, with the first syllable indicating the order, such as *alf* for Alfisol and *oll* for Mollisol.

Soil Great Groups and Subgroups

Suborders are divided into great groups that are defined largely by the presence or absence of diagnostic horizons and the arrangements of those horizons. Great group names are coined by prefixing one or more additional formative elements to the appropriate suborder name. More than 300 great groups have been identified. Subgroups are subdivisions of great groups. Subgroup names indicate to what extent the central concept of the great group is expressed. Typic Fragiaqualf is a soil that is typical for the Fragiaqualf great group.

Soil Families and Series

The family category of classification is based on features that are important to plant growth such as texture, particle size, mineralogical class, and depth. Terms such as *clayey, sandy, loamy*, and others are used to identify textural classes. Terms used to describe mineralogical classes include *mixed, oxidic, carbonatic*, and others. For temperature classes, terms such as *hypothermic, frigid, cryic*, and others are used. The soil series (subdivided from soil family) gets down to the individual soil, and the name is that of a natural feature or place near where the soil was first recognized. Familiar series names include Amarillo (Texas), Carlsbad (New Mexico), and Fresno (California). In the United States, there are more than 18,000 soil series.

THE BOTTOM LINE

Land subsidence is all about soil. More specifically, land subsidence is related to what is in or not in the soil. Dry soil without any moisture (desert areas, in particular) does not experience subsidence because of a lack of water; rather, the pore spaces within the soil normally occupied by water and air collapse, leading to land subsidence. This is important because much of the uninformed literature and many uneducated opinions have formed around the idea that it is the soil itself that compacts in land subsidence events. No. It is the vacant pore spaces that collapse or compact, not necessarily the soil mineral components. To understand land subsidence, we must understand not only soil but also soil mechanics. The topic of soil mechanics is discussed in the next chapter.

REFERENCES AND RECOMMENDED READING

ASTM. (1969). *Manual on Water.* Philadelphia, PA: American Society for Testing and Materials.

Beazley, J. (1992). *The Way Nature Works.* New York: Macmillan.

Brady, N.C. and Weil, R.R. (2007). *The Nature and Properties of Soils,* 14th ed. Upper Saddle River, NJ: Prentice Hall.

Carson, R. (1962). *Silent Spring.* Boston, MA: Houghton Mifflin.

Ciardi, J. (1997). In the stoneworks. In: *The Collected Poems of John Ciardi* (Cifelli, E.M., Ed.). Fayetteville: University of Arkansas Press.

Davis, G.H. and Pollock, G.L. (2003). Geology of Bryce Canyon National Park, Utah. In: *Geology of Utah's Parks and Monuments,* 2nd ed. (Anderson, P.B., Chidsey, T.C., and Sprinkel, D.A., Eds.), pp. 37–60. Salt Lake City: Utah Geological Association.

Davis, M.L. and Cornwell, D.A. (1991). *Introduction to Environmental Engineering.* New York: McGraw-Hill.

Eswaran, H. (1993). Assessment of global resources: current status and future needs. *Pedologie,* 43: 19–39.

Foth, H.D. (1978). *Fundamentals of Soil Science,* 6th ed. New York: John Wiley & Sons.

Franck, I. and Brownstone, D. (1992). *The Green Encyclopedia.* New York: Prentice Hall.

GLOBE Program. (2017). *Pedosphere (Soil),* https://www.globe.gov/do-globe/globe-teachers -guide/soil-pedosphere.

Illich, I., Groeneveld, S., and Hoinacki, L. (1991). *Declaration on Soil.* Kassel, Germany: University of Kassel.

Kemmer, F.N. (1979). *Water: The Universal Solvent.* Oak Brook, IL: Nalco.

Konigsburg, E.M. (1996). *The View from Saturday.* New York: Scholastic.

MacKay, D., Shiu, W.Y., and Ma, K.-C. (1997). *Illustrated Handbook of Physical–Chemical Properties and Environmental Fate for Organic Chemicals.* Boca Raton, FL: CRC Press.

Morris, D. (1991). As if materials mattered. *The Amicus Journal,* 13(4): 17–21.

NPS. (2008). *The Hoodoo.* Washington, DC: National Park Service.

Robotham, M. and Hart, J. (2003). *Soil Sampling for Home Gardens and Small Acreages,* EC 628. Corvallis: Oregon State University Extension Service (http://extension.oregon-state.edu/catalog/html/ec/ec628/).

Soil Survey Staff. (1990). *Keys to Soil Taxonomy,* 4th ed. Blacksburg, VA: Virginia Polytechnic Institute.

Spellman, F.R. (1998). *The Science of Environmental Pollution.* Boca Raton, FL: CRC Press.

Spellman, F.R. (2008). *The Science of Water,* 2nd ed. Boca Raton, FL: CRC Press.

Spellman, F.R. (2009). *The Science of Environmental Pollution,* 2nd ed. Boca Raton, FL: CRC Press.

Spellman, F.R. (2015). *The Science of Water,* 3rd ed. Boca Raton, FL: CRC Press.

Spellman, F.R. (2017). *The Science of Environmental Pollution,* 3rd ed. Boca Raton, FL: CRC Press.

Spellman, F.R. and Whiting, N.E. (2006). *Environmental Science and Technology,* 2nd ed. Rockville, MD: Government Institutes.

Sposito, G. (2008). *The Chemistry of Soils*. New York: Oxford University Press.

SSSA. (2008). *Soil Science Glossary*. Madison, WI: Soil Science Society of America.

Tomera, A.N. (1989). *Understanding Basic Ecological Concepts*. Portland, ME: J. Weston Walch, Publisher.

USDA. (1975). *Soil Taxonomy: A Basic System of Soil Classification for Making and Interpreting Soil Surveys*. Washington, DC: USDA Natural Resources Conservation Service.

USDA. (1999). *Soil Taxonomy: A Basic System of Soil Classification for Making and Interpreting Soil Surveys,* 2nd ed. Washington, DC: USDA Natural Resources Conservation Service.

USDA. (2009). *Urban Soil Primer*. Washington, DC: U.S. Department of Agriculture.

Winegardner, D. (1996). *An Introduction to Soils for Environmental Professionals*. Boca Raton, FL: CRC Press.

3 The Science of Soil Mechanics

Soil mechanics is the application of laws of mechanics and hydraulics to engineering problems dealing with sediments and other unconsolidated accumulations of solid particles produced by the mechanical and chemical disintegration of rocks regardless of whether or not they contain an admixture of organic constituent.

—Karl Terzaghi, Austrian civil engineer and geologist (1883–1963)

Why does the Leaning Tower of Pisa lean? The tower leans because it was built on a nonuniform consolidation of a clay layer beneath the structure. This process is ongoing (by about 1/25 of inch per year) and may eventually lead to failure of the tower. The factors that caused the Leaning Tower of Pisa to lean (and are relevant to using soil as a foundational and building material) are what this chapter is all about. The mechanics and physics of soil are important factors in making the determination as to whether a particular building site is viable for building. Simply put, these two factors can help to answer the question of whether or not the soils present will support buildings. This concerns us because wherever humans build the opportunity for land subsidence follows.

SOIL MECHANICS

The mechanics of soil are physical factors important to engineers because their focus is on the suitability of soil as a construction material. Simply put, the engineer must determine the response of a particular volume of soil to internal and external mechanical forces. Obviously, this is important in determining the suitability of the soil to withstand the load applied by structures of various types. By studying soil survey maps and reports and checking with soil scientists and other engineers familiar with the region and the soil types in that region, an engineer can determine the suitability of a particular soil for whatever purpose. Conducting field sampling to ensure that the soil product possesses the soil characteristics for its intended purpose is also essential. The soil characteristics important for engineering purposes include soil texture, kinds of clay present, depth to bedrock, soil density, erodibility, corrosivity, surface geology, plasticity, content of organic matter, salinity, and depth to seasonal water table. Engineers will also want to know the space and volume (weight–volume) relationship of the soil, stress–strain, slope stability, and compaction. Because these concepts are also of paramount importance to determining the fate of materials that are carried through soil, we present these concepts in this section.

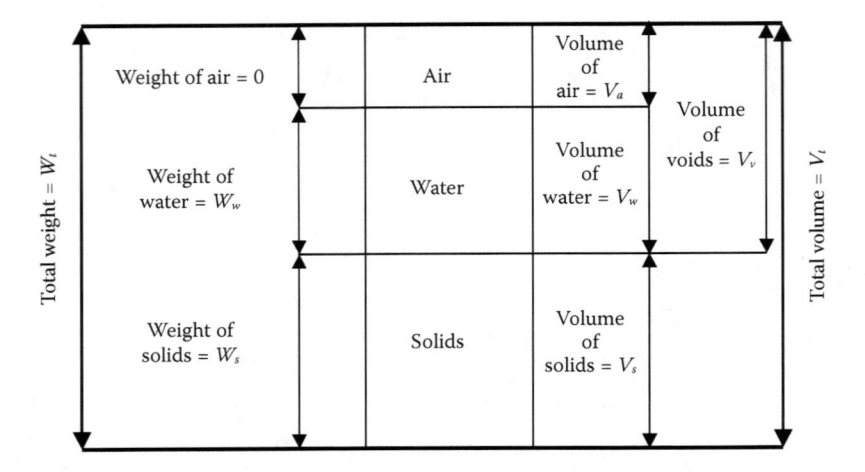

FIGURE 3.1   Weight–volume relationship of soil.

Weight–Volume or Space and Volume Relationships

All natural soil consists of at least three primary components (or phases): solids, water, and air (within void spaces between the solid particles). Examining the physical relationships (for soils in particular) between these phases is essential (see Figure 3.1). For convenience and clarity, in Figure 3.1, the mass of soil is represented as a block diagram. Each phase shown in the diagram is a separate block, and each major component has been reduced to a concentrated commodity within a unit volume. Note that the proportions of the components shown in Figure 3.1 will vary (sometimes widely) between and within various soil types. Remember that all water that is not chemically attached acts as a void filler. The relationship between free water and void spaces depends on available water (moisture). *Note:* This is an important point with regard to land subsidence. Remember, it is not the soil that compacts because of itself but instead it is the lack of or withdrawal of pore water that enables the soil to collapse or compact.

The volume of the soil mass is the sum of the volumes of the three components, or

$$V_T = V_a + V_w + V_s \tag{3.1}$$

The volume of the voids is the sum of V_a and V_w; however, the weight of the voids is only W_w, the weight of the water. Because weighing air in soil voids must be done within the Earth's atmosphere, the weight of air in the soil is factored in at zero. The total weight is expressed as the sum of the weights of the soil solids and water:

$$W_T = W_s + W_w \tag{3.2}$$

The relationship between weight and volume can be expressed as

$$W_m = V_m G_m Y_w \tag{3.3}$$

where

W_m = Weight of the material (solid, liquid, or gas).
V_m = Volume of the material.
G_m = Specific gravity of the material (dimensionless).
Y_w = Unit weight of water.

We can solve a few useful problems with the relationships described above. More importantly, this information about a particular location's soil allows engineers to mechanically adjust the proportions of the three major components by reorienting the mineral grains through compaction or tilling. In remediation, a decision to blend soil types to alter the proportions (such as increasing or decreasing the percentage of void space) may be part of a site cleanup process.

Relationships between volumes of soil and voids are described by the *void ratio* (*e*) and *porosity* (η). We must first determine the void ratio, which is the ratio of the void volume (V_v) to the volume of solids (V_s):

$$e = V_v/V_s \qquad (3.4)$$

The first step is to determine the ratio of the volume of void spaces to the total volume. We do this by determining the porosity (η) of the soil, which is the ratio of void volume to total volume. Porosity is usually expressed as a percentage:

$$\eta = V_v/V_t \times 100\% \qquad (3.5)$$

The terms *moisture content* (*w*) and *degree of saturation* (*S*) relate the water content of the soil and the volume of the water in the void space to the total void volume:

$$w = W_w/W_s \times 100\% \qquad (3.6)$$

and

$$S = V_w/V_v \times 100\% \qquad (3.7)$$

SOIL PARTICLE CHARACTERISTICS

The size and shape of particles in the soil, along with density and other characteristics, give the engineer information on shear strength, compressibility, and other aspects of soil behavior. These index properties are used to create engineering classifications of soil. Simple classification tests are used to measure index properties (see Table 3.1) in the lab or the field.

From the engineering point of view, the separation of *cohesive* (fine-grained) from *incohesive* (coarse-grained) soils is an important distinction. Let's take a closer look at these two terms. The level of cohesion of a soil describes the tendency of the soil particles to stick together. Cohesive soils contain silt and clay, which, along with water content, make these soils hold together through the attractive forces between individual clay and water particles. Because the clay particles so strongly influence cohesion, the index properties of cohesive soils are more complicated than

---

TABLE 3.1
Index Properties of Soils

Soil Type	Index Property
Cohesive (fine-grained)	Water content
	Sensitivity
	Type and amount of clay
	Consistency
	Atterburg limits
Incohesive (coarse-grained)	Relative density
	In-place density
	Particle-size distribution
	Clay content
	Shape of particles

Source: Adapted from Kehew, A.E., *Geology for Engineers and Environmental Scientists*, 2nd ed., Prentice Hall, Englewood Cliffs, NJ, 1995.

---

the index properties of cohesionless soils. The *consistency* of the soil—the arrangement of clay particles—describes the resistance of soil at various moisture contents to mechanical stresses or manipulations and is the most important characteristic of cohesive soils.

Sensitivity (the ratio of unconfined compressive strength in the undisturbed state to strength in the remolded state; see Equation 3.8) is another important index property of cohesive soils. Soils with high sensitivity are highly unstable.

$$S_t = \text{(Strength in undisturbed condition)/(Strength in remolded condition)} \quad (3.8)$$

As we described earlier, soil water content also influences soil behavior. Water content values of soil—the *Atterburg limits*, a collective designation of the so-called limits of consistency of fine-grained soils determined with simple laboratory tests—are usually presented as the liquid limit (LL), plastic limit (PL), and the plasticity index (PI). Plasticity is exhibited over a range of moisture contents referred to as *plasticity limits*. The lower plastic limit is the water level at which soil begins to be malleable in the semisolid state but the molded pieces still crumble easily with a little applied pressure. When the volume of the soil becomes nearly constant with further decreases in water content, the soil reaches the shrinkage state. The upper plasticity limit (or *liquid limit*) is reached when the water content in a soil–water mixture changes from a liquid to a semifluid or plastic state and tends to flow when jolted. Obviously, a soil that tends to flow when wet presents special problems for both engineering purposes and remediation of contamination. The range of water content over which the soil is plastic, called the *plasticity index*, provides the difference between the liquid limit and the plastic limit. Soils with the highest plasticity

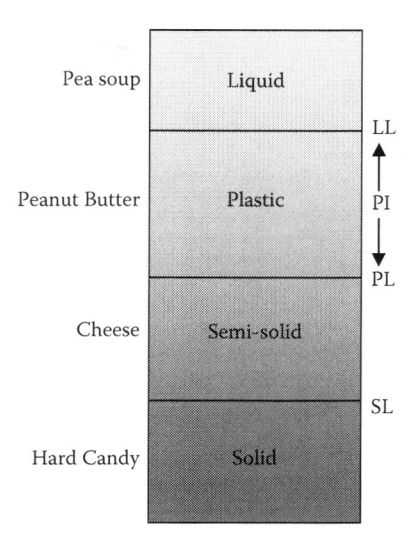

FIGURE 3.2 Atterburg limits and key definitions.

indices are unstable in bearing loads—a key point to remember. Soil consistency limits are important parameters (Atterburg limits) related to land subsidence due to groundwater withdrawal (see Figure 3.2).

The best known and probably the most useful system of the several systems designed for classifying the stability of soil materials, is called the *Unified System of Classification*. This system gives each soil type (14 classes) a two-letter designation based on particle-size distribution, liquid limit, and plasticity index.

Cohensionless coarse-grained soils are classified by index properties including the size and distribution of particles in the soil. Other index properties (including particle shape, in-place density, and relative density) are important in describing cohesionless soils because they relate how closely particles can be packed together.

Soil Stress

Just as water pressure increases as you go deeper into water, the pressure within soil increases as the depth increases. A soil with a unit weight of 75 pounds per cubic feet (lb/ft^3) exerts a pressure of 75 psi at a depth of 1 foot and 225 psi at 3 feet. Of course, as the pressure on a soil unit increases, soil particles reorient themselves to support the cumulative load. This is critically important information to remember, because a soil sample retrieved from beneath the load may not be truly representative once delivered to the surface. Representative samples are essential. The response of a soil to pressure (*stress*), as when a load is applied to a solid object, is transmitted throughout the material. The load puts the material under pressure, which equals the amount of load divided by the surface area of the external face of the object over which it is applied. The response to this pressure or stress is called *displacement* or *strain*. Stress (like pressure) at any point within the object can be defined as force per unit area.

SOIL COMPRESSIBILITY

Compressibility, the tendency of soil to decrease in volume under load, is most significant in clay soils because of inherently high porosity. The actual evaluation process for these properties relies on the *consolidation test*. This test subjects a soil sample to an increasing load. The change in thickness is measured after the application of each load increment.

SOIL COMPACTION

Compaction reduces the void ratio and increases the soil density, which affects how materials (including pollutants) travel through soil. Compaction is accomplished by working the soil to reorient the soil grains into a more compact state. Water content sufficient to lubricate particle movement is critical to obtaining efficient compaction.

SOIL FAILURE

Soil structure affects natural processes such as frost heave (which could damage septic systems, disturb improperly set footings, or shift soils under an improperly seated underground storage tank and its piping) as well as changes applied to soils during remediation efforts (e.g., when excavating to mitigate a hazardous materials spill). When soil cannot support a load, *soil failure* occurs, which can include events as diverse as foundation overload, collapse of the sides of an excavation, or slope failure on the sides of a dike or hill. Because of the safety factors involved, the structural stability of a soil is critically important.

Classifying the type of soil before excavating involves determining the soil type. Finding a combination of soil types at an excavation site is common. Soil types include the following:

- *Stable rock* is generally stable but may lose stability when excavated; it is natural solid mineral material that can be excavated with vertical sides and will remain intact while exposed.
- *Type A soil* is the most stable soil; it includes clay, silty clay, sandy clay, clay loam, and sometimes silty clay loam and sandy clay loam.
- *Type B soil* is moderately stable soil that includes silt, silt loam, sandy loam, and sometimes silty clay loam and sandy clay loam.
- *Type C soil* is the least stable soil; it includes granular soils such as gravel, sand, loamy sand, and submerged soil, as well as soil from which water is freely seeping, and unstable submerged rock.

Both visual and manual tests are used to classify soil for excavation. Visual soil testing concerns soil particle size and type. In a mixture of soils, if the soil clumps when dug, it could be clay or silt. The presence of cracks in walls and spalling (breaking up into chips or fragments) may indicate Type B or C soil. Standing water or water seeping through trench walls automatically classifies the soil as Type C.

Manual soil testing includes the sedimentation test, wet shaking test, thread test, and ribbon test. A sample taken from soil should be tested, onsite or offsite, as soon as possible to preserve its natural moisture. A *sedimentation test* determines how much silt and clay are in sandy soil. Saturated sandy soil is placed in a straight-sided jar with about 5 inches of water. After the sample is thoroughly mixed (by shaking it) and allowed to settle, the percentage of sand is visible. A sample containing 80% sand, for example, will be classified as Type C. The *wet shaking test* is another way to determine the amount of sand vs. clay and silt in a soil sample. A saturated sample is shaken by hand to gauge soil permeability based on the following facts: (1) shaken clay resists water movement through it, and (2) water flows freely through sand and less freely through silt. The *thread test* is used to determine cohesion (remember, cohesion relates to stability, or how well the grains hold together). A representative soil sample is rolled between the palms of the hands to about 1/8-inch diameter and several inches in length. The rolled piece is placed on a flat surface, then picked up. If the sample holds together for 2 inches, it is considered cohesive. The *ribbon test* is used as a backup for the thread test. It also determines cohesion. A representative soil sample is rolled out (using the palms of your hands) to 3/4 inch in diameter and several inches in length. The sample is then squeezed between the thumb and fore-finger into a flat unbroken ribbon 1/8 to 1/4 inch thick, which is allowed to fall freely over the fingers. If the ribbon does not break off before several inches are squeezed out, the soil is considered cohesive.

When soil has been properly classified, the necessary measures for safe excavation can be chosen, based on both soil classification and site restrictions. The two standard protective systems include sloping or benching and shoring or shielding.

SOIL WATER

As a dynamic, heterogeneous body, soil is not isotropic—it does not have the same properties in all directions. Because soil properties vary directionally, various physical processes are always active in soil, as Winegardner (1996, p. 63) made clear: "All of the factors acting on a particular soil, in an established environment, at a specified time, are working from some state of imbalance to achieve a balance." Soil practitioners must understand the factors involved in the physical processes that are active in soil. These include physical interactions related to soil water, soil grains, organic matter, soil gases, and soil temperature. Because the subject of this text is land subsidence due to groundwater withdrawal, our focus here is on soil water.

WATER AND SOIL

Not only is water a vital component of every living being, but it is also essential to plant growth and to the microorganisms that live in the soil, in addition to being important in the weathering process, which involves the breakdown of rocks and minerals to form soil and release plant nutrients. In this section, we focus on soil water and its importance in soil, but first we need to take a closer look at water—what it is and its physical properties. Water exists as a liquid between 0° and 100°C

(32° to 212°F), as a solid at or below 0°C (32°F), and as a gas at or above 100°C (212°F). One gallon of water weighs 8.33 pounds (3.778 kilograms) and is equal to 3.785 liters. One cubic foot of water equals 7.5 gallons (28.35 liters). One ton of water equals 240 gallons. One acre-foot of water equals 43,560 cubic feet (325,900 gallons). Earth's rate of rainfall equals 340 cubic miles per day (16 million tons per second). Finally, water is dynamic (constantly in motion). It evaporates from seas, lakes, and the soil; is transported through the atmosphere; falls to Earth; runs across the land; and filters downward into and through the soil to flow along rock strata.

Water: What Is It?[*]

Water is often assumed to be one of the simplest compounds known on Earth, but water is not simple. Nowhere in nature is absolutely simple (pure) water to be found. Here on Earth, with a geologic origin dating back over 3 to 5 billion years, water found in even its purest form is composed of many constituents. Along with H_2O molecules, hydrogen (H^+), hydroxyl (OH^-), sodium, potassium, and magnesium, other ions and elements are present. Water contains additional dissolved compounds, including various carbonates, sulfates, silicates, and chlorides. Rainwater (often assumed to be the equivalent of distilled water) is not immune to contamination, which it collects as it descends through the atmosphere. The movement of water across the face of land contributes to its contamination, as it takes up dissolved gases such as carbon dioxide and oxygen and a multitude of organic substances and minerals leached from the soil.

Water Physical Properties

In soil physics, the physical properties of water (which are also a function of the chemical structure of water) that concern us are density, viscosity, surface tension, and capillary action. Let's take a closer look at each of these physical properties. *Density* is a measure of the mass per unit volume. The number of water molecules occupying the space of a unit volume determines the magnitude of the density. As temperature (which measures internal energy) increases or decreases, the molecules vibrate more or less strongly and frequently (which changes the distance between them), expanding or diminishing the volume occupied by the molecules. As discussed previously, liquid water reaches its maximum density at 4°C and its minimum at 100°C. In soil science work, the density may be considered to be a unit weight (62.4 lb/ft³ or 1 g/cm³).

Viscosity is the measure of the internal flow resistance of a liquid or gas. Stated differently, viscosity is ease of flow of a liquid, or the capacity of a fluid to convert energy of motion (kinetic energy) into heat energy. Viscosity is the result of the cohesion between fluid particles and the interchange of molecules between layers of different viscosities. High-viscosity fluids flow slowly, while low viscosity fluids flow freely. Viscosity decreases as temperature rises for liquids.

---

[*] Much of the following information is adapted from Spellman, F.R., *The Science of Water*, 3rd ed., CRC Press, Boca Raton, FL, 2015.

Have you ever wondered why a needle can float on water? Or why some insects can stand on water? The reason is *surface tension*. Surface tension (or cohesion) is the property that causes the surface of a liquid to behave as if it were covered with a weak elastic skin. It is caused by the tendency of the exposed surface to contract to the smallest possible area because of unequal cohesive forces between molecules at the surface. What does surface tension have to do with soil? The surface tension property of water markedly influences the behavior of water in soils. Consider an example you may be familiar with, one that will help you understand surface tension and the other important physical properties of the water–soil interface. Water commonly rises in clays, fine silts, and other soils, and surface tension plays a major role. The rise of water through clays, silts, and other soils is its *capillarity* or *capillary action* (the property of the interaction of the water with a solid), and the two primary factors of capillary rise are surface tension (cohesion) and adhesion (the attraction of water for the solid walls of channels through which it moves).

Why does the water rise? Because the water molecules are attracted to the sides of the tube (or soil pores) and start moving up the tube in response to this attraction. The cohesive force between individual water molecules ensures that water not directly in contact with the side walls is also pulled up the tube (or soil pores). This action continues until the weight of water in the tube counterbalances the cohesive and adhesive forces. Keep in mind that for water in soil the rate of movement and the rise in height of soil water are less than one might expect on the basis of soil pore size, because soil pores are not straight like glass tubes nor are the openings uniform. Also, many soil pores are filled with air, which may prevent or slow down the movement of water by capillarity. A final word on capillarity—keep in mind that capillarity means movement in any direction, not just upward. Because the attractions between soil pores and water are as effective with horizontal pores as with vertical ones, water movement in any direction occurs.

WATER CYCLE (HYDROLOGIC CYCLE)

The importance of the water cycle cannot be overstated; thus, we discuss the cycle again here to emphasize this point. Water is never stationary; it is constantly in motion. Again, this phenomenon occurs because of the water or hydrological cycle. In simple terms, the water cycle can be explained as follows: The sun helps transfer water from lakes and oceans to the land. As the sun shines on the Earth, the surface water is heated and evaporates, forming an invisible gas that mixes with the air. This gas is water vapor; it is pure water without any minerals or bacteria in it. Water vapor rises in the air, then cools and condenses into tiny drops of water that form clouds. Further cooling may form drops large enough to fall as rain. In this way, water is brought from the oceans to the land, where it reappears in springs and wells, soaks into the ground, or runs off again through streams and rivers back to the ocean. Of course, the actual movement of water on Earth is much more complex. Three different methods of transport are involved in this water movement: evaporation, precipitation, and runoff.

Evaporation of water is a major factor in hydrologic systems. Evaporation is a function of temperature, wind velocity, and relative humidity. Evaporation (or vaporization) is, as the name implies, the formation of vapor. Dissolved constituents (such as salts) remain behind when water evaporates. Evaporation of the surface water of oceans provides most water vapor, although water can also vaporize through plants, especially from leaf surfaces in a process known as evapotranspiration. Ice can also vaporize without melting first; however, this sublimation process is slower than the vaporization of liquid water.

Precipitation includes all forms in which atmospheric moisture descends to Earth—rain, snow, sleet, and hail. Before precipitation can occur, the water that enters the atmosphere by vaporization must first condense into a liquid (clouds and rain) or solid (snow, sleet, and hail) before it can fall. This vaporization process absorbs energy, which is released in the form of heat when the water vapor condenses. You can best understand this phenomenon when you compare it to what occurs when water evaporates from your skin; this absorbs heat, making you feel cold. Note that the annual evaporation from ocean and land areas is the same as the annual precipitation.

Runoff is the flow back to the oceans of the precipitation that falls on land. This journey to the oceans is not always unobstructed—flow back may be intercepted by vegetation (from which it later evaporates), a portion of the precipitation is held in depressions, and some infiltrates into the ground. A part of the infiltrated water is taken up by plant life and returned to the atmosphere through evapotranspiration, while the remainder either moves through the ground or is held by capillary action. Eventually, water drips, seeps, and flows its way back into ponds, lakes, streams, rivers, and the oceans.

WATER MOVEMENT IN SOIL

Have you ever wondered what happens to water after it enters the soil? Maybe not, but if you are to work in the soil science field the answer to this question is one that you definitely must know and must also have a full and complete understanding of. Water that enters the soil has (in simple terms) four ways it may go:

1. It may move on through the soil and percolate out of the root zone, eventually reaching the water table.
2. It may be drawn back to the surface and evaporate.
3. It may be taken up (transpired, or used) by plants.
4. It may be held in storage in the water profile.

What determines how much water ends up in each of these categories? It depends. Climate and the properties of the particular soil and the requirements of the plants growing in that soil all have an impact on how much water ends up in each of the categories. But, don't forget the influence of anthropogenic actions (what we like to call the heavy hand of man). People alter the movement of water, not only by irrigation and stream diversion practices and by building but also in their choice of which crops to plant and the types of tillage practices employed.

REFERENCES AND RECOMMENDED READING

ASTM. (1969). *Manual on Water.* Philadelphia, PA: American Society for Testing and Materials.

Brady, N.C. and Weil, R.R. (2007). *The Nature and Properties of Soils*, 14th ed. Upper Saddle River, NJ: Prentice Hall.

Carson, R. (1962). *Silent Spring.* Boston, MA: Houghton Mifflin.

Ciardi, J. (1997). In the stoneworks. In: *The Collected Poems of John Ciardi* (Cifelli, E.M., Ed.). Fayetteville: University of Arkansas Press.

Davis, G.H. and Pollock, G.L. (2003). Geology of Bryce Canyon National Park, Utah. In: *Geology of Utah's Parks and Monuments*, 2nd ed. (Anderson, P.B., Chidsey, T.C., and Sprinkel, D.A., Eds.), pp. 37–60. Salt Lake City: Utah Geological Association.

Davis, M.L. and Cornwell, D.A. (1991). *Introduction to Environmental Engineering.* New York: McGraw-Hill.

Eswaran, H. (1993). Assessment of global resources: current status and future needs. *Pedologie*, 43: 19–39.

Foth, H.D. (1978). *Fundamentals of Soil Science*, 6th ed. New York: John Wiley & Sons.

Franck, I. and Brownstone, D. (1992). *The Green Encyclopedia.* New York: Prentice Hall.

GLOBE Program. (2017). *Pedosphere (Soil)*, https://www.globe.gov/do-globe/globe-teachers -guide/soil-pedosphere.

Illich, I., Groeneveld, S., and Hoinacki, L. (1991). *Declaration on Soil.* Kassel, Germany: University of Kassel.

Kemmer, F.N. (1979). *Water: The Universal Solvent.* Oak Brook, IL: Nalco.

MacKay, D., Shiu, W.Y., and Ma, K.-C. (1997). *Illustrated Handbook of Physical–Chemical Properties and Environmental Fate for Organic Chemicals.* Boca Raton, FL: CRC Press.

Morris, D. (1991). As if materials mattered. *The Amicus Journal*, 13(4): 17–21.

Soil Survey Staff. (1960). *Soil Classification: A Comprehensive System*, 7th approximation. Washington, DC: U.S. Department of Agriculture.

Spellman, F.R. (2008). *The Science of Water*, 2nd ed. Boca Raton, FL: CRC Press.

Spellman, F.R. (2015). *The Science of Water*, 3rd ed. Boca Raton, FL: CRC Press.

Spellman, F.R. (2017). *The Science of Environmental Pollution*, 3rd ed. Boca Raton, FL: CRC Press.

Spellman, F.R. and Whiting, N.E. (2006). *Environmental Science and Technology*, 2nd ed. Rockville, MD: Government Institutes.

USDA. (1975). *Soil Taxonomy: A Basic System of Soil Classification for Making and Interpreting Soil Surveys.* Washington, DC: USDA Natural Resources Conservation Service.

USDA. (1999). *Soil Taxonomy: A Basic System of Soil Classification for Making and Interpreting Soil Surveys,* 2nd ed. Washington, DC: USDA Natural Resources Conservation Service.

USDA. (2009). *Urban Soil Primer.* Washington, DC: U.S. Department of Agriculture.

Winegardner, D. (1996). *An Introduction to Soils for Environmental Professionals.* Boca Raton, FL: CRC Press.

4 Basic Water Hydraulics

Note: The practice and study of water hydraulics are not new. Even in medieval times, water hydraulics was not new, as medieval Europe inherited a highly developed range of Roman hydraulic components (Magnusson, 2001). The basic conveyance technology, based on low-pressure systems of pipe and channels, was already established. When studying "modern" water hydraulics, it is important to remember that, as Magnusson put it, the science of water hydraulics is the direct result of two immediate and enduring problems: "the acquisition of freshwater and access to a continuous strip of land with a suitable gradient between the source and the destination."

WHAT IS WATER HYDRAULICS?

The word "hydraulics" is derived from the Greek words *hydro* ("water") and *aulis* ("pipe"). Originally, the term referred only to the study of water at rest and in motion (flowing through pipes or channels). Today, it is taken to mean the flow of any liquid in a system. What is a liquid? In terms of hydraulics, a liquid can be either oil or water. In fluid power systems used in modern industrial equipment, the hydraulic liquid of choice is oil. Some common examples of hydraulic fluid power systems include automobile braking and power steering systems, hydraulic elevators, and hydraulic jacks or lifts. Probably the most familiar hydraulic fluid power systems in water/wastewater operations are those in dump trucks, front-end loaders, graders, and earth-moving and excavation equipment. In this text, we are concerned with liquid water. Many find the study of water hydraulics difficult and puzzling (especially the licensure examination questions), but we know it is not mysterious or incomprehensible. It is the function or output of practical applications of the basic principles of water physics.

BASIC CONCEPTS

Air pressure (at sea level) = 14.7 pounds per square inch (psi)

This relationship is important because our study of hydraulics begins with air. A blanket of air many miles thick surrounds the Earth. The weight of this blanket on a given square inch of the Earth's surface will vary according to the thickness of the atmospheric blanket above that point. As shown above, at sea level the pressure exerted is 14.7 pounds per square inch (psi). On a mountain top, air pressure decreases because the blanket is not as thick.

$$1 \text{ ft}^3 \text{ H}_2\text{O} = 62.4 \text{ lb}$$

The relationship shown above is also important; note that both cubic feet and pounds are used to describe a volume of water. A defined relationship exists between these two methods of measurement. The specific weight of water is defined relative to a cubic foot. One cubic foot of water weighs 62.4 lb (see Figure 4.1). This relationship is true only at a temperature of 4°C and at a pressure of 1 atmosphere, conditions that are referred to as *standard temperature and pressure* (STP).

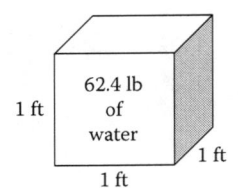

FIGURE 4.1 One cubic foot of water weighs 62.4 lb.

Note that 1 atmosphere = 14.7 lb/in.^{2} at sea level and 1 ft^3 of water contains 7.48 gal. The weight varies so little that, for practical purposes, this weight is used for temperatures ranging from 0 to 100°C. One cubic inch of water weighs 0.0362 lb. Water 1 ft deep will exert a pressure of 0.43 lb/in.^{2} on the bottom area (12 in. × 0.0362 lb/in.^{3}). A column of water 2 ft high exerts 0.86 psi (2 ft × 0.43 psi/ft); one 10 ft high exerts 4.3 psi (10 ft × 0.43 psi/ft); and one 55 ft high exerts 23.65 psi (55 ft × 0.43 psi/ft). A column of water 2.31 feet high will exert 1.0 psi (2.31 ft × 0.43 psi/ft). To produce a pressure of 50 psi requires a 115.5-ft water column (50 psi × 2.31 ft/psi).

Remember the important points being made here:

1. 1 ft^3 H$_2$O = 62.4 lb (see Figure 4.1).
2. A column of water 2.31 ft high will exert 1.0 psi.

Another relationship is also important:

$$1 \text{ gal H}_2\text{O} = 8.34 \text{ lb}$$

At standard temperature and pressure, 1 ft^3 of water contains 7.48 gal. With these two relationships, we can determine the weight of 1 gal of water:

$$\text{Weight of 1 gal of water} = 62.4 \text{ lb} \div 7.48 \text{ gal} = 8.34 \text{ lb/gal}$$

Thus,

$$1 \text{ gal H}_2\text{O} = 8.34 \text{ lb}$$

Note: Further, this information allows cubic feet to be converted to gallons by simply multiplying the number of cubic feet by 7.48 gal/ft^3.

■ **EXAMPLE 4.1**

Problem: Find the number of gallons in a reservoir that has a volume of 855.5 ft^3.

Solution:

$$855.5 \text{ ft}^3 \times 7.48 \text{ gal/ft}^3 = 6399 \text{ gal (rounded)}$$

Note: The term *head* is used to designate water pressure in terms of the height of a column of water in feet; for example, a 10-ft column of water exerts 4.3 psi. This can be referred to as 4.3-psi pressure or 10 ft of head.

STEVIN'S LAW

Stevin's law deals with water at rest. Specifically, it states: "The pressure at any point in a fluid at rest depends on the distance measured vertically to the free surface and the density of the fluid." Stated as a formula, this becomes

$$p = w \times h \qquad (4.1)$$

where
p = Pressure in pounds per square foot (lb/ft^2 or psf).
w = Density in pounds per cubic foot (lb/ft^3).
h = Vertical distance in feet.

■ EXAMPLE 4.2

Problem: What is the pressure at a point 18 ft below the surface of a reservoir?

Solution: To calculate this, we must know that the density of the water (w) is 62.4 lb/ft^3.

$$p = w \times h = 62.4 \text{ lb/ft}^3 \times 18 \text{ ft} = 1123 \text{ lb/ft}^2 \text{ (psf)}$$

Waterworks operators generally measure pressure in pounds per square inch rather than pounds per square foot; to convert, divide by 144 in.^{2}/ft^2 (12 in. × 12 in. = 144 in.^{2}):

$$p = 1123 \text{ psf} \div 144 \text{ in.}^2/\text{ft}^2 = 7.8 \text{ lb/in.}^2 \text{ (psi)}$$

DENSITY AND SPECIFIC GRAVITY

Table 4.1 shows the relationships among temperature, specific weight, and density of water. When we say that iron is heavier than aluminum, we say that iron has a greater density than aluminum. In practice, what we are really saying is that a given volume of iron is heavier than the same volume of aluminum.

TABLE 4.1
Water Properties (Temperature, Specific Weight, and Density)

Temperature (°F)	Specific Weight (lb/ft³)	Density (slugs/ft³)	Temperature (°F)	Specific Weight (lb/ft³)	Density (slugs/ft³)
32	62.4	1.94	130	61.5	1.91
40	62.4	1.94	140	61.4	1.91
50	62.4	1.94	150	61.2	1.90
60	62.4	1.94	160	61.0	1.90
70	62.3	1.94	170	60.8	1.89
80	62.2	1.93	180	60.6	1.88
90	62.1	1.93	190	60.4	1.88
100	62.0	1.93	200	60.1	1.87
110	61.9	1.92	210	59.8	1.86
120	61.7	1.92			

Note: What is *density*? Density is the *mass per unit volume* of a substance.

Suppose you have a tub of lard and a large box of cold cereal, each having a mass of 600 g. The density of the cereal would be much less than the density of the lard because the cereal occupies a much larger volume than the lard occupies.

The density of an object can be calculated by using the following formula:

$$\text{Density} = \text{Mass} \div \text{Volume} \qquad (4.2)$$

In water treatment operations, perhaps the most common measures of density are pounds per cubic foot (lb/ft^3) and pounds per gallon (lb/gal):

- 1 ft^3 of water weighs 62.4 lb; density = 62.4 lb/ft^3.
- 1 gal of water weighs 8.34 lb; density = 8.34 lb/gal.

The density of a dry material, such as cereal, lime, soda, or sand, is usually expressed in pounds per cubic foot. The density of a liquid, such as liquid alum, liquid chlorine, or water, can be expressed either as pounds per cubic foot or as pounds per gallon. The density of a gas, such as chlorine gas, methane, carbon dioxide, or air, is usually expressed in pounds per cubic foot.

As shown in Table 4.1, the density of a substance like water changes slightly as the temperature of the substance changes. This occurs because substances usually increase in volume (size) by expanding as they become warmer. Because of this expansion with warming, the same weight is spread over a larger volume, so the density is lower when a substance is warm than when it is cold.

Note: What is *specific gravity*? Specific gravity is the weight (or density) of a substance compared to the weight (or density) of an equal volume of water. The specific gravity of water is 1.

This relationship is easily seen when a cubic foot of water, which weighs 62.4 lb, is compared to a cubic foot of aluminum, which weighs 178 lb. Aluminum is 2.8 times heavier than water.

It is not that difficult to find the specific gravity of a piece of metal. All you have to do is weigh the metal in air, then weigh it under water. The loss of weight is the weight of an equal volume of water. To find the specific gravity, divide the weight of the metal by its loss of weight in water.

$$\text{Specific gravity} = \text{Weight of substance} \div \text{Weight of equal volume of water} \qquad (4.3)$$

■ EXAMPLE 4.3

Problem: Suppose a piece of metal weighs 150 lb in air and 85 lb under water. What is the specific gravity?

Solution:

$$150 \text{ lb} - 85 \text{ lb} = 65 \text{ lb loss of weight in water}$$

$$\text{Specific gravity} = 150 \text{ lb} \div 65 \text{ lb} = 2.3$$

Note: When calculating specific gravity, it is *essential* that the densities be expressed in the same units.

As stated earlier, the specific gravity of water is 1, which is the standard, the reference against which all other liquid or solid substances are compared. Specifically, any object that has a specific gravity greater than 1 will sink in water (e.g., rocks, steel, iron, grit, floc, sludge). Substances with specific gravities of less than 1 will float (e.g., wood, scum, gasoline). Considering the total weight and volume of a ship, its specific gravity is less than 1; therefore, it can float.

The most common use of specific gravity in water treatment operations is in gallon-to-pound conversions. In many cases, the liquids being handled have a specific gravity of 1 or very nearly 1 (between 0.98 and 1.02), so 1 may be used in the calculations without introducing significant error. For calculations involving a liquid with a specific gravity of less than 0.98 or greater than 1.02, however, the conversions from gallons to pounds must consider specific gravity. The technique is illustrated in the following example.

■ EXAMPLE 4.4

Problem: A basin contains 1455 gal of a liquid. If the specific gravity of the liquid is 0.94, how many pounds of liquid are in the basin?

Solution: Normally, for a conversion from gallons to pounds, we would use the factor 8.34 lb/gal (the density of water) if the specific gravity of the substance is between 0.98 and 1.02. In this instance, however, the substance has a specific gravity outside this range, so the 8.34 factor must be adjusted by multiplying 8.34 lb/gal by the specific gravity to obtain the adjusted factor:

$$8.34 \text{ lb/gal} \times 0.94 = 7.84 \text{ lb/gal (rounded)}$$

Then convert 1455 gal to pounds using the correction factor:

$$1455 \text{ gal} \times 7.84 \text{ lb/gal} = 11,407 \text{ lb (rounded)}$$

FORCE AND PRESSURE

Water exerts force and pressure against the walls of its container, whether it is stored in a tank or flowing in a pipeline. Force and pressure are different, although they are closely related. *Force* is the push or pull influence that causes motion. In the English system, force and weight are often used in the same way. The weight of 1 ft^3 of water is 62.4 lb. The force exerted on the bottom of a 1-ft cube is 62.4 lb (see Figure 4.1). If we stack two 1-ft cubes on top of one another, the force on the bottom will be 124.8 lb. *Pressure* is the force per unit of area. In equation form, this can be expressed as

$$P = F \div A \tag{4.4}$$

where

P = Pressure.

F = Force.

A = Area over which the force is distributed.

Earlier we pointed out that pounds per square inch (lb/in.^{2} or psi) or pounds per square foot (lb/ft^2 or psf) are common expressions of pressure. The pressure on the bottom of the cube is 62.4 lb/ft^2 (see Figure 4.1). It is normal to express pressure in pounds per square inch. This is easily accomplished by determining the weight of 1 in.^{2} of a 1-ft cube. If we have a cube that is 12 in. on each side, the number of square inches on the bottom surface of the cube is $12 \times 12 = 144$. Dividing the weight by the number of square inches determines the weight on each square inch:

$$62.4 \text{ lb/ft} \div 144 \text{ in.}^2 = 0.433 \text{ psi/ft}$$

This is the weight of a column of water 1 in. square and 1 ft tall. If the column of water were 2 ft tall, the pressure would be 2 ft $\times$ 0.433 psi/ft = 0.866.

Note: 1 foot of water = 0.433 psi. To convert feet of head to psi, multiply the feet of head by 0.433 psi/ft.

■ EXAMPLE 4.5

Problem: A tank is mounted at a height of 90 ft. Find the pressure at the bottom of the tank.

Solution:
$$90 \text{ ft} \times 0.433 \text{ psi/ft} = 39 \text{ psi (rounded)}$$

Note: To convert psi to feet, divide the psi by 0.433 psi/ft.

■ EXAMPLE 4.6

Problem: Find the height of water in a tank if the pressure at the bottom of the tank is 22 psi.

Solution:
$$\text{Height} = 22 \text{ psi} \div 0.433 \text{ psi/ft} = 51 \text{ ft (rounded)}$$

Note: One of the problems encountered in a hydraulic system is storing the liquid. Unlike air, which is readily compressible and is capable of being stored in large quantities in relatively small containers, a liquid such as water cannot be compressed. It is not possible to store a large amount of water in a small tank, as 62.4 lb of water occupies a volume of 1 ft^3, regardless of the pressure applied to it.

HYDROSTATIC PRESSURE

Figure 4.2 shows a number of differently shaped, connected, open containers of water. Note that the water level is the same in each container, regardless of the shape or size of the container. This occurs because pressure is developed within a liquid by

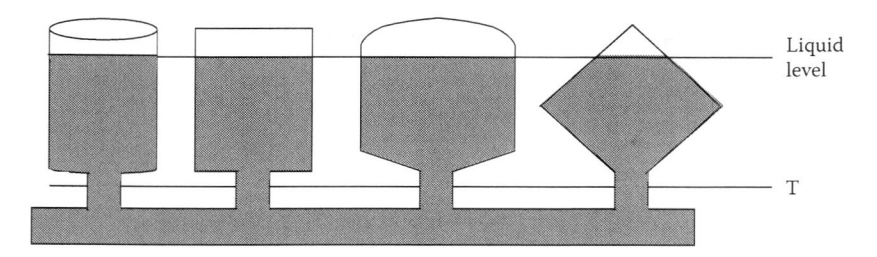

FIGURE 4.2 Hydrostatic pressure.

the weight of the liquid above. If the water level in any one container is momentarily higher than that in any of the other containers, the higher pressure at the bottom of this container would cause some water to flow into the container having the lower liquid level. In addition, the pressure of the water at any level (such as line T) is the same in each of the containers. Pressure increases because of the weight of the water. The farther down from the surface, the more pressure is created. This illustrates that the weight, not the volume, of water contained in a vessel determines the pressure at the bottom of the vessel. Some very important principles that always apply for hydrostatic pressure are listed below (Nathanson, 1997):

1. The pressure depends only on the depth of water above the point in question (not on the water surface area).
2. The pressure increases in direct proportion to the depth.
3. The pressure in a continuous volume of water is the same at all points that are at the same depth.
4. The pressure at any point in the water acts in all directions at the same depth.

Effects of Water Under Pressure

Water under pressure and in motion can exert tremendous forces inside a pipeline (Hauser, 1993). One of these forces, called *hydraulic shock* or *water hammer*, is the momentary increase in pressure that occurs due to a sudden change of direction or velocity of the water. When a rapidly closing valve suddenly stops water from flowing in a pipeline, pressure energy is transferred to the valve and pipe wall. Shock waves are set up within the system. Waves of pressure move in a horizontal yo-yo fashion—back and forth—against any solid obstacles in the system. Neither the water nor the pipe will compress to absorb the shock, which may result in damage to pipes and valves and shaking of loose fittings.

Another effect of water under pressure is called *thrust*, which is the force that water exerts on a pipeline as it rounds a bend. As shown in Figure 4.3, thrust usually acts perpendicular (90°) to the inside surface it pushes against. It affects not only bends in a pipe but also reducers, dead ends, and tees. Uncontrolled, the thrust can cause movement in the fitting or pipeline, which will lead to separation of the pipe coupling away from both sections of pipeline or at some other nearby coupling upstream or downstream of the fitting.

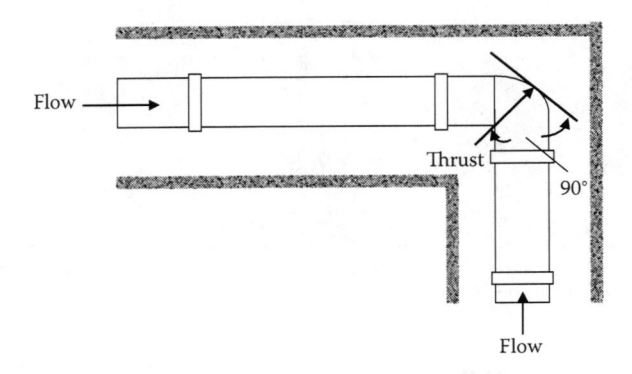

FIGURE 4.3   Direction of thrust in a pipe in a trench (viewed from above).

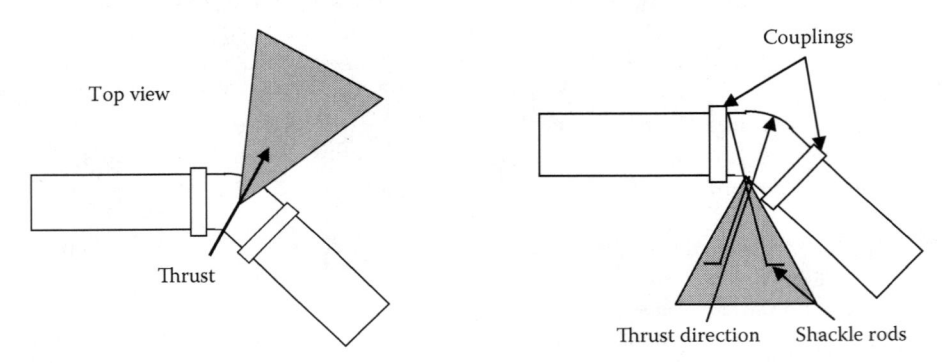

FIGURE 4.4   Thrust block.   **FIGURE 4.5**   Thrust anchor.

Two types of devices commonly used to control thrust in larger pipelines are thrust blocks and thrust anchors. A *thrust block* is a mass of concrete cast in place onto the pipe and around the outside bend of the turn. An example is shown in Figure 4.4. Thrust blocks are used for pipes with tees or elbows that turn left or right or slant upward. The thrust is transferred to the soil through the larger bearing surface of the block. A *thrust anchor* is a massive block of concrete, often a cube, cast in place below the fitting to be anchored (see Figure 4.5). As shown in Figure 4.5, imbedded steel shackle rods anchor the fitting to the concrete block, effectively resisting upward thrusts. The size and shape of a thrust control device depend on pipe size, type of fitting, water pressure, water hammer, and soil type.

HEAD

Head is defined as the vertical distance water must be lifted from the supply tank to the discharge or as the height a column of water would rise due to the pressure at its base. A perfect vacuum plus atmospheric pressure of 14.7 psi would lift the water 34 ft. When the top of the sealed tube is open to the atmosphere and the reservoir is enclosed, the pressure in the reservoir is increased; the water will rise in the tube.

Because atmospheric pressure is essentially universal, we usually ignore the first 14.7 psi of actual pressure measurements and measure only the difference between the water pressure and the atmospheric pressure; we call this *gauge pressure*. Consider water in an open reservoir subjected to 14.7 psi of atmospheric pressure; subtracting this 14.7 psi leaves a gauge pressure of 0 psi, indicating that the water would rise 0 feet above the reservoir surface. If the gauge pressure in a water main were 120 psi, the water would rise in a tube connected to the main:

$$120 \text{ psi} \times 2.31 \text{ ft/psi} = 277 \text{ ft (rounded)}$$

The *total head* includes the vertical distance the liquid must be lifted (static head), the loss to friction (friction head), and the energy required to maintain the desired velocity (velocity head):

$$\text{Total head} = \text{Static head} + \text{Friction head} + \text{Velocity head} \qquad (4.5)$$

STATIC HEAD

Static head is the actual vertical distance the liquid must be lifted:

$$\text{Static head} = \text{Discharge elevation} - \text{Supply elevation} \qquad (4.6)$$

■ EXAMPLE 4.7

Problem: The supply tank is located at elevation 118 ft. The discharge point is at elevation 215 ft. What is the static head in feet?

Solution:
$$\text{Static head} = 215 \text{ ft} - 118 \text{ ft} = 97 \text{ ft}$$

FRICTION HEAD

Friction head is the equivalent distance of the energy that must be supplied to overcome friction. Engineering references include tables showing the equivalent vertical distance for various sizes and types of pipes, fittings, and valves. The total friction head is the sum of the equivalent vertical distances for each component:

$$\text{Friction head} = \text{Energy losses due to friction} \qquad (4.7)$$

VELOCITY HEAD

Velocity head is the equivalent distance of the energy consumed to achieve and maintain the desired velocity in the system:

$$\text{Velocity head} = \text{Energy losses to maintain velocity} \qquad (4.8)$$

TOTAL DYNAMIC HEAD (TOTAL SYSTEM HEAD)

$$\text{Total dynamic head = Static head + Friction head + Velocity head} \qquad (4.9)$$

PRESSURE AND HEAD

The pressure exerted by water or wastewater is directly proportional to its depth or head in the pipe, tank, or channel. If the pressure is known, the equivalent head can be calculated:

$$\text{Head (ft) = Pressure (psi)} \times 2.31 \text{ (ft/psi)} \qquad (4.10)$$

■ EXAMPLE 4.8

Problem: The pressure gauge on the discharge line from the influent pump reads 72.3 psi. What is the equivalent head in feet?

Solution:

$$\text{Head} = 72.3 \times 2.31 \text{ ft/psi} = 167 \text{ ft}$$

HEAD AND PRESSURE

If the head is known, the equivalent pressure can be calculated by

$$\text{Pressure (psi) = Head (ft)} \div 2.31 \text{ ft/psi} \qquad (4.11)$$

■ EXAMPLE 4.9

Problem: A tank is 22 ft deep. What is the pressure in psi at the bottom of the tank when it is filled with water?

Solution:

$$\text{Pressure} = 22 \text{ ft} \div 2.31 \text{ ft/psi} = 9.52 \text{ psi (rounded)}$$

FLOW AND DISCHARGE RATES: WATER IN MOTION

The study of fluid flow is much more complicated than that of fluids at rest, but it is important to have an understanding of these principles because the water in a waterworks system is nearly always in motion. *Discharge* (or flow) is the quantity of water passing a given point in a pipe or channel during a given period. Stated another way for open channels, the flow rate through an open channel is directly related to the velocity of the liquid and the cross-sectional area of the liquid in the channel:

$$Q = A \times V \qquad (4.12)$$

where
Q = Flow or discharge (cfs).
A = Cross-sectional area of the pipe or channel (ft^2).
V = Water velocity (fps or ft/sec).

■ EXAMPLE 4.10

Problem: A channel is 6 ft wide and the water depth is 3 ft. The velocity in the channel is 4 fps. What is the discharge or flow rate in cubic feet per second?

Solution:
$$\text{Flow} = 6\text{ ft} \times 3\text{ ft} \times 4\text{ ft/sec} = 72\text{ cfs}$$

Discharge or flow can be recorded as gal/day (gpd), gal/min (gpm), or cubic feet per second (cfs). Flows treated by many waterworks plants are large and are often referred to in million gallons per day (MGD). The discharge or flow rate can be converted from cubic feet per second to other units such as gpm or MGD by using appropriate conversion factors.

■ EXAMPLE 4.11

Problem: A 12-in.-diameter pipe has water flowing through it at 10 fps. What is the discharge in (a) cfs, (b) gpm, and (c) MGD?

Solution: Before we can use the basic formula, we must determine the area (A) of the pipe. The formula for the area of a circle is

$$\text{Area } (A) = \pi \times (D^2/4) = \pi \times r^2 \tag{4.13}$$

where
 π = Constant value 3.14159, or simply 3.14.
 D = Diameter of the circle in feet.
 r = Radius of the circle in feet.

Therefore, the area of the pipe is

$$A = \pi \times (D^2/4) = 3.14 \times (1\text{ ft}^2/4) = 0.785\text{ ft}^2$$

(a) Now we can determine the discharge in cfs:

$$Q = V \times A = 10\text{ ft/sec} \times 0.785\text{ ft}^2 = 7.85\text{ ft}^3/\text{sec (cfs)}$$

(b) Because 1 cfs is 449 gpm, then 7.85 cfs × 449 gpm/cfs = 3525 gpm (rounded).

(c) 1 million gallons per day is 1.55 cfs, so

$$7.85\text{ cfs} \div 1.55\text{ cfs/MGD} = 5.06\text{ MGD}$$

Note: Flow may be *laminar* (i.e., streamline; see Figure 4.6) or *turbulent* (see Figure 4.7). Laminar flow occurs at extremely low velocities. The water moves in straight parallel lines, called *streamlines* or *laminae*, which slide upon each other as they travel, rather than mixing up. Normal pipe flow is turbulent flow, which occurs because of friction encountered on the inside of the pipe. The outside layers of flow are thrown into the inner layers; the result is that all of the layers mix and are moving in different directions and at different velocities; however, the direction of flow is forward.

FIGURE 4.6 Laminar (streamline) flow.

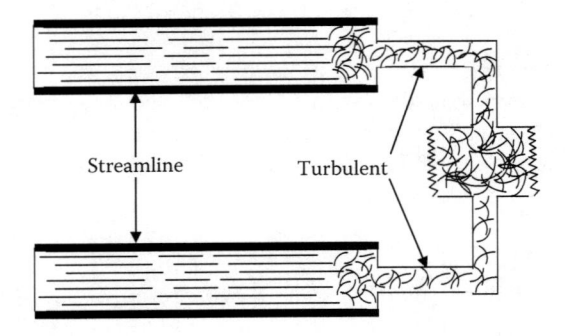

FIGURE 4.7 Turbulent flow.

Note: Flow may be steady or unsteady. For our purposes, most of the hydraulic calculations in this manual assume steady-state flow.

AREA AND VELOCITY

The *law of continuity* states that the discharge at each point in a pipe or channel is the same as the discharge at any other point (if water does not leave or enter the pipe or channel). That is, under the assumption of steady-state flow, the flow that enters the pipe or channel is the same flow that exits the pipe or channel. In equation form, this becomes

$$Q_1 = Q_2 \quad \text{or} \quad A_1 \times V_1 = A_2 \times V_2 \tag{4.14}$$

Note: With regard to the area/velocity relationship, Equation 4.14 also makes clear that, for a given flow rate, the velocity of the liquid varies indirectly with changes in cross-sectional area of the channel or pipe. This principle provides the basis for many of the flow measurement devices used in open channels (weirs, flumes, and nozzles).

■ EXAMPLE 4.12

Problem: A pipe 12 inches in diameter is connected to a 6-in.-diameter pipe. The velocity of the water in the 12-in. pipe is 3 fps. What is the velocity in the 6-in. pipe?

Solution: Using the equation $A_1 \times V_1 = A_2 \times V_2$, we need to determine the area of each pipe.

- 12-inch pipe:

$$A = \pi \times (D^2/4) = 3.14 \times (1 \text{ ft}^2/4) = 0.785 \text{ ft}^2$$

- 6-inch pipe:

$$A = \pi \times (D^2/4) = 3.14 \times (0.5\ \text{ft}^2/4) = 0.196\ \text{ft}^2$$

The continuity equation now becomes

$$0.785\ \text{ft}^2 \times 3\ \text{ft/sec} = 0.196\ \text{ft}^2 \times V_2$$

Solving for V_2,

$$V_2 = \frac{0.785\ \text{ft}^2 \times 3\ \text{ft/sec}}{0.196\ \text{ft}^2} = 12\ \text{ft/sec (fps)}$$

PRESSURE AND VELOCITY

In a closed pipe flowing full (under pressure), the pressure is indirectly related to the velocity of the liquid. This principle, when combined with the principle discussed in the previous section, forms the basis for several flow measurement devices (e.g., Venturi meters, rotameters):

$$\text{Velocity}_1 \times \text{Pressure}_1 = \text{Velocity}_2 \times \text{Pressure}_2 \qquad (4.15)$$

or

$$V_1 \times P_1 = V_2 \times P_2$$

PIEZOMETRIC SURFACE AND BERNOULLI'S THEOREM

> They will take your hand and lead you to the pearls of the desert, those secret wells swallowed by oyster crags of wadi, underground caverns that bubble salty rust water you would sell your own mothers to drink.
>
> **Holman (1998)**

Most applications of hydraulics in water treatment systems involve water in motion—in pipes under pressure or in open channels under the force of gravity. The volume of water flowing past any given point in the pipe or channel per unit time is called the *flow rate* or *discharge*—or just *flow*. The *continuity of flow* and the *continuity equation* have already been discussed (see Equation 4.14). Along with the continuity of flow principle and continuity equation, the law of conservation of energy, piezometric surface, and Bernoulli's theorem (or principle) are also important to our study of water hydraulics.

CONSERVATION OF ENERGY

Many of the principles of physics are important to the study of hydraulics. When applied to problems involving the flow of water, few of the principles of physical science are more important and useful to us than the *law of conservation of energy*.

Simply, the law of conservation of energy states that energy can be neither created nor destroyed, but it can be converted from one form to another. In a given closed system, the total energy is constant.

ENERGY HEAD

In hydraulic systems, two types of energy (kinetic and potential) and three forms of mechanical energy (potential energy due to elevation, potential energy due to pressure, and kinetic energy due to velocity) exist. Energy is measured in units of foot-pounds (ft-lb). It is convenient to express hydraulic energy in terms of *energy head* in feet of water. This is equivalent to foot-pounds per pound of water (ft-lb/lb = ft).

PIEZOMETRIC SURFACE

We have seen that when a vertical tube, open at the top, is installed onto a vessel of water the water will rise in the tube to the water level in the tank. The water level to which the water rises in a tube is the *piezometric surface*. That is, the piezometric surface is an imaginary surface that coincides with the level of the water to which water in a system would rise in a *piezometer* (an instrument used to measure pressure). In groundwater, piezometric surface is a synonym for *potentiometric surface*, which is an imaginary surface that defines the level to which water in a confined aquifer would rise were it completely bored with wells (Younger, 2007).

The surface of water that is in contact with the atmosphere is known as *free water surface*. Many important hydraulic measurements are based on the difference in height between the free water surface and some point in the water system. The piezometric surface is used to locate this free water surface in a vessel where it cannot be observed directly. To understand how a piezometer actually measures pressure, consider the following example. If a clear, see-through pipe is connected to the side of a clear glass or plastic vessel, the water will rise in the pipe to indicate the level of the water in the vessel. Such a see-through pipe—a piezometer—allows us to see the level of the top of the water in the pipe; this is the piezometric surface. In practice, a piezometer is connected to the side of a tank or pipeline. If the water-containing vessel is not under pressure (as is the case in Figure 4.8), the piezometric surface will be the same as the free water surface in the vessel, just as when a drinking straw (the piezometer) is left standing in a glass of water.

In pressurized tank and pipeline systems, as they often are, the pressure will cause the piezometric surface to rise above the level of the water in the tank. The greater the pressure, the higher the piezometric surface (see Figure 4.9). An increased pressure in a water pipeline system is usually obtained by elevating the water tank.

> **Note:** In practice, piezometers are not installed on water towers because water towers are hundreds of feet high, or on pipelines. Instead, pressure gauges are used that record pressure in feet of water or in psi.

Water only rises to the water level of the main body of water when it is at rest (static or standing water). The situation is quite different when water is flowing. Consider, for example, an elevated storage tank feeding a distribution system pipeline. When

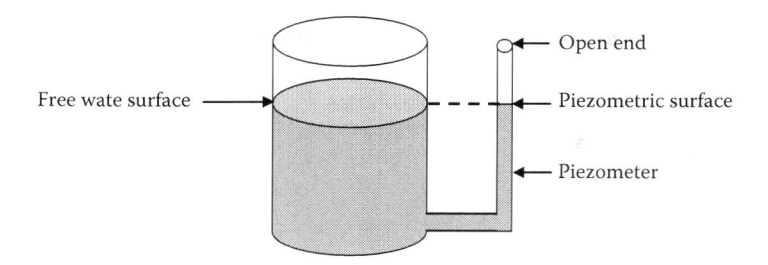

FIGURE 4.8 A container not under pressure; the piezometric surface is the same as the free water surface in the vessel.

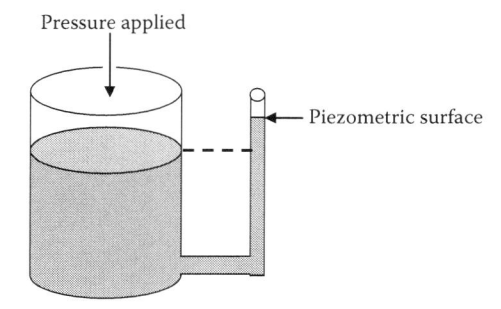

FIGURE 4.9 A container under pressure; the piezometric surface is above the level of the water in the tank.

the system is at rest, with all of the valves closed, all of the piezometric surfaces are the same height as the free water surface in storage. On the other hand, when the valves are opened and the water begins to flow, the piezometric surface changes. This is an important point because, as water continues to flow down a pipeline, less and less pressure is exerted. This happens because some pressure is lost (used up) to keep the water moving over the interior surface of the pipe (friction). The pressure that is lost is called *head loss*.

HEAD LOSS

Head loss is best explained by example. Figure 4.10 shows an elevated storage tank feeding a distribution system pipeline. When the valve is closed (Figure 4.10A), all the piezometric surfaces are the same height as the free water surface in storage. When the valve opens and water begins to flow (Figure 4.10B), the piezometric surfaces *drop*. The farther along the pipeline, the lower the piezometric surface, because some of the pressure is used up keeping the water moving over the rough interior surface of the pipe. Thus, pressure is lost and is no longer available to push water up in a piezometer; this, again, is the head loss.

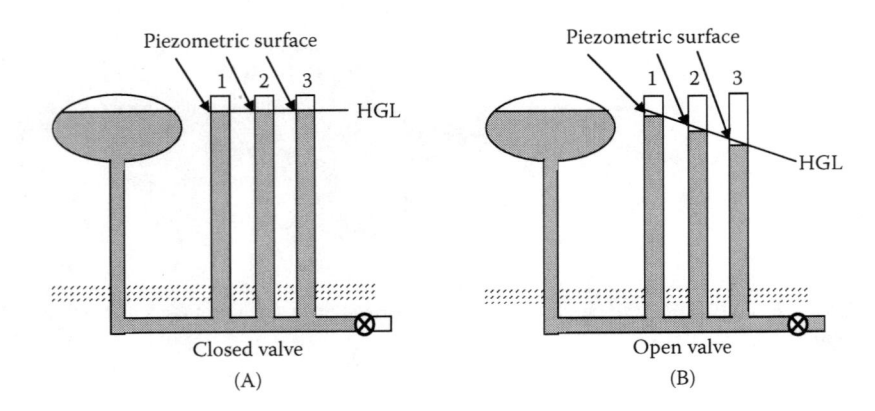

FIGURE 4.10 Head loss and piezometric surface changes when water is flowing.

HYDRAULIC GRADE LINE

When the valve shown in Figure 4.10 is opened, flow begins with a corresponding energy loss due to friction. The pressures along the pipeline can measure this loss. In Figure 4.10B, the difference in pressure heads between sections 1, 2, and 3 can be seen in the piezometer tubes attached to the pipe. A line connecting the water surface in the tank with the water levels at sections 1, 2, and 3 shows the pattern of continuous pressure loss along the pipeline. This is the *hydraulic grade line* (HGL), or *hydraulic gradient*, of the system.

> *Note:* It is important to point out that in a static water system the HGL is always horizontal. The HGL is a very useful graphical aid when analyzing pipe flow problems.

> *Note:* Changes in the piezometric surface occur when water is flowing.

BERNOULLIS'S THEOREM

Swiss physicist and mathematician Samuel Bernoulli developed the calculation for the total energy relationship from point to point in a steady-state fluid system in the 1700s. Before discussing Bernoulli's energy equation, it is important to understand the basic principle behind Bernoulli's equation. Water (and any other hydraulic fluid) in a hydraulic system possesses two types of energy—kinetic and potential. *Kinetic energy* is present when the water is in motion. The faster the water moves, the more kinetic energy is used. *Potential energy* is a result of the water pressure. The *total energy* of the water is the sum of the kinetic and potential energy. Bernoulli's principle states that the total energy of the water (fluid) always remains constant; therefore, when the water flow in a system increases, the pressure must decrease. When water starts to flow in a hydraulic system, the pressure drops. When the flow stops, the pressure rises again. The pressure gauges shown in Figure 4.11 illustrate this balance more clearly.

> *Note:* This discussion of Bernoulli's equation ignores friction losses from point to point in a fluid system employing steady-state flow.

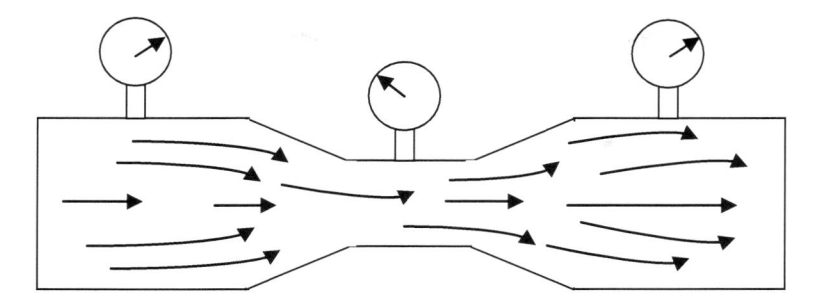

FIGURE 4.11 Demonstration of Bernoulli's principle.

Bernoulli's Equation

In a hydraulic system, total energy head is equal to the sum of three individual energy heads. This can be expressed as

Total head = Elevation head + Pressure head + Velocity head

where elevation head is the pressure due to the elevation of the water, pressure head is the height of a column of water that a given hydrostatic pressure in a system could support, and velocity head is the energy present due to the velocity of the water. This can be expressed mathematically as

$$E = z + \frac{p}{w} + \frac{V^2}{2g} \tag{4.16}$$

where
E = Total energy head.
z = Height of the water above a reference plane (ft).
p = Pressure (psi).
w = Unit weight of water (62.4 lb/ft³).
V = Flow velocity (ft/sec).
g = Acceleration due to gravity (32.2 ft/sec²).

Consider the constriction in the section of pipe shown in Figure 4.11. We know, based on the law of energy conservation, that the total energy head at section A (E_1) must equal the total energy head at section B (E_2). Using Equation 4.16, we get Bernoulli's equation:

$$z_A + \frac{p_A}{w} + \frac{V_A^2}{2g} = z_B + \frac{p_B}{w} + \frac{V_B^2}{2g} \tag{4.17}$$

The pipeline system shown in Figure 4.11 is horizontal; therefore, we can simplify Bernoulli's equation because $z_A = z_B$. Because they are equal, the elevation heads cancel out from both sides, leaving

$$\frac{p_A}{w} + \frac{V_A^2}{2g} = \frac{p_B}{w} + \frac{V_B^2}{2g} \qquad (4.18)$$

As water passes through the constricted section of the pipe (section B), we know from continuity of flow that the velocity at section B must be greater than the velocity at section A, because of the smaller flow area at section B. This means that the velocity head in the system increases as the water flows into the constricted section; however, the total energy must remain constant. For this to occur, the pressure head, and therefore the pressure, must drop. In effect, pressure energy is converted into kinetic energy in the constriction. The fact that the pressure in the narrower pipe section (constriction) is less than the pressure in the bigger section seems to defy common sense; however, it does follow logically from continuity of flow and conservation of energy. The fact that there is a pressure difference allows measurement of flow rate in the closed pipe.

■ **EXAMPLE 4.13**

Problem: In Figure 4.12, the diameter at section A is 8 in., and at section B it is 4 in. The flow rate through the pipe is 3.0 cfs and the pressure at section A is 100 psi. What is the pressure in the constriction at section B?

Solution: Compute the flow area at each section, as follows:

$$A_A = \pi \times (D^2/4) = 3.14 \times (0.666 \text{ ft}^2/4) = 0.349 \text{ ft}^2$$

and

$$A_B = \pi \times (D^2/4) = 3.14 \times (0.333 \text{ ft}^2/4) = 0.087 \text{ ft}^2$$

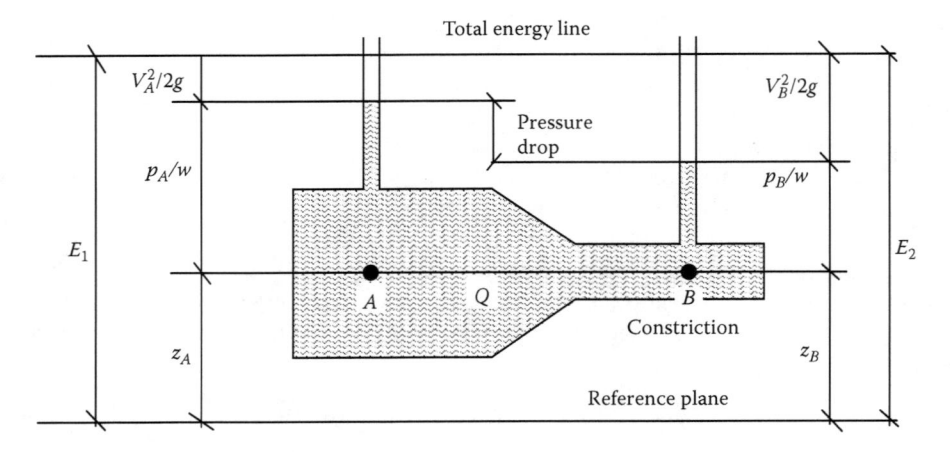

FIGURE 4.12 The result of the law of conservation of energy. Because the velocity and kinetic energy of the water flowing in the constricted section must increase, the potential energy may decrease. This is observed as a pressure drop in the constriction. (Adapted from Nathanson, J.A., *Basic Environmental Technology: Water Supply, Waste Management, and Pollution Control*, 2nd ed., Prentice Hall, Upper Saddle River, NJ, 1997, p. 29.)

From $Q = A \times V$ or $V = Q/A$, we get

$$V_A = 3.0 \text{ ft}^3/\text{sec} \div 0.349 \text{ ft}^2 = 8.6 \text{ ft/sec}$$

and

$$V_B = 3.0 \text{ ft}^3/\text{sec} \div 0.087 \text{ ft}^2 = 34.5 \text{ ft/sec}$$

Applying Equation 4.18, we get

$$\frac{p_A}{w} + \frac{V_A^2}{2g} = \frac{p_B}{w} + \frac{V_B^2}{2g}$$

$$\frac{100 \times 144}{62.4} + \frac{(8.6)^2}{2 \times 32.2} = \frac{p_B \times 144}{62.4} + \frac{(34.5)^2}{2 \times 32.2}$$

$$231 + 1.15 = 2.3 p_B + 18.5$$

$$(231 + 1.15) - 18.5 = 2.3 p_B$$

$$p_B = \frac{232.2 - 18.5}{2.3}$$

$$p_B = \frac{213.7}{2.3}$$

$$p_B = 93 \text{ psi}$$

Note: The pressures are multiplied by 144 in.^{2}/ft^2 to convert from psi to lb/ft^2 to be consistent with the units for w; the energy head terms are in feet of head.

REFERENCES AND RECOMMENDED READINGS

AWWA. (1995). *Basic Science Concepts and Applications: Principles and Practices of Water Supply Operations Series*, 2nd ed. Denver, CO: American Water Works Association.

Grant, D.M. (1991). Open channel flow measurement. In: *Flow Measurement*, 2nd ed. (Spitzer, D.W., Ed.), pp. 252–290. Research Triangle Park, NC: Instrument Society of America.

Hauser, B.A. (1993). *Hydraulics for Operators*. Boca Raton, FL: Lewis Publishers.

Hauser, B.A. (1996). *Practical Hydraulics Handbook*, 2nd ed. Boca Raton, FL: CRC Press.

Holman, S. (1998). *A Stolen Tongue*. New York: Anchor Press.

Kawamura, S. (2000). *Integrated Design and Operation of Water Treatment Facilities*, 2nd ed. New York: John Wiley & Sons.

Lindeburg, M.R. (1986). *Civil Engineering Reference Manual*, 4th ed. San Carlos, CA: Professional Publications.

Magnusson, R.J. (2001). *Water Technology in the Middle Ages*. Baltimore, MD: The Johns Hopkins University Press.

McGhee, T.J. (1991). *Water Supply and Sewerage*, 2nd ed. New York: McGraw-Hill.

Nathanson, J.A. (1997). *Basic Environmental Technology: Water Supply Waste Management, and Pollution Control*, 2nd ed. Upper Saddle River, NJ: Prentice Hall.

Spellman, F.R. (2008). *The Science of Water*, 2nd ed. Boca Raton, FL: CRC Press.

Spellman, F.R. (2015). *The Science of Water*, 3rd ed. Boca Raton, FL: CRC Press.

Spellman, F.R. and Drinan, J. (2001). *Water Hydraulics*. Boca Raton, FL: CRC Press.

USEPA. (1991). *Flow Instrumentation: A Practical Workshop on Making Them Work*. Sacramento, CA: Water & Wastewater Instrumentation Testing Association.

Viessman, Jr., W. and Hammer, M.J. (1998). *Water Supply and Pollution Control*, 6th ed. Menlo Park, CA: Addison–Wesley.

Younger, P. (2007). *Groundwater in the Environment*. Oxford, U.K.: Blackwell.

5 Groundwater Hydraulics

Groundwater is not a nonrenewable resource, such as a mineral or petroleum deposit, nor is it completely renewable in the same manner and time frame as solar energy.

GROUNDWATER

Unbeknownst to most of us, our Earth possesses an unseen ocean—a hidden resource. This ocean, unlike the surface oceans that cover most of the globe, is freshwater, the groundwater contained in aquifers beneath Earth's crust. This gigantic water source forms a reservoir that feeds all the natural fountains and springs of Earth. But, how does water travel into the aquifers that lie under the Earth's surface?

Groundwater sources are replenished from a percentage of the average approximately 3 feet of water that falls to Earth each year on every square foot of land. Water falling to Earth as precipitation follows three courses. Some runs off directly to rivers and streams (roughly 6 inches of that 3 feet), eventually working its way back to the sea. Evaporation and transpiration through vegetation take up about 2 feet. The remaining 6 inches of water seep into the ground, entering and filling every interstice, each hollow and cavity. Gravity pulls water toward the center of the Earth. That means that water on the surface will try to seep into the ground below it. Although groundwater accounts for only one-sixth of the total 1,680,000 miles of water, if we could spread out this water over the land it would blanket it to a depth of 1000 feet.

The science of groundwater hydraulics is concerned with evaluating the occurrence, availability, and quality of groundwater. In particular, groundwater hydraulics is concerned with the natural or induced movement of water through permeable rock formations. To understand groundwater hydraulics, it is important to understand the operation of the natural plumbing system within Earth's geologic framework—flow through permeable rock formations. To get even close to understanding basic groundwater hydraulics, an understanding of the fundamental properties of unconfined and confined aquifers is necessary.

Part of the precipitation that falls on land infiltrates the land surface, percolates downward through the soil under the force of gravity, and becomes groundwater. Groundwater, like surface water, is extremely important to the hydrologic cycle and to our water supplies. Almost half of the people in the United States drink public water from groundwater supplies. Overall, more water exists as groundwater than surface water in the United States, including the water in the Great Lakes, but sometimes pumping it to the surface is not economical, and in recent years pollution of groundwater supplies from improper disposal has become a significant problem.

We find groundwater in saturated layers called *aquifers* under the Earth's surface. Three types of aquifers exist: unconfined, confined, and springs. Aquifers are made up of a combination of solid material such as rock and gravel and open spaces called *pores*. Regardless of the type of aquifer, the groundwater in the aquifer is in a constant state of motion. This motion is caused by gravity or by pumping.

The actual amount of water in an aquifer depends upon the amount of space available between the various grains of material that make up the aquifer. The amount of space available is called *porosity*. The ease of movement through an aquifer is dependent on how well the pores are connected; for example, clay can hold a lot of water and has high porosity, but the pores are not connected so water moves through the clay with difficulty. The ability of an aquifer to allow water to infiltrate is called *permeability*.

UNCONFINED AQUIFERS

The unconfined aquifer that lies just under the Earth's surface is called the *zone of saturation* (Figure 5.1). This type of aquifer is composed of granular materials, such as mixtures of clay, silt, sand, and gravel. The top of the zone of saturation is the *water table*. An unconfined aquifer is only contained on the bottom and is dependent on local precipitation for recharge. Unconfined aquifers are a primary source of shallow well water (Figure 5.1). Because these wells are shallow they are not desirable as public drinking water sources. When a well is sunk a few feet into an unconfined aquifer, the water level remains, for a time, the same as when it was first reached by drilling, but of course this level may fluctuate later in response to many factors. They are subject to local contamination from hazardous and toxic materials, such as fuel and oil, as well as septic tank and agricultural runoff that provides increased levels of nitrates and microorganisms. These wells may be classified as groundwater under the direct influence of surface water (GUDISW) and therefore require treatment for control of microorganisms.

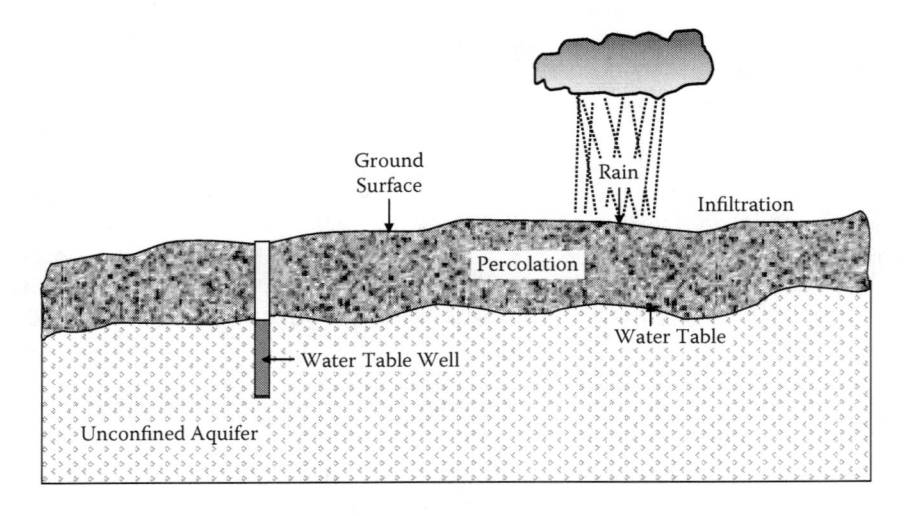

FIGURE 5.1   Unconfined aquifer.

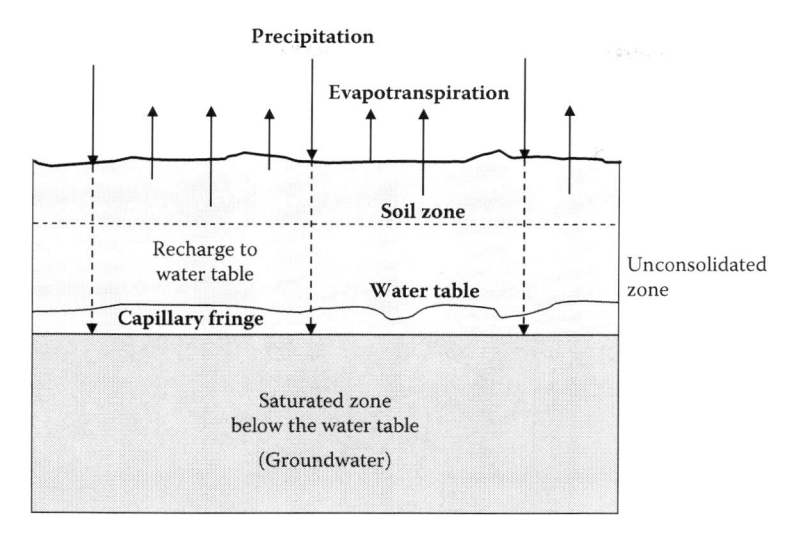

FIGURE 5.2   The unsaturated zone, capillary fringe, water table, and saturated zone.

The water level in wells sunk to greater depths in unconfined aquifers may stand at, above, or below the water table, depending on whether the well is in the discharge or recharge area of the aquifer. This type of aquifer is often called a *water table aquifer*. The unconfined groundwater below the water table is under pressure greater than atmospheric. Unconfined aquifers consist of an unsaturated and a saturated zone (see Figure 5.2). In the unsaturated zone, the spaces between particle grains and the cracks in rocks contain both air and water. Although a considerable amount of water can be present in the unsaturated zone, this water cannot be pumped by wells because capillary forces hold it too tightly. In contrast to the unsaturated zone, the voids in the saturated zone are completely filled with water.

SATURATED ZONE

The approximate upper surface of the saturated zone is referred to as the water table. Water in the saturated zone below the water table is referred to as groundwater. Below the water table, the water pressure is high enough to allow water to enter a well as the water level in the well is lowered by pumping, thus permitting groundwater to be withdrawn for use. Between the unsaturated zone and the water table is a transition zone known as the capillary fringe. In this zone, the voids are saturated or almost saturated with water that is held in place by capillary forces (USDI, 1975).

CAPILLARY FRINGE

The capillary fringe (see Figure 5.2) acts like a sponge sucking water up from the water table. The capillary fringe ranges in thickness from a small fraction of an inch in coarse gravel to more than 5 ft in silt. The lower part is completely saturated, like the material below the water table, but it contains water under less than atmospheric pressure, so the water in it normally does not enter a well. The capillary fringe rises and declines with fluctuations of the water table, and may change in thickness as it moves through materials of different grain sizes.

Unsaturated Zone

The unsaturated zone contains water in the gas phase under atmospheric pressure, water temporarily or permanently under less than atmospheric pressure, and air or other gases. The fine-grained materials may be temporarily or permanently saturated with waste under less than atmospheric pressure, but the coarse-grained materials are unsaturated and generally contain liquid water only in rings surrounding the contacts between grains.

Capillarity

The rise of water in the interstices in rocks or soil may be considered to be caused by (1) the molecular attraction (adhesion) between the solid material and the fluid, and (2) the surface tension of the fluid, an expression of the attraction (cohesion) between the molecules of the fluid. Have you ever noticed that water will wet and adhere to a clean floor whereas it will remain in drops without wetting a dust-covered floor? The molecular attraction between a solid material and a fluid varies with the cleanliness of the solid material. In addition, the height of capillary rise is also governed by the size of the opening. The surface of water resists considerable tension without losing its continuity. Thus, a carefully placed greased needle floats on water, as do certain insects having greasy pads on their feet or water-resistant hair on their underbodies. A good insect example is the water strider ("Jesus bug"; Order: Hemiptera), which rides the top of the water, with only their feet making dimples in the surface film. Like all insects, a water strider has a three-part body (head, thorax, and abdomen), six jointed legs, and two antennae. It has a long, dark, narrow body (see Figure 5.3).

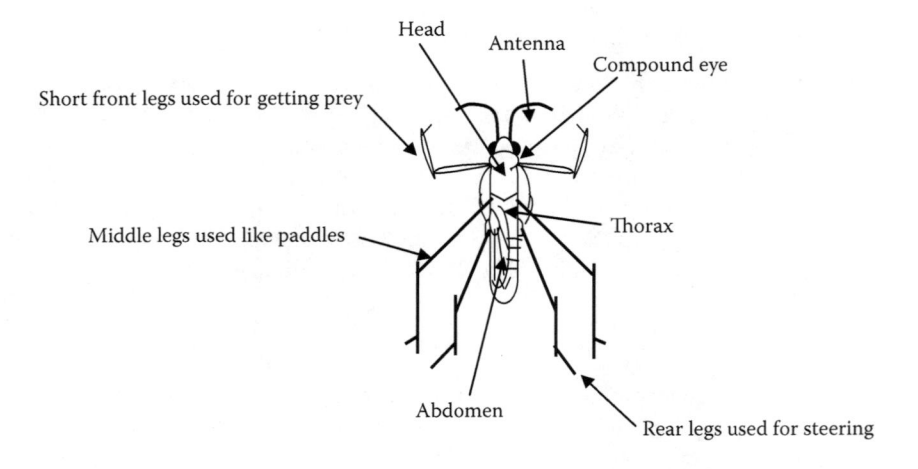

FIGURE 5.3   Water strider.

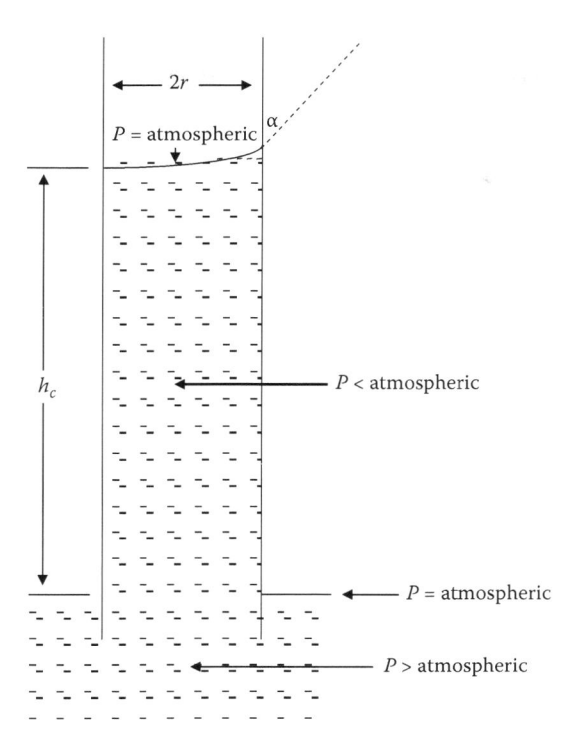

FIGURE 5.4 Capillary rise of water in a tube (exaggerated).

In Figure 5.4, the water has risen to h_c in a tube of radius r immersed in a container of water. The relations shown in Figure 5.4 may be expressed as

$$\pi r^2 \rho g h_c = 2\pi r T \cos\alpha \tag{5.1}$$

where
  r = Radius of capillary tube.
  ρ = Density of fluid.
  g = Acceleration due to gravity.
  h_c = Height of capillary rise.
  T = Surface tension of fluid.
  α = Angle between meniscus and tube.

Note that, according to Equation 5.1, weight equals lift by surface tension. Solving Equation 5.1 for h_c,

$$h_c = (2T/r\rho g)/\cos\alpha \tag{5.2}$$

For pure water in a clean glass, $\alpha = 0$ and $\cos\alpha = 1$. At 20°C, $T = 72.8$ dyne/cm, ρ may be taken as 1 g/cm³, and $g = 980.665$ cm/sec²; thus,

$$h_c = (0.15)/r \tag{5.3}$$

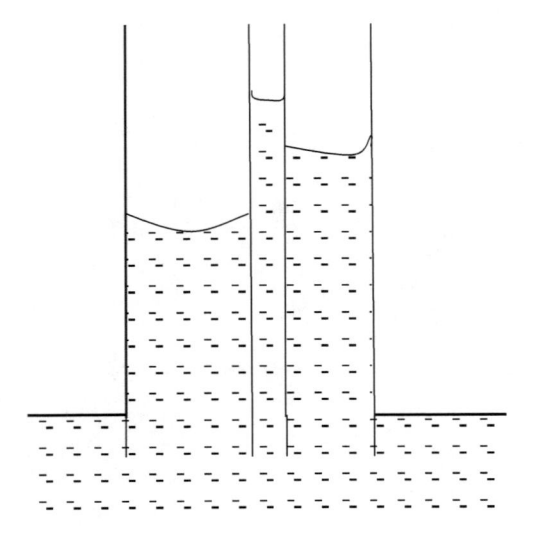

FIGURE 5.5 Rise in capillary tubes of different diameters (exaggerated).

Surface tension is sometimes given in grams per centimeter; for pure water in contact with air at 20°C, that value is 0.074 g/cm. To express dynes per centimeter in grams per centimeter, we must divide the 72.8 dyne/cm by g, the standard acceleration of gravity; thus, 72.8 dyne/cm ÷ 980.665 cm/sec = 0.074 g/cm. From Equation 5.3, it can be seen that the *height of capillary rise* in tubes is inversely proportional to the radius of the tube. The rise of water in interstices of various sizes in the capillary fringe may be likened to the rise of water in a bundle of capillary tubes of various diameters, as shown in Figure 5.5. In Table 5.1, note that the capillary rise is nearly inversely proportional to the grain size.

TABLE 5.1
**Capillary Rise in Samples Having Virtually
the Same Porosity, 41%, after 72 Days**

Material	Grain Size (mm)	Capillary Rise (cm)
Fine gravel	5-2	2.5
Very coarse sand	2-1	6.5
Coarse sand	1-0.5	13.5
Medium sand	0.5-0.2	24.6
Fine sand	0.2–0.1	42.8
Silt	0.1–0.05	105.5
Silt	0.05–0.02	200

Source: Lohman, S.W., *Ground-Water Hydraulics*, U.S. Geological Survey Professional Paper 708, U.S. Government Printing Office, Washington, DC, 1972.

HYDROLOGIC PROPERTIES OF WATER-BEARING MATERIALS

POROSITY

Porosity is defined as the ratio of (1) the volume of the void spaces to (2) the total volume of the rock or soil mass. Stated differently, the porosity of a rock or solid is simply its property of containing interstices. It can be expressed quantitatively as the ratio of the volume of the interstices to the total volume, and may be expressed as decimal fraction or as a percentage. Thus,

$$\theta = \frac{v_i}{V} = \frac{v_w}{V} = \frac{V - v_m}{V} = 1 - \frac{v_m}{V} \tag{5.4}$$

where
θ = Porosity, as a decimal fraction.
v_i = Volume of interstices.
V = Total volume.
v_w = Volume of water (in a saturated sample).
v_m = Volume of mineral particles.

Porosity may be expressed also as

$$\theta = \frac{\rho_m - \rho_d}{\rho_m} = 1 - \frac{\rho_d}{\rho_m} \tag{5.5}$$

where
ρ_m = Mean density of mineral particles (grain density).
ρ_d = Density of dry sample (bulk density).

Multiplying the right-hand sides of Equations 5.4 and 5.5 by 100 gives the porosity as a percentage.

Primary Porosity

Primary porosity is porosity associated with the original depositional texture of the sediment. In soil and sedimentary rocks, the primary interstices are the spaces between grains or pebbles. In intrusive igneous rocks, the few primary interstices result from cooling and crystallization. Extrusive igneous rocks may have large openings and high porosity resulting from the expansion of gas, but the openings may or may not be connected. With time, metamorphism of igneous or sedimentary rocks generally reduces the primary porosity and may virtually obliterate it.

Secondary Porosity

Secondary porosity is porosity that developed after the deposition and burial of sediment in the sedimentary basin. Fractures such as joints, faults, and openings along planes of bedding or schistosity in consolidated rocks having low primary porosity and permeability may afford appreciable secondary porosity.

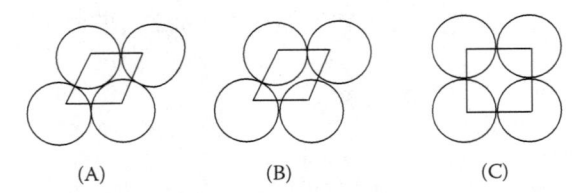

(A)   (B)   (C)

FIGURE 5.6   Sections of four contiguous spheres of equal size. (A) most compact arrange-ment, lowest porosity; (B) less compact arrangement, higher porosity; (C) least compact arrangement, highest porosity. (Adapted from Slichter, C.S., *Theoretical Investigations of the Motion of Ground Waters*, 19th Annual Report, U.S. Geological Survey, Washington, DC, 1899.)

Controlling Porosity of Granular Materials

When describing conditions that control the porosity of granular materials, it is con-venient to use the time-proven approach originally developed by Slichter in 1899. He explained that, if a hypothetical granular material were composed of spherical par-ticles of equal size, then the porosity would be independent of particle size (whether the particles were the size of silt or the size of the Earth) but would vary with the packing arrangement of the particles. Slichter explained that the lowest porosity of 25.95% (~26%) would result from the most compact rhombohedral arrangement (Figure 5.6A), and the highest porosity of 47.64% (~48%) would result from the least compact cubical arrangement (Figure 5.6C). The porosity of the other arrangements, such as that shown in Figure 5.6B, would be between these limits.

In addition to the arrangement of grains (as shown in Figure 5.6) having an impact on the porosity of granular materials, the shape of the grains (i.e., their angular-ity) and their degree of assortment (i.e., range in particle size) also have an impact on porosity. The angularity of particles causes wide variations in porosity and may increase or decrease it, according to whether the particles tend to bridge openings or pack together like pieces of a mosaic. The greater the range in particle size, the lower the porosity, as the small particles occupy the voids between the larger ones.

Void Ratio

The void ratio of a rock or soil is the ratio of the volume of its interstices to the vol-ume of its mineral particles. It may be expressed as

$$\text{Void ratio} = \frac{v_i}{v_m} = \frac{v_w}{v_m} = \frac{\theta}{1-\theta} \tag{5.6}$$

where
   v_i = Volume of interstices.
   v_m = Volume of mineral particles.
   v_w = Volume of water (in a saturated sample).
   θ = Porosity, as a decimal fraction.

PERMEABILITY

The permeability of a rock or soil is a measure of how easily a fluid (e.g., water) can travel through it, under a hydropotential gradient. Soil and loose sediments, such as sand and gravel, are porous and permeable. They can hold a lot of water, and it flows easily through them. The permeability is approximately proportional to the square of the mean grain diameter:

$$k \approx Cd^2 \tag{5.7}$$

where
  k = Intrinsic permeability.
  C = Dimensionless constant depending on porosity, range and distribution of particle size, shape of grains, and other factors.
  d = Mean grain diameter of some workers and the effective grain diameter of others.

Intrinsic Permeability

Intrinsic permeability, as defined by the U.S. Geological Survey, is an intensive property (not a spatial average of a heterogeneous block of material). It is a function of the material structure only, not of the fluid, and the value is explicitly distinguished from that of relative permeability. Intrinsic permeability may be expressed as

$$k = \frac{qv}{g\left(dh/dl\right)} = -\frac{qv}{\left(dp/dl\right)} \tag{5.8}$$

where
  k = Intrinsic permeability.
  q = Rate of flow per unit area = Q/A.
  v = Kinematic viscosity.
  g = Acceleration of gravity.
  dh/dl = Gradient, or unit change in head per unit length of flow.
  dp/dl = Potential gradient, or unit change in potential per unit length of flow.

From Equation 5.8 it may be stated that a porous medium has an intrinsic permeability of one unit of length squared if it will transmit in unit time a unit volume of fluid of unit kinematic viscosity though a cross-section of unit area, measured at right angles to the flow direction, under a unit potential gradient.

DID YOU KNOW?

Although clay and shale are porous and can hold a lot of water, the pores in these fine-grained materials are so small that water flows very slowly through them. Clay has a low permeability.

HYDRAULIC CONDUCTIVITY

Hydraulic conductivity, which is represented as K, is a property of soils and rocks that describes the ease with which water can move through pore spaces or fractures. A medium has a hydraulic conductivity of unit length per unit time if it will transmit in unit time a unit volume of groundwater at the prevailing viscosity through a cross-section of unit area, measured at right angles to the direction of flow, under a hydraulic gradient of unit change in head through unit length of flow. The suggested units are

$$K = -\frac{q}{dh/dl} \tag{5.9}$$

where q is the rate of flow per unit area (Q/A), and dh/dl is the gradient.

$$K = -\frac{\text{ft}^3}{\text{ft}^2 \text{ day} \left(-\text{ft ft}^{-1}\right)} = \text{ft day}^{-1} \tag{5.10}$$

or

$$K = -\frac{\text{m}^3}{\text{m}^2 \text{ day} \left(-\text{m m}^{-1}\right)} = \text{m day}^{-1} \tag{5.11}$$

Note that the minus signs in these equations result from the fact that the water moves in the direction of decreasing head.

TRANSMISSIVITY

The transmissivity (T) is the rate at which water of the prevailing kinematic viscosity flows horizontally through an aquifer, such as to a pumping well.

WATER YIELDING AND RETAINING CAPACITY

SPECIFIC YIELD

Specific yield (S_y) is defined as the ratio of (1) the volume of water that a saturated rock or soil will yield by gravity to (2) the total volume of the rock or soil. This may be expressed as

$$S_y = v_g/V \tag{5.12}$$

where
    S_y = Specific yield, as a decimal fraction.
    v_g = Volume of water drained by gravity.
    V = Total volume.

> **DID YOU KNOW?**
>
> Specific yield is usually expressed as a percentage. The value is not definitive, because the quantity of water that will drain by gravity depends on variables such as duration of drainage, temperature, mineral composition of the water, and various physical characteristics of the rock or soil under consideration. Values of specific yield, nevertheless, offer a convenient means by which hydrologists can estimate the water-yielding capacities of earth materials and, as such, are very useful in hydrologic studies.

SPECIFIC RETENTION

The specific retention of a rock or soil with respect to water was defined by Meinzer (1923) as the ratio of (1) the volume of water, which, after being saturated, the rock or soil will retain against the pull of gravity to (2) the volume of the rock or soil:

$$S_r = v_r/V = \theta - S_y \tag{5.13}$$

where

S_r = Specific retention, as a decimal fraction.
v_r = Volume of water retained against gravity, mostly by molecular attraction.
V = Total volume.
S_y = Specific yield, as a decimal fraction.

CONFINED AQUIFERS

A confined aquifer is sandwiched between two impermeable layers that block the flow of water. The water in a confined aquifer is under hydrostatic pressure, and it does not have a free water table (see Figure 5.7). Confined aquifers are called

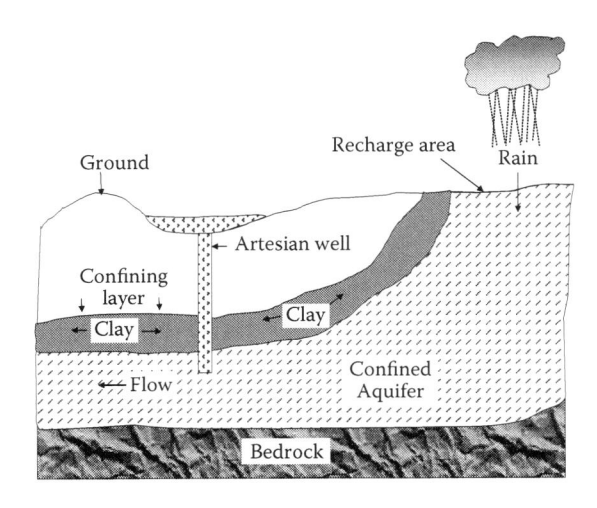

FIGURE 5.7 Confined aquifer.

artesian aquifers. Wells drilled into artesian aquifers are referred to as *artesian wells* and commonly yield large quantities of high-quality water. An artesian well is any well where the water in the well casing would rise above the saturated strata. Wells in confined aquifers are deep wells and are not generally affected by local hydrological events. A confined aquifer is recharged by rain or snow in the mountains where the aquifer lies close to the surface of the Earth. Because the recharge area is some distance from areas of possible contamination, the possibility of contamination is usually very low. However, once contaminated, confined aquifers may take centuries to recover. Groundwater naturally exits the Earth's crust in areas called *springs*. The water in a spring can originate from a water table aquifer or from a confined aquifer. Only water from a confined spring is considered desirable for a public water system.

STEADY FLOW OF GROUNDWATER

For steady flow of groundwater through permeable material, there is no change in head with time. Mathematically, this statement is symbolized by $dh/dt = 0$, which says that the change in head, dh, with respect to the change in time, dt, equals zero. Note that steady flow generally does not occur in nature, but it is a very useful concept in that steady flow can be closely approached in nature and in aquifer tests, and this condition may be symbolized $dh/dt \rightarrow 0$. Figure 5.8 shows a hypothetical example of true steady radial flow. Here, steady radial flow will be reached and maintained when all of the recoverable groundwater in the cone of depression has been drained by gravity into the well discharging at constant rate Q.

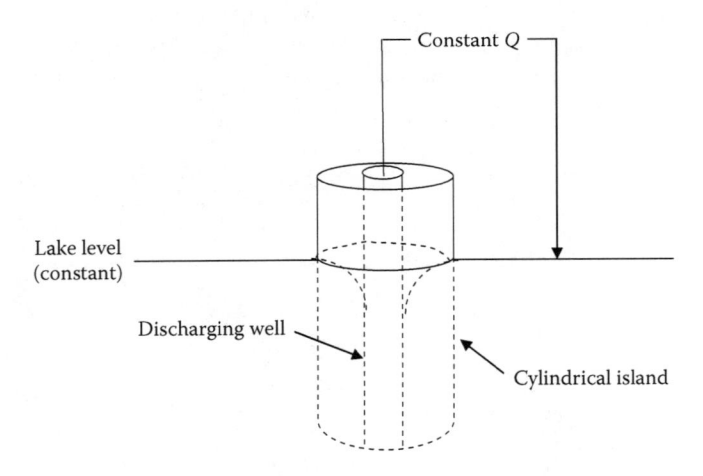

FIGURE 5.8 Hypothetical example of steady flow; the well is discharging at constant rate Q from a cylindrical island in a lake of constant level. (From Lohman, S.W., *Ground-Water Hydraulics*, U.S. Geological Survey Professional Paper 708, U.S. Government Printing Office, Washington, DC, 1972.)

DARCY'S LAW

Hagen (1797–1884) and Poiseuille (1799–1869) found that the rate of flow through capillary tubes is proportional to the hydraulic gradient, but in 1856 Henry Darcy experimented with the flow of water through sand and determined that the rate of laminar (viscous) flow of water through sand is also proportional to the hydraulic gradient. His discovery came to be known as *Darcy's law*, and it is generally expressed, by rewriting Equation 5.10, as

$$q = Q/A = (Kdh)/(dl) \qquad (5.14)$$

Note that K, the constant of proportionality in Darcy's law, is the hydraulic conductivity.

To illustrate the use of Equation 5.14 (Darcy's law), assume that we wish to compute the total rate of groundwater movement in a valley where A, the cross-sectional area, is 100 ft deep by 1 mile wide; where $K = 500$ ft/day; and where $dh/dl = 5$ ft per mile. Then,

$$Q = -(100 \text{ ft})(5280 \text{ ft})(500 \text{ ft/day})(-5 \text{ ft/5280 ft}) = 250{,}000 \text{ ft}^3/\text{day}$$

VELOCITY

Because the hydraulic conductivity, K, has the dimensions of velocity, some might mistake this for the particle velocity of the water, whereas, as may be seen from Equations 5.10 and 5.11, K is actually a measure of the volume rate of flow through unit cross-sectional area. For the average particle velocity, $\bar{v}$, we must also know the porosity of the material. Thus,

$$Q = \bar{v}A\theta = -KA(dh/dl) \qquad (5.15)$$

where
$\bar{v}$ = Average velocity, in feet per day.
θ = Porosity, as a decimal fraction.

GROUNDWATER FLOW AND EFFECTS OF PUMPING

Water pumped from a groundwater system causes the water table to lower, alters the direction of groundwater movement, leads to land subsidence and can have a debilitating impact on the quality of the water within an aquifer. Generally, groundwater is of high chemical, bacteriological, and physical quality. When pumped from an aquifer composed of a mixture of sand and gravel and when not directly influenced by surface water, groundwater is often used without filtration. It can also be used without disinfection if it has a low coliform count; however, as mentioned, groundwater can become contaminated. Septic systems fail, saltwater intrudes, improper disposal of wastes occurs, stockpiled chemicals leach, underground storage tanks leak, hazardous materials spill, fertilizers and pesticides are misplaced, and mines are recklessly abandoned. To understand how an underground aquifer becomes contaminated, it is important to understand what occurs when pumping is taking place within the well. When groundwater is removed from its underground source (the water-bearing stratum) via a well, water flows toward the center of the well. In a water table aquifer, this movement causes the water table to sag toward the well. This sag is the *cone of depression*. The shape and size of the cone depend on the relationship between the pumping rate and the rate at which water can move toward the well. If the rate is high, the cone is shallow and its growth stabilizes. The area that is included in the cone of depression is the *cone of influence* and any contamination in this zone will be drawn into the well.

TYPES OF WELLS

Water supply wells may be characterized as shallow or deep. In addition, wells are classified as follows:

- Class I, cased and grouted to 100 ft
- Class II A, cased to a minimum of 100 ft and grouted to 20 ft
- Class II B, cased and grouted to 50 ft

Note: During the well development process, mud/silt forced into the aquifer during the drilling process is removed, allowing the well to produce the best-quality water at the highest rate from the aquifer.

SHALLOW WELLS

Shallow wells are those that are less than 100 ft deep. Such wells are not particularly desirable for municipal supplies because the aquifers they tap are likely to fluctuate considerably in depth, making the yield somewhat uncertain. Municipal wells in such aquifers cause a reduction in the water table (or phreatic surface) that affects nearby private wells, which are more likely to utilize shallow strata. Such interference with private wells may result in damage suits against the community. Shallow wells may be dug, bored, or driven:

- *Dug wells*—Dug wells are the oldest type of well and date back many centuries; they are dug by hand or by a variety of unspecialized equipment. They range in size from approximately 4 to 15 ft in diameter and are usually about 20 to 40 ft deep. Such wells are usually lined or cased with concrete or brick. Dug wells are prone to failure from drought or heavy pumpage. They are vulnerable to contamination and are not acceptable as a public water supply in many locations.
- *Driven wells*—Driven wells consist of a pipe casing terminating in a point slightly greater in diameter than the casing. The pointed well screen and the lengths of pipe attached to it are pounded down or driven in the same

manner as a pile, usually with a drop hammer, to the water-bearing strata. Driven wells are usually 2 to 3 inches in diameter and are used only in unconsolidated materials. This type of shallow well is not acceptable as a public water supply.

- *Bored wells*—Bored wells range from 1 to 36 inches in diameter and are constructed in unconsolidated materials. The boring is accomplished with augers (either hand or machine driven) that fill with soil and then are drawn to the surface to be emptied. The casing may be placed after the well is completed (in relatively cohesive materials) but must advance with the well in noncohesive strata. Bored wells are not acceptable as a public water supply.

DEEP WELLS

Deep wells are the usual source of groundwater for municipalities. Deep wells tap thick and extensive aquifers that are not subject to rapid fluctuations in water level (remember that the *piezometric surface* is the height to which water will rise in a tube penetrating a confined aquifer) and that provide a large and uniform yield. Deep wells typically yield water of more consistent quality than shallow wells, although the quality is not necessarily better. Deep wells are constructed by a variety of techniques; we discuss two of these techniques below:

- *Jetted wells*—Jetted well construction commonly employs a jetting pipe with a cutting tool. This type of well cannot be constructed in clay or hardpan or where boulders are present. Jetted wells are not acceptable as a public water supply.
- *Drilled wells*—Drilled wells are usually the only type of well allowed for use in most public water supply systems. Several different methods of drilling are available, all of which are capable of drilling wells of extreme depth and diameter. Drilled wells are constructed using a drilling rig that creates a hole into which the casing is placed. Screens are installed at one or more levels when water-bearing formations are encountered.

COMPONENTS OF A PRODUCTION WELL

The components that make up a well system include the well itself, the building and the pump, and related piping system. In this section, we focus on the components that make up the well itself. Many of these components are shown in Figure 5.9.

Well Casing

A well is a hole in the ground called the *borehole*. To prevent collapse, a casing is placed inside the borehole. The well casing prevents the walls of the hole from collapsing and prevents contaminants (either surface or subsurface) from entering the water source. The casing also provides a column of stored water and housing for the pump mechanisms and pipes. Well casings constructed of steel or plastic material are acceptable. The well casing must extend a minimum of 12 inches above grade.

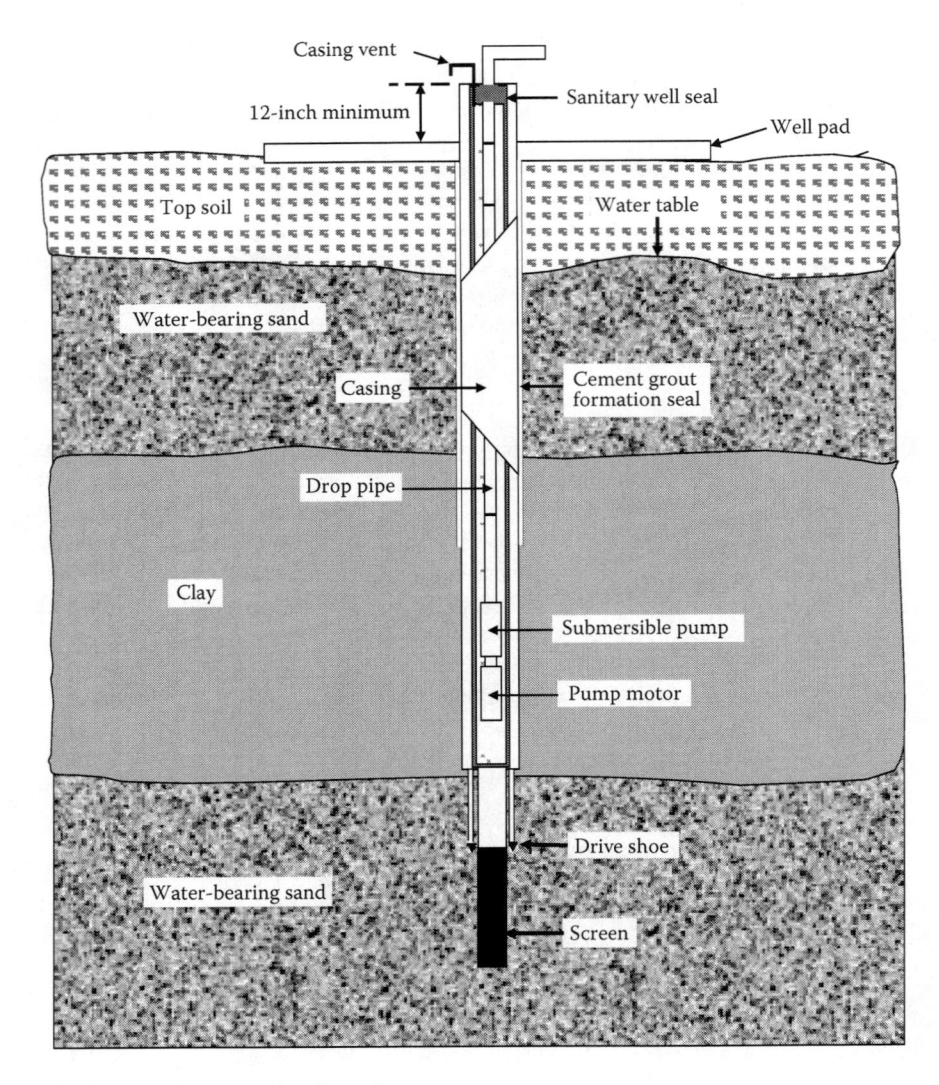

FIGURE 5.9 Components of a well.

Grout

To protect the aquifer from contamination, the casing is sealed to the borehole near the surface and near the bottom where it passes into the impermeable layer with grout. This sealing process keeps the well from being polluted by surface water and seals out water from water-bearing strata that have undesirable water quality. Sealing also protects the casing from external corrosion and restrains unstable soil and rock formations. Grout consists of near cement that is pumped into the annular space (it is completed within 48 hours of well construction); it is pumped under continuous pressure starting at the bottom and progressing upward in one continuous operation.

Well Pad

The well pad provides a ground seal around the casing. The pad is constructed of reinforced concrete 6 ft by 6 ft (6 inches thick) with the well head located in the middle. The well pad prevents contaminants from collecting around the well and seeping down into the ground along the casing.

Sanitary Seal

To prevent contamination of the well, a sanitary seal is placed at the top of the casing. The type of seal varies depending on the type of pump used. The sanitary seal contains openings for power and control wires, pump support cables, a drawdown gauge, discharge piping, pump shaft, and air vent, while providing a tight seal around them.

Well Screen

Screens can be installed at the intake points on the end of a well casing or on the end of the inner casing on a gravel pack well. These screens perform two functions: (1) support the borehole, and (2) reduce the amount of sand that enters the casing and the pump. They are sized to allow the maximum amount of water while preventing the passage of sand, sediment, and gravel.

Casing Vent

The well casing must have a vent to allow air into the casing as the water level drops. The vent terminates 18 inches above the floor with a return bend pointing downward. The opening of the vent must be screened with No. 24 mesh stainless steel to prevent entry of vermin and dust.

Drop Pipe

The drop pipe or riser is the line leading from the pump to the well head. It provides adequate support so an aboveground pump does not move and so a submersible pump is not lost down the well. This pipe is either steel or polyvinylchloride (PVC). Steel is the most desirable.

WELL HYDRAULICS

When the source of water for a water distribution system is a groundwater supply, operator knowledge of well hydraulics is important. In this section, basic well hydraulics terms are presented and defined, and they are related pictorially (see Figure 5.10).

DID YOU KNOW?

For confined aquifers, 80 to 90% of the thickness of the water-bearing zone should be screened. Best results are obtained by centering the screen section in the aquifer. For unconfined aquifers, maximum specific capacity is obtained by using the longest screen possible, but more available drawdown results from using the shortest screen possible. These factors are optimized by screening the bottom 30 to 50% of the aquifer (Driscoll, 1986).

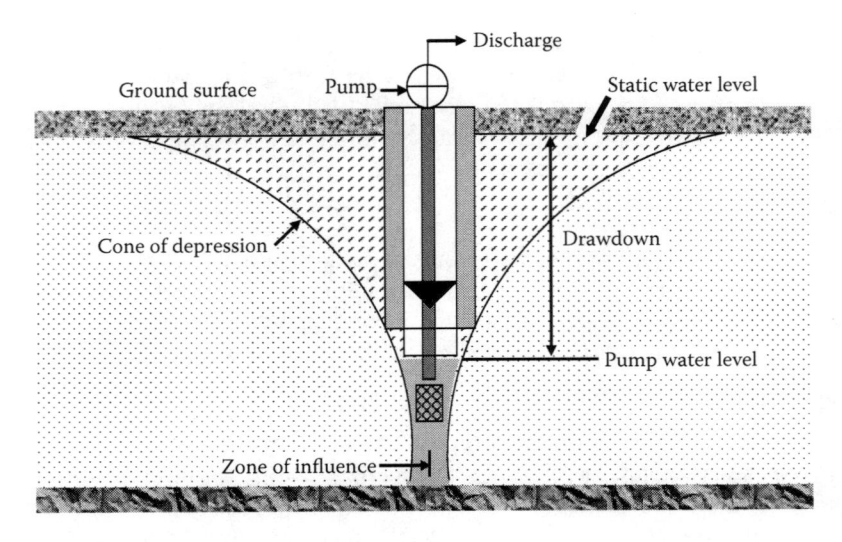

FIGURE 5.10   Hydraulic characteristics of a well.

Basic Well Hydraulics

- *Static water level*—The water level in a well when no water is being taken from the groundwater source (i.e., the water level when the pump is off; see Figure 5.10). Static water level is normally measured as the distance from the ground surface to the water surface. This is an important parameter because it is used to measure changes in the water table.
- *Pumping water level*—The water level when the pump is off. When water is pumped out of a well, the water level usually drops below the level in the surrounding aquifer and eventually stabilizes at a lower level; this is the pumping level (see Figure 5.10).
- *Drawdown*—The difference, or the drop, between the static water level and the pumping water level, measured in feet. Simply, it is the distance the water level drops when pumping begins (see Figure 5.10).
- *Cone of depression*—In unconfined aquifers, water flows in the aquifer from all directions toward the well during pumping. The free water surface in the aquifer then takes the shape of an inverted cone or curved funnel line. The curve of the line extends from the pumping water level to the static water level at the outside edge of the zone (or radius) of influence (see Figure 5.10).

Note: The shape and size of the cone of depression are dependent on the relationship between the pumping rate and the rate at which water can move toward the well. If the rate is high, the cone will be shallow and its growth will stabilize. If the rate is low, the cone will be sharp and continue to grow in size.

- *Zone (or radius) of influence*—The distance between the pump shaft and the outermost area affected by drawdown. The distance depends on the porosity of the soil and other factors. This parameter becomes important in well fields with many pumps. If wells are set too closely together, the zones of influence will overlap, increasing the drawdown in all wells. Obviously, pumps should be spaced apart to prevent this from happening.

Two important parameters not shown in Figure 5.10 are well yield and specific capacity:

1. *Well yield* is the rate of water withdrawal that a well can supply over a long period, or, alternatively, the maximum pumping rate that can be achieved without increasing the drawdown. The yield of small wells is usually measured in gallons per minute (liters per minute) or gallons per hour (liters per hour). For large wells, it may be measured in cubic feet per second (cubic meters per second).
2. *Specific capacity* is the pumping rate per foot of drawdown (gpm/ft), or

$$\text{Specific capacity} = \text{Well yield} \div \text{Drawdown} \tag{5.16}$$

■ **EXAMPLE 5.1**

Problem: If the well yield is 300 gpm and the drawdown is measured to be 20 ft, what is the specific capacity?

Solution:

$$\text{Specific capacity} = 300 \div 20 = 15 \text{ gpm per ft of drawdown}$$

Specific capacity is one of the most important concepts in well operation and testing. The calculation should be made frequently in the monitoring of well operation. A sudden drop in specific capacity indicates problems such as pump malfunction, screen plugging, or other problems that can be serious. Such problems should be identified and corrected as soon as possible.

Well Drawdown Calculations

As shown in Figure 5.9, drawdown is the drop in the level of water in a well when water is being pumped. Drawdown is usually measured in feet or meters. One of the most important reasons for measuring drawdown is to make sure that the source water is adequate and not being depleted. The data collected to calculate drawdown can indicate if the water supply is slowly declining. Early detection can give the system time to explore alternative sources, establish conservation measures, or obtain any special funding that may be needed to get a new water source. Well drawdown is the difference between the pumping water level and the static water level:

$$\text{Drawdown (ft)} = \text{Pumping water level (ft)} - \text{Static water level (ft)} \tag{5.17}$$

■ **EXAMPLE 5.2**

Problem: The static water level for a well is 70 ft. If the pumping water level is 90 ft, what is the drawdown?

Solution:

$$\text{Drawdown} = \text{Pumping water level (ft)} - \text{Static water level (ft)} = 90 \text{ ft} - 70 \text{ ft} = 20 \text{ ft}$$

■ EXAMPLE 5.3

Problem: The static water level of a well is 122 ft. The pumping water level is determined using the sounding line. The air pressure applied to the sounding line is 4.0 psi, and the length of the sounding line is 180 ft. What is the drawdown?

Solution: Calculate the water depth in the sounding line and the pumping water level:

1. Water depth in sounding line = 4.0 psi × 2.31 ft/psi = 9.2 ft
2. Pumping water level = 180 ft − 9.2 ft = 170.8 ft

Then calculate drawdown as usual:

$$\text{Drawdown} = \text{Pumping water level (ft)} - \text{Static water level (ft)}$$
$$\text{Drawdown} = 170.8 \text{ ft} - 122 \text{ ft} = 48.8 \text{ ft}$$

Well Yield Calculations

Well yield is the volume of water per unit of time that is produced from the well pumping. Usually, well yield is measured in terms of gallons per minute (gpm) or gallons per hour (gph). Sometimes, large flows are measured in cubic feet per second (cfs). Well yield is determined by using the following equation:

$$\text{Well yield (gpm)} = \text{Gallons produced} \div \text{Duration of test (min)} \qquad (5.18)$$

■ EXAMPLE 5.4

Problem: When the drawdown level of a well was stabilized, it was determined that the well produced 400 gal during a 5-min test. What was the well yield?

Solution:

Well yield = Gallons produced ÷ Duration of test (min) = 400 gal ÷ 5 min = 80 gpm

■ EXAMPLE 5.5

Problem: During a 5-min test for well yield, a total of 780 gal was removed from the well. What was the well yield in gpm? In gph?

Solution:

Well yield = Gallons produced ÷ Duration of test (min) = 780 gal ÷ 5 min = 156 gpm

Then convert gpm flow to gph flow:

$$156 \text{ gpm} \times 60 \text{ min/hr} = 9360 \text{ gph}$$

Specific Yield Calculations

Specific yield is the discharge capacity of the well per foot of drawdown. The specific yield may range from 1 gpm/ft drawdown to more than 100 gpm/ft drawdown for a properly developed well. Specific yield is calculated using Equation 5.16:

$$\text{Specific yield (gpm/ft)} = \text{Well yield (gpm)} \div \text{Drawdown (ft)} \qquad (5.19)$$

■ EXAMPLE 5.6

Problem: A well produces 260 gpm. If the drawdown for the well is 22 ft, what is the specific yield in gpm/ft?

Solution:

Specific yield = Well yield (gpm) ÷ Drawdown (ft) = 260 gpm ÷ 22 ft = 11.8 gpm/ft

■ EXAMPLE 5.7

Problem: The yield for a particular well is 310 gpm. If the drawdown for this well is 30 ft, what is the specific yield in gpm/ft?

Solution:

Specific yield = Well yield (gpm) ÷ Drawdown (ft) = 310 gpm ÷ 30 ft = 10.3 gpm/ft

DEPLETING THE GROUNDWATER BANK ACCOUNT

Normally, as shown in Figure 5.11, water is recharged to the groundwater system by percolation of water from precipitation and then flows to a stream or other water body through the groundwater system. Where surface water, such as lakes and rivers, is scarce or inaccessible, groundwater provides many of the hydrologic needs of people everywhere. In the United States, it is the source of drinking water for about half of the total population and nearly all of the rural population, and it provides over 50 billion gallons per day for agricultural needs. Water stored in the ground for use can be compared to money in a bank account. However, water pumped from the groundwater system causes the water table to lower and alters the direction of groundwater movement. Some water that used to flow to a stream or other water body no longer

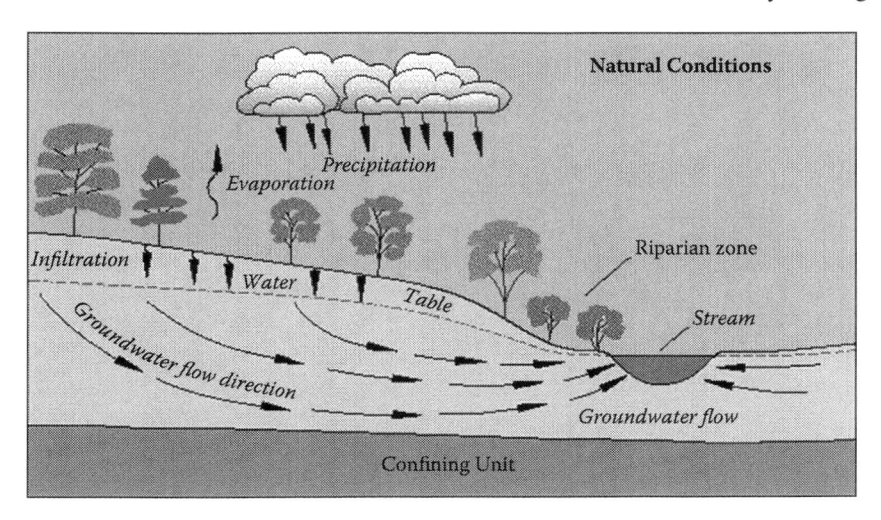

FIGURE 5.11 Normal groundwater conditions. (From USGS, *Groundwater Flow and Effects of Pumping*, U.S. Geological Survey, Washington, DC, 2016.)

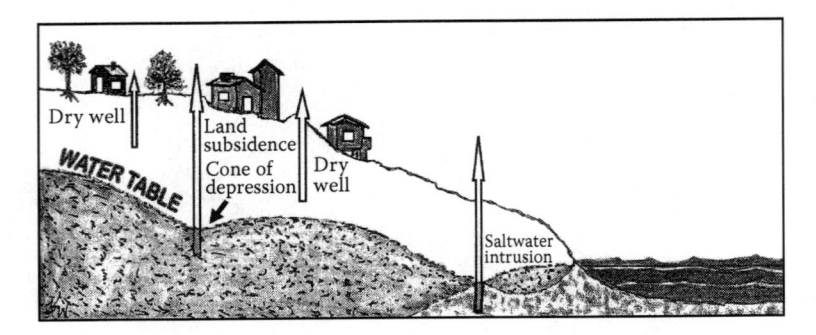

FIGURE 5.12 Impacts of overpumping groundwater and groundwater depletion. (Illustration by F.R. Spellman and Kathern Welsh, adapted from USGS, *Groundwater Flow and Effects of Pumping*, U.S. Geological Survey, Washington, DC, 2016.)

does so, and some water may be drawing from the steam into the groundwater system, thereby reducing the amount of streamflow. *Groundwater depletion*, or long-term declines in water level, is a key issue associated with groundwater use. Many areas of the United States are experiencing groundwater depletion (USGS, 2016a).

OVERDRAWING THE GROUNDWATER BANK ACCOUNT

If you withdraw money at a faster rate than you deposit new money in your bank account, you will eventually begin to have money supply problems. Likewise, if water is pumped out of the ground faster than it is being replenished over the long term, then similar problems will develop. The volume of groundwater in storage is decreasing in many areas of the United States in response to pumping. Groundwater depletion is primarily caused by sustained groundwater pumping. Some of the negative effects of groundwater depletion are listed below and depicted in Figure 5.12:

- Drying up of wells
- Reduction of water in streams and lakes
- Deterioration of water quality
- Increased pumping costs
- Land subsidence

Although the negative effects of groundwater depletion listed above and shown in Figure 5.12 are of major concern to those affected by depleted groundwater bank accounts, in this book our major concern related to groundwater depletion is the lowering of the water table and land subsidence. The next chapter focuses on land subsidence and all of its implications.

THE BOTTOM LINE

The absolute value of quality groundwater and the ever-increasing need for it cannot be overstated. The bottom line on groundwater is simple: Without groundwater, Planet Earth and all of its inhabitants could not survive.

REFERENCES AND RECOMMENDED READING

AWWA. (1995). *Basic Science Concepts and Applications: Principles and Practices of Water Supply Operations*, 2nd ed. Denver, CO: American Water Works Association.

Driscoll, F. (1986). *Groundwater and Wells*. St. Paul, MN: Johnson Screens.

Grant, D.M. (1991). Open channel flow measurement. In: *Flow Measurement* (Spitzer, D.W., Ed.), pp. 252–290. Research Triangle Park, NC: Instrument Society of America.

Hauser, B.A. (1993). *Hydraulics for Operators*. Boca Raton, FL: Lewis Publishers.

Hauser, B.A. (1996). *Practical Hydraulics Handbook*, 2nd ed. Boca Raton, FL: Lewis Publishers.

Holman, S. (1998). *A Stolen Tongue*. New York: Anchor.

Johnson, A.I. (1967). *Specific Yield: Compilation of Specific Yields for Various Materials*, Water-Supply Paper 1662-D. Washington, DC: U.S. Geological Survey.

Kawamura, S. (2000). *Integrated Design and Operation of Water Treatment Facilities*, 2nd ed. New York: John Wiley & Sons.

Lindeburg, M.R. (1986). *Civil Engineering Reference Manual*, 4th ed. San Carlos, CA: Professional Publications.

Lohman, S.W. (1972). *Ground-Water Hydraulics*, U.S. Geological Survey Professional Paper 708. Washington, DC: U.S. Government Printing Office.

Magnusson, R.J. (2001). *Water Technology in the Middle Ages*. Baltimore, MD: The Johns Hopkins University Press.

McGhee, T.J. (1991). *Water Supply and Sewerage*, 2nd ed. New York: McGraw-Hill.

Meinzer, O.E. (1923). *Outline of Groundwater Hydrology, with Definitions*, Water-Supply Paper 494. Washington, DC: U.S. Geological Survey.

Slichter, C.S. (1899). *Theoretical Investigations of the Motion of Ground Waters*, 19th Annual Report. Washington, DC: U.S. Geological Survey.

Spellman, F.R. (2008). *The Science of Water*, 2nd ed. Boca Raton, FL: CRC Press.

Spellman, F.R. (2015). *The Science of Water*, 3rd ed. Boca Raton, FL: CRC Press.

Spellman, F.R. and Drinan, J. (2001). *Water Hydraulics*. Boca Raton, FL: CRC Press.

Terzaghi, K. (1942). Soil moisture and capillary phenomena in soils. In: *Hydrology* (Meinzer, O.E., Ed.), pp. 331–363. New York: McGraw-Hill.

USEPA. (1991). *Flow Instrumentation: A Practical Workshop on Making Them Work*. Sacramento, CA: Water and Wastewater Instrumentation Testing Association.

USGS. (2016a). *Groundwater Flow and Effects of Pumping*. Washington, DC: U.S. Geological Survey (https://water.usgs.gov/edu/earthgwdecline.html).

USGS. (2016b). *Groundwater Depletion*. Washington, DC: U.S. Geological Survey (http://water.usgs.gov/edu/gwdepletion.html).

Viessman, Jr., W. and Hammer, M.J. (1998). *Water Supply and Pollution Control*, 6th ed. Menlo Park, CA: Addison-Wesley.

6 Land Subsidence

I came where the river
Ran over stones;
My ears knew
An early joy.
And all the waters
Of all the streams
Sang in my veins
That summer day.

Roethke (1948)

The science of hydrology would be relatively simple if water were unable to penetrate below the earth's surface.

—Harold E. Thomas, U.S. Geological Survey

Some hydrologists believe that a predevelopment water budget for a groundwater system (that is, a water budget for the natural conditions before humans used the water) can be used to calculate the amount of water available for consumption (or the safe yield). This concept has been referred to as the "Water-Budget Myth."

Alley et al. (1999)

VANISHING LAND*

From the Lower Chesapeake Bay Region (Hampton Roads or Tidewater) of Virginia to San Francisco Bay/Delta to the Florida Everglades and from upstate New York to Houston, people are dealing with a common problem in these diverse locations—vanishing land as a result of land subsidence due to the withdrawal of groundwater. Vanishing land due to subsidence is not an isolated problem: an area of more than 17,000 square miles in 45 states, an area roughly the size of New Hampshire and Vermont combined, has been directly affected by land subsidence. More than 80% of the identified subsidence in this country is a consequence of our exploitation of underground water. Moreover, it seems certain that the increasing development of land and water resources threatens to exacerbate existing land-subsidence problems and initiate new ones (Galloway et al., 1999). This chapter focuses on three principal processes causing land subsidence: compaction of aquifer systems, oxidation of organic soils, and the collapse of cavities in carbonate and evaporite rocks. Several examples are presented to illustrate the impacts of land subsidence, past and present, and, most importantly, Chapter 7 presents the value of applying science and engineering innovations to effectively mitigate or limit damages from land subsidence.

---

* Adapted from Leatherman, S.P. et al., *Vanishing Lands: Sea Level, Society and Chesapeake Bay*, U.S. Fish and Wildlife Service, Annapolis, MD, 1995.

> **DID YOU KNOW?**
>
> Water budgets provide a means for evaluating the availability and sustainability of a water supply. A water budget simply states that the rate of change in water stored in an area, such as a watershed, is balanced by the rate at which water flows into and out of the area.

One thing is certain: scientific understanding is critical to the formulation of balanced decisions about the management of land and water resources. When scientific information is presented in plain English, and in a conversational way, understanding flows like unimpeded groundwater.

WITHDRAWALS FROM THE GROUNDWATER BANK ACCOUNT

Permanent subsidence can occur when water stored beneath the Earth's surface is removed by pumping. The reduction of fluid pressure in the pores and cracks of aquifer systems, especially in unconsolidated rocks, is inevitably accompanied by some deformation of the aquifer system. Because the granular structure—the so-called "skeleton"—of the aquifer system is not rigid, a shift in the balance of support for the overlying material causes the skeleton to deform slightly. Both the aquifers and aquitards that constitute the aquifer system undergo deformation, but to different degrees. During the typically slow process of aquitard drainage (when the irreversible compression or consolidation of aquitards occurs) is when almost all of the permanent subsidence takes place (Tolman and Poland, 1940). This concept, known as the *aquitard drainage model*, has formed the theoretical basis of many successful subsidence investigations.

EFFECTIVE STRESS

The principle of effective stress was first proposed by Terzaghi (1925). According to this principle, when the support provided by fluid pressure is reduced, such as when groundwater levels are lowered, support previously provided by the pore-fluid pressure is transferred to the skeleton of the aquifer system, which compresses to a degree. On the other hand, when the pore-fluid pressure is increased, such as when groundwater recharges the aquifer system, support previously provided by the skeleton is transferred to the fluid and the skeleton expands. In this way, the skeleton

> **DID YOU KNOW?**
>
> Studies of subsidence in the Santa Clara Valley in California established the theoretical and field application of the laboratory-derived principle of effective stress and theory of hydrodynamic consolidation to the drainage and compaction of aquitards (Green, 1964; Poland and Green, 1962; Poland and Ireland, 1988; Tolman and Poland, 1940).

alternatively undergoes compression and expansion as the pore-fluid pressure fluctuates with aquifer-system discharge and recharge. When the load on the skeleton remains less than any previous maximum load, the fluctuations create only a small elastic deformation of the aquifer system and small displacement of land surface. This fully recoverable deformation occurs in all aquifer systems, commonly resulting in seasonal, reversible displacements in land surface of up to 1 inch or more in response to the seasonal changes in groundwater pumpage (Galloway et al., 1999).

PRECONSOLIDATION STRESS

The maximum level of past stressing of a skeletal element is the preconsolidation stress. When the load on the aquitard skeleton exceeds the preconsolidation stress, the aquitard skeleton may undergo significant, permanent rearrangement, resulting in irreversible compaction. Because the skeleton defines the pore structure of the aquitards, this results in a permanent reduction of pore volume as the pore fluid is squeezed out of the aquitards into the aquifers. In confined aquifer systems subject to large-scale overdraft, the volume of water derived from irreversible aquitard compaction is essentially equal to the volume of subsidence and can typically range from 10 to 30% of the total volume of water pumped. This represents a one-time mining of stored groundwater and a small permanent reduction in the storage capacity of the aquifer system. Alternative names for preconsolidation stress are preconsolidation pressure, precompression stress, precompaction stress, and preload stress (Dawidowski and Koolen, 1994).

ROLE OF AQUITARDS IN COMPACTION

In recent decades, increasing recognition has been given to the critical role of aquitards in the intermediate and long-term response of alluvial systems to groundwater pumpage. Aquitard systems play an important role in compaction. In many such systems, interbedded layers of silts and clays, once dismissed as non-water yielding, comprise the bulk of the groundwater storage capacity of the confined aquifer system. This is the case because of their substantially greater porosity and compressibility and, in many cases, their greater aggregate thickness compared to the more transmissive, coarser-grained sand and gravel layers.

Aquitards are less permeable than aquifers; thus, the vertical drainage of aquitards into adjacent pumped aquifers may proceed very slowly and thus lag far behind the changing water levels in adjacent aquifers. The lagged response within the inner portions of a thick aquitard may be largely isolated from the higher frequency seasonal fluctuations and more influenced by lower frequency, longer term trends in groundwater levels. Because the migration of increased internal stress into the aquitard accompanies its drainage, as more fluid is squeezed from the interior of the aquitard, larger and larger internal stresses propagate farther into the aquitard (Galloway et al., 1999).

When the preconsolidation stress is exceeded by the internal stresses, the compressibility increases dramatically, typically by a factor of 20 to 100 times, and the resulting compaction is largely nonrecoverable. At stresses greater than the preconsolidation stress, the lag in aquitard drainage increases by comparable factors, and concomitant

DID YOU KNOW?

Responses to changing water levels following several decades of groundwater development suggest that stresses directly driving much of the compaction are somewhat insulated from the changing stresses caused by short-term water-level variations in the aquifers.

DID YOU KNOW?

Hydrocompaction—compaction due to wetting—is a near-surface phenomenon that produces land-surface subsidence through a mechanism entirely different from compaction of deep, overpumped aquifer systems.

compaction may require decades or centuries to approach completion. The theory of hydrodynamic consolidation (Terzaghi, 1925)—an essential element of the *aquitard drainage model*—describes the delay involved in draining aquitards when heads are lowered in adjacent aquifers, as well as the residual compaction that may continue long after drawdowns in the aquifers have essentially stabilized. Numerical modeling based on Terzaghi's theory has successfully simulated complex histories of compaction observed in response to measure water-level fluctuations (Helm, 1975). Hydrodynamic lag, which is a delay in the propagation fluid-pressure changes between the aquifers and aquitards, can be seen in the Antelope Valley, Mojave Desert, California.

SUBSIDENCE MODEL OF ANTELOPE VALLEY[*]

PROJECT

In cooperation with the Los Angeles County Department of Public Works, Antelope Valley–East Kern Water Agency, Palmdale Water District, and Edwards Air Force Base, the U.S. Geological Survey (USGS) updated a three-layer version of the Antelope Valley groundwater-flow and land-subsidence model, based on MODFLOW, the USGS modular three-dimensional groundwater-flow model. In this scientific investigations study, physical processes were added in the model simulation to obtain an improved representation to an improved representation of the aquifer system.

STUDY PROCESS

In their project report, Siade et al. (2014) described Antelope Valley, California, as a topographically closed basin in the western part of the Mojave Desert, approximately 50 miles northeast of Los Angeles. The Antelope Valley groundwater basin is about 940 square miles and is separated from the northern part of Antelope Valley by faults

---

[*] Adapted from Siade, A.J. et al., *Groundwater-Flow and Land-Subsidence Model of Antelope Valley, California*, Report 2014-5166, U.S. Geological Survey, Reston, VA, 2014.

and low-lying hills. Prior to 1972, more than 90% of the total water supply in the valley was provided by groundwater; after 1972, it has provided between 50 and 90% (the rest of the water supplies have been imported). The Antelope Valley groundwater basin, which includes the growing cities of Lancaster and Palmdale, is the source of most of the groundwater pumping in the valley. Declines of more than 270 feet in the groundwater level have resulted in an increase in pumping lifts, reduced well efficiency, and land subsidence of more than 6 feet in some areas. Limited supplies of imported water and continued urban growth may increase reliance on groundwater.

In 2011, the Los Angeles County Superior Court of California ruled that the Antelope Valley groundwater basin was in overdraft—that is, groundwater extractions were in excess of the Court-defined safe yield of the groundwater basin (110,000 acre-ft/yr). Natural recharge is an important component of total recharge in Antelope Valley; however, the exact quantity and distribution of natural recharge, primarily in the form of mountain-front recharge (i.e., contributions of mountains to recharge), are uncertain.

Antelope Valley's groundwater-flow system consists of three aquifers: the upper, middle, and lower aquifers. The three aquifers, which were identified on the basis of the hydrologic properties, age, and depth of the unconsolidated deposits, consist of gravel, sand, silt, and clay alluvial deposits and clay and silt clay lacustrine (lake) deposits. Prior to groundwater development in the valley, recharge was primarily the infiltration of runoff from the surrounding mountains. Groundwater flowed from the recharge areas to discharge areas around the playas (flat, dried-up land) where it discharged from the aquifer system from springs or via evapotranspiration. Faults, which are partial barriers to horizontal groundwater flow, have been identified in the groundwater basin. Because water levels have declined due to groundwater development, natural sources of discharge have been eliminated and pumping for agricultural and urban uses has become the primary source of discharge from the groundwater system. Agricultural irrigation return flow has become an important source of recharge to the aquifer system.

When devising and implementing the model for groundwater flow and land subsidence in Antelope Valley, the basin was discretized horizontally into a grid of 130 rows and 118 columns of square cells 1 kilometer (0.621 mile) on a side, and vertically into four layers representing the upper (two layers), middle (one layer), and lower (one layer) aquifers. The model simulated faults that were thought to act as horizontal-flow barriers. The model was calibrated to simulate steady-state conditions, represented by 1915 water levels and transient-state conditions, represented by 1915–95, by using water-level and subsidence data. Data from a previously published numerical model of the Antelope Valley groundwater basin were used to provide initial estimates of the aquifer system properties and stresses; estimates also were obtained from recently collected hydrologic data and from results of simulations of groundwater-flow and land-subsidence models of the Edwards Air Force Base area. Researchers modified some of these initial estimates during model calibration. Groundwater pumpage for agriculture was estimated on the basis of irrigated crop acreage and crop consumptive-use data. Pumpage for public supply, which is metered, was compiled and entered into a database used for this study. Estimated annual agricultural pumpage peaked at 395,000 acre-ft in 1951 and then declined because of declining agricultural production. Recharge from irrigation return flows was assumed to be 30% of agricultural pumpage; delays associated with return flow moving though the unsaturated zone were also simulated. Estimates from previous studies were used to approximate the annual quantity of mountain-front recharge. The model was calibrated using the PEST software suite; prior information from the area was incorporated through the use of Tikhonov regularization. During model calibration, the estimated mountain-front recharge was reduced from the previously estimated 30,300 acre-ft/yr to 29,150 acre-ft/yr.

RESULTS

In their study, the investigators reported that the results of the simulations using the calibrated model indicated that simulated groundwater pumpage exceeded recharge in most years, resulting in an estimated cumulative depletion in groundwater storage of 8,700,000 acre-ft during the transient-simulation period (1915–2005). About 15,000,000 acre-ft of cumulative groundwater pumpage was simulated during the transient-simulation period (1915–2005), reaching a maximum rate of about 400,000 acre-ft/yr in 1951. Compared to 1915 conditions in agricultural areas, groundwater pumpage resulted in simulated hydraulic heads declining by more than 150 feet. Depletion of groundwater storage caused the decline in hydraulic head in the groundwater basin. In turn, the simulated decline in hydraulic head in the groundwater basin has resulted in the decrease in natural discharge from the basin and has caused compaction of aquitards, resulting in land subsidence. After about 90 years of groundwater development, the areal distribution of total simulated land subsidence for 2005 indicates that land subsidence occurred through almost the entire Lancaster subbasin, with a maximum of approximately 9.4 feet in the central and eastern parts of the subbasin.

An important objective of this study was to systematically address the uncertainty in estimates of natural recharge and related aquifer parameters by using the groundwater-flow and land-subsidence model with observational data and expert knowledge. After the model was calibrated to the observations and a reasonable parameter set obtained, the parameter null space—parameter values that do not appreciably affect the model calibration but may have importance for production—was identified. The Null Space Monte Carlo method was used to address the effect of parameter uncertainty on the estimation of mountain-front recharge. To portray the reasonableness of larger natural-recharge rates, the Pareto trade-off method of visualizing uncertainty was also employed. The results indicated that the total mountain-front recharge likely ranges between 28,000 and 44,000 acre-ft/yr, which is appreciably less than published estimates of 60,000 acre-ft/yr. Moreover, expected errors associated with agricultural pumpage estimates used in this study were found to have relatively little effect on the estimates of mountain-front recharge, reflecting the difficulty in increasing recharge through manipulation of other components of the water budget. The calibrated model was used to simulate the response of the aquifer to potential future pumping scenarios:

- No change in the distribution of pumpage, or *status quo*
- Redistribution of pumpage
- Artificial recharge

All of these scenarios specify a total pumpage throughout the Antelope Valley of 110,000 acre-ft/yr according to the safe yield value ruled by the Los Angeles County Superior Court of California. This reduction of groundwater pumpage is assumed to be uniform throughout the basin, based on a 10% reduction of the total pumpage in 2005 to achieve the110,000 acre-ft/yr level. The calibrated Antelope Valley groundwater-flow and land-subsidence model was used to simulate the hydrologic effects of the three groundwater management scenarios during a 50-year period by using the reduced, temporally constant, pumpage distribution.

- *First scenario*—Results indicated that the total drawdown observed since predevelopment would continue, with values exceeding 325 feet near Palmdale; consequently, land subsidence would also continue, with additional subsidence (since 2005) exceeding 3 feet in the central part of the Lancaster subbasin.
- *Second scenario*—This scenario evaluated redistributing pumpage from areas in the Lancaster subbasin where simulated hydraulic head declines were the greatest to areas where declines were smallest. Neither a formal optimization algorithm nor water-rights allocations were considered when redistributing the pumpage. Results indicated that hydraulic heads near Palmdale, where the pumpage was reduced, would recover by about 200 feet compared to 2005 conditions, with only 30 feet of additional drawdown in the northwestern part of the Lancaster subbasin, where the pumpage was increased. The magnitude of the simulated addition land subsidence decreased slightly compared to the first, *status quo* scenario, but land subsidence continued to be simulated throughout most of the northern part of the Lancaster subbasin.
- *Third scenario*—This scenario consisted of two artificial-recharge simulations along the Upper Amargosa Creek channel and at a site located north of Antelope Buttes. Results indicated that applying artificial recharge at these sites would yield continued drawdowns and associated land subsidence. However, the magnitudes of drawdown and subsidence would be smaller than those simulated in the *status quo* scenario, indicating that artificial-recharge operations in the Antelope Valley could be expected to reduce the magnitude and extent of continued water-level declines and associated land subsidence.

SUBSIDENCE IN SILICON VALLEY[*]

Silicon Valley, the birthplace of the global electronics industry, is located in the Santa Clara Valley, California, and is part of a structured trough extending about 90 miles southeast from San Francisco. The northern third of the trough is occupied by the San Francisco Bay, the central third by the Santa Clara Valley, and the southern third by the San Benito Valley. The northern Santa Clara Valley, roughly from Palo Alto to the Coyote Narrows (10 miles southeast of downtown San Jose), is now densely populated.

[*] Adapted from Ingebritsen, S.E. and Jones, D.R., in *Land Subsidence in the United States*, Galloway, D., Jones, D.R., and Ingebritsen, S.E., Eds., U.S. Geological Survey Circular 1182, U.S. Department of Interior, Reston, VA, 1999, pp. 15–22.

From the early 1900s to prior to World War II, the Santa Clara Valley region was a premier fruit-growing region that was intensely cultivated; it was an area of extensive orchards of apricots, plums, cherries, and pears. In the post-World War II era (circa 1945–1970), rapid population growth was associated with the transition from an agriculturally based economy to an industrial and urban economy. Because of the changing land and water use and importation of surface water to support the growing population, land subsidence in the Santa Clara Valley became a fact of life; it was the first area in the United States where land subsidence due to groundwater withdrawal was recognized (Tolman and Poland, 1940).

With regard to the recognition and detection of land subsidence in the Santa Clara Valley area, let's begin with the substantial land subsidence that occurred in the northern Santa Clara Valley as a result of the massive groundwater overdrafts. Detectable subsidence of the land surface (greater than 0.1 feet) took place over much of the area. The maximum subsidence occurred in downtown San Jose, where land-surface elevations decreased from about 98 feet about sea level in 1910 to about 84 feet above sea level in 1995.

By 1969, lands adjacent to the southern end of San Francisco Bay had sunk from 2 to 8 feet, putting 17 square miles of dry land below the high-tide level. The southern end of the Bay is now ringed with dikes to prevent landward movement of saltwater, and flood-control levees have been built to control the bayward-ends of stream channels. The stream channels must now be maintained well above the surrounding land in order to provide a gradient for flow to the Bay. To prevent widespread flooding during local storms in the land that has sunk below the high-tide level, storm discharge must be captured and pumped over levees.

The real eye-opener in the Santa Clara Valley region occurred in 1933, when benchmarks in San Jose that were established in 1912 were resurveyed and found to have subsided 4 feet. This finding motivated the U.S. Coast and Geodetic Survey to establish benchmarks tied to stable bedrock on the edges of the Valley. Between 1934 and 1967, the benchmark network was remeasured many times and forms the basis for mapping subsidence.

Regional leaders in the Santa Clare Valley realized that the subsidence had to be stopped. In 1935 and 1936, the Santa Clara Valley Water District built five storage dams on local streams to capture storm flows. This permitted controlled release to increase groundwater recharge through streambeds. Wet years in the early 1940s enhanced both natural and artificial recharge. Although subsidence was briefly arrested during World War II, these measures proved inadequate to halt water-level declines over the long term, and, between 1950 and 1965, subsidence resumed at an accelerated rate. In 1965, increased imports of surface water allowed the Santa Clara Valley Water District to greatly expand its program of groundwater recharge, leading to substantial recovery of groundwater levels, and there has been little additional subsidence since about 1969.

More significantly, as of 1995, water levels in the USGS monitoring well in downtown San Jose were only 35 feet below land surface, the highest levels observed since the early 1920s. A series of relatively wet years in the mid-1990s even caused a return to artesian conditions in some areas near San Francisco Bay. Some capped and long-forgotten wells near the Bay began to leak and were thereby rediscovered!

> **DID YOU KNOW?**
>
> Hydrologists use the term *acre-foot* to describe a volume of water. One acre-foot is the volume of water that will cover an area of 1 acre to a depth of 1 foot. The term is especially useful where large volumes of water are being described. One acre-foot is equivalent to 43,560 cubic feet, or about 325,829 gallons.

The bottom line is that subsidence in the Santa Clara Valley was caused by the decline of artesian pressures and the resulting increase in the effective overburden load on the water-bearing sediments. The sediments compacted under the increasing stress and the land surface sank. Most of the compaction occurred in fined-grained clay (aquitards), which are more compressible, though less permeable, than coarser-grained sediments. The low permeability of the clay layers retards and smoothes the compaction of the aquifer system relative to the water-level variations in the permeable aquifers. Since 1969, despite water-level recoveries, a small amount of additional residual compaction and subsidence has occurred. The total subsidence has been large and chiefly permanent, but future subsidence can be controlled if groundwater levels are maintained safely above their subsidence thresholds. More significantly, the Santa Clara Valley region was the first area where organized remedial action was undertaken, and subsidence was effectively arrested by about 1969.

MINING GROUNDWATER IN THE SAN JOAQUIN VALLEY[*]

Today, the San Joaquin Valley is the backbone of California's modern and highly technological agricultural industry. California ranks as the largest agricultural producing state in the nation, producing 11% of the total U.S. agricultural value. The Central Valley of California, which includes the San Joaquin Valley, the Sacramento Valley, and the Sacramento–San Joaquin Delta, produces about 25% of the nation's table food on only 1% of the country's farmland (Cone, 1997).

After reading about how valuable California is to the production of foodstuffs for the nation, many might think that only in America could we produce such a bountiful food output. But at what cost? Every benefit has a cost. The plain truth is that any kind of successful agricultural production process requires the right ingredients. Along with fertile soil and accommodating climate conditions, water is required. It is the mining of groundwater for agriculture that has enabled the San Joaquin Valley of California to become one of the world's most productive agricultural regions, while simultaneously contributing to one of the single largest alterations of land surface attributed to humankind. For example, in 1970, when the last comprehensive surveys of land subsidence were made, subsidence in excess of 1 foot had affected more than 5200 square miles of irrigable land—one-half of the entire San Joaquin Valley (Poland et al., 1975). The maximum subsidence, near Mendota, was more than 28 feet (see Figure 6.1).

---

[*] Adapted from Galloway, D. and Riley, F.S., in *Land Subsidence in the United States*, Galloway, D., Jones, D.R., and Ingebritsen, S.E., Eds., U.S. Geological Survey Circular 1182, U.S. Department of Interior, Reston, VA, 1999, pp. 23–34.

FIGURE 6.1   Type of sign typically observed in land subsidence regions.

Since the early 1970s, land subsidence has continued in some locations but has generally slowed due to reductions in groundwater pumpage and the accompanying recovery of groundwater levels made possible by supplemental use of surface water diverted principally from the Sacramento–San Joaquin Delta and the San Joaquin, the Kings, Kern, and Feather Rivers for irrigation. When two droughts occurred in 1975, surface-water deliveries in the valley were sharply curtailed and demonstrated the valley's vulnerability to continued land subsidence when groundwater pumpage increased.

The history of land subsidence in the San Joaquin Valley is integrally linked to the development of agriculture and the availability of water for irrigation. Further agricultural development without accompanying subsidence is dependent on the con- tinued availability of surface water, which is subject to uncertainties due to climatic variability and pending regulatory decisions.

Throughout the valley region, groundwater occurs in shallow, unconfined (water table) or partially confined aquifers. Such aquifers are particularly important near the margins of the valley and near the toes of younger alluvial fans. A laterally extensive lacustrine clay known as the Corcoran Clay is distributed throughout the central and western valley. The Corcoran Clay, which varies in thickness from a feather edge to about 160 feet beneath the current bed of Tulare Lake, confines a deeper aquifer system comprised of fine-grained aquitards interbedded with coarser aquifers. Most of the subsidence measured in the valley has been correlated with the distribution of groundwater pumpage and the reduction of water levels in the deep confined aquifer system.

Before human intervention and development, groundwater in the alluvial sediments was replenished primarily by infiltration through stream channels near the valley margins. The eastern-valley streams carrying runoff from the Sierra Nevada provided most of the recharge for valley aquifers. Some recharge also occurred from precipitation falling directly on the valley floor and from stream and lake seepage occurring there. Over the long term, natural replenishment was dynamically balanced by natural depletion through groundwater discharge, which occurred primarily through evapotranspiration and contributions to streams flowing into the Delta. The areas of natural discharge in the valley generally corresponded with the areas of flowing, artesian wells mapped in an early USGS investigation (Mendenhall et al., 1916). Direct groundwater outflow to the Delta is thought to have been negligible.

Today, nearly 150 years since water was first diverted at Peoples Weir on the Kings River and more than 120 years after the first irrigation colonies were established in the valley, intensive development of groundwater resources for agricultural uses has drastically altered the valley's water budget. The natural replenishment of the aquifer system has remained about the same, but more water has discharged than recharged the aquifer system; the deficit may have amounted to as much as 800,000 acre-ft/yr during the late 1960s (Williamson et al., 1989). Most of the surface water now being imported is transpired by crops or evaporated from the soil. The amount of surface-water outflow from the valley has actually been reduced compared to predevelopment conditions. Groundwater in the San Joaquin Valley has generally been depleted and redistributed from the deeper aquifer system to the shallow aquifer system. This has created problems of groundwater quality and drainage in the shallow aquifer system, which is infiltrated by excess irrigation water that has been exposed to agricultural chemicals and natural salts concentrated by evapotranspiration.

The history of the development of a reliable water supply for the San Joaquin Valley began right after the 1849 Gold Rush. In 1857, the California Legislature passed an act that promoted the drainage and reclamation of river-bottom lands (Manning, 1967). By 1900, the entire flow of the Kings River and much of the flow of the Kern River had been diverted through canals and ditches to irrigate lands throughout the southern part of the valley (Nady and Larragueta, 1983). Because no significant storage facilities accompanied these earliest diversions, the agricultural water supply, and hence crop demand, was largely limited by the summer low-flows. The development of groundwater resources was prompted by the restrictions imposed by the need for constant surface-water flows, coupled with a drought occurring around 1880 and the fact that, by 1910, nearly all of the available surface-water supply in the San Joaquin Valley had been diverted.

Initially, development of the groundwater resource occurred in regions where shallow groundwater was plentiful and particularly where flowing wells were commonplace, near the central part of the valley around the old lake basins. With time and increasing usage, however, the yields of flowing wells diminished as water levels were reduced, and it became necessary to install pumps in wells to sustain flow rates. Around 1930, the development of an improved deep-well turbine pump (see Figure 6.2) and rural electrification enabled additional groundwater development

DID YOU KNOW?

The *turbine pump* consists of a motor, drive shaft, a discharge pipe of varying lengths, and one or more impeller-bowl assemblies. It is normally a vertical assembly in which the water enters at the bottom, passes axially through the impeller–bowl assembly where the energy transfer occurs, then moves upward through additional impeller–bowl assemblies to the discharge pipe. The length of this discharge pipe will vary with the distance from the wet well to the desired point of discharge (see Figure 6.2). The two basic types of turbine pumps are *line shaft turbines* and *can turbines* (for dry well installations). Due to the construction of the turbine pump, the major applications have traditionally been for pumping relatively clean water. The line shaft turbine pump has been used extensively for drinking water pumping, especially where water is withdrawn from deep wells.

for irrigation. The groundwater resource had been established as a reliable, stable water supply for irrigation. Similar histories were repeated in many other basins in California and throughout the Southwest, where surface water was limited and groundwater was readily available.

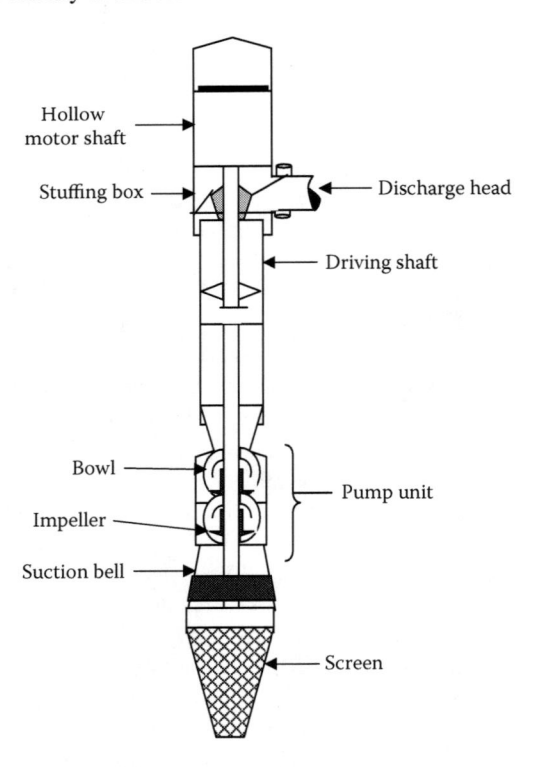

FIGURE 6.2 Vertical turbine pump.

APPLICATION

In 1951, shortly after the U.S. Bureau of Reclamation had completed the Delta–Mendota Canal, subsidence caused by withdrawal of groundwater in the northern San Joaquin valley had begun to raise concerns, largely because of the impending threat to the canal and the specter of remedial repairs. Because of this threat to the canal and to help plan other major canals and engineering proposed for construction in the subsiding areas, the USGS, in cooperation with the California Department of Water Resources, began an intensive investigation into land subsidence in the San Joaquin Valley. The objectives were to determine the causes, rates, and extent of land subsidence and to develop scientific criteria for the estimation and control of subsidence. The USGS concurrently began a federally funded research project to determine the physical principles and mechanisms governing the expansion and compaction of aquifer systems resulting from changes in aquifer hydraulic heads.

In 1955, about one-fourth (almost 8 million acre-ft) of the total groundwater extracted for irrigation in the United States was pumped in the San Joaquin Valley. The maximum changes in water levels occurred in the deep confined aquifer systems in the western and southern portions of the valley. More than 400 ft of water-level decline occurred in some west-side areas in the deep aquifer system. Until 1968, irrigation water in these areas was supplied almost entirely by groundwater. As of 1960, water levels in the deep aquifer system were declining at a rate of about 10 feet per year. Western and southern portions of the valley generally experienced more than 100 feet of water-level decline in the deep aquifer system. Water levels in the southeastern and eastern portions of the valley were generally less affected because some surface water was also available for irrigation. In the water table aquifer, few areas exceeded 100 feet of water-level decline, but a large portion of the southern valley did experience declines of more than 40 feet. In some areas on the northwest side, the water table aquifer rose up to 40 feet due to infiltration of excess irrigation water.

About 75% of the land subsidence in the San Joaquin Valley was caused by accelerated groundwater pumpage and water-level declines, principally in the deep aquifer system during the 1950s and 1960s. By the late 1960s, surface water was being diverted to agricultural interests from the Sacramento–San Joaquin Delta and the San Joaquin River through federal reclamation projects and from the Delta through the newly completed, massive State (California) Water Project. Groundwater levels began a dramatic period of recovery, and subsidence slowed or was arrested over a large part of the affected area. Water levels in the deep aquifer system recovered as much as 200 feet in the 6 years from 1967 to 1974 (Ireland et al., 1984).

Aquifer-system compaction and land subsidence began to abate when water levels began to recover in the deep aquifer system; however, many areas continued to subside, although at a lesser rate. During the period from 1968 to 1974, water levels measured in an observation well near Cantua Creek recovered more than 200 feet, while another 2 feet of subsidence continued to accrue. This apparent contradiction is the result of the time delay in the compaction of the aquitards in the aquifer system. This delay is caused by the time that it takes for pore-fluid pressures in the aquitards to equilibrate with the pressure changes occurring in the aquifers, which are much more responsive to the current volume of groundwater being pumped (or

DID YOU KNOW?

The economic impacts of land subsidence in the San Joaquin Valley are not well known. Damages directly related to subsidence have been identified, and some have been quantified. Other damages indirectly related to subsidence, such as flooding and long-term environmental effects, merit additional assessment.

not pumped) from the aquifer system. The time needed for pressure equilibrium depends largely on the thickness and permeability of the aquitards. Typically, as in the San Joaquin Valley, centuries will be required for most of the pressure equilibration to occur and therefore the ultimate compaction to be realized. Swanson (1998) stated that, "Subsidence is continuing in all historical subsidence areas ..., but at lower rates than before."

Since 1974, land subsidence has been greatly slowed or largely arrested but remains poised to resume. In fact, during the severe droughts in California in 1976–77 and 1987–91, diminished deliveries of imported water prompted some water agencies and farmers, especially in the western valley, to refurbish old pumping plants, drill new wells and begin pumping groundwater to make up for cutbacks in the imported water supply. The decisions to renew groundwater pumpage were encouraged by the fact that groundwater levels had recovered nearly to predevelopment levels. During the 1976–77 drought, after only one-third of the peak annual pumpage volumes of the 1960s had been produced, groundwater levels rapidly declined more than 150 feet over a large area and subsidence resumed. Nearly half a foot of subsidence was measured in 1977 near Cantua Creek. This scenario was repeated during the 1987–91 drought, thus underscoring the sensitive dependence between subsidence and the dynamic state of imported-water availability and use.

The bottom line is that a relatively small amount of renewed pumping caused a rapid decline in water levels, reflecting the reduced groundwater storage capacity (i.e., lost pore space) caused by aquifer-system compaction. This demonstrates the nonrenewable nature of the resource embodied in the "water of compaction" and emphasizes the fact that extraction of this resource, available only on the first cycle of large-scale drawdown, must be viewed, like more traditional forms of mining, in terms of not only its obvious economic return but also its less readily identifiable costs.

COASTAL SUBSIDENCE IN HOUSTON–GALVESTON AREA*

Houston, Texas, and land subsidence are almost synonymous and becoming much more so with each passing day. Notwithstanding the progression of sea-level rise and land subsidence in the Hampton Roads area of Virginia, possibly no other location in the United States has been more adversely affected by land subsidence than the Houston area. Extensive subsidence, caused mainly by groundwater pumping but

---

* Adapted from Coplin, L.S. and Galloway, D., in *Land Subsidence in the United States*, Galloway, D., Jones, D.R., and Ingebritsen, S.E., Eds., U.S. Geological Survey Circular 1182, U.S. Department of Interior, Reston, VA, 1999, pp. 35–48.

also by oil and gas extraction, has increased the frequency of flooding; caused extensive damage to industrial and transportation infrastructure; motivated major investments in levees, reservoirs, and surface-water distribution facilities; and caused substantial loss of wetland habitat.

In some localities near Houston the effects of subsidence are quite evident. In this low-lying coastal environment, as much as 10 feet of subsidence has shifted the position of the coastline and changed the distribution of wetlands and aquatic vegetation. In fact, the San Jacinto Battleground State Historic Site, site of the battle that won Texas independence, is now partly submerged. This park, about 20 miles east of downtown Houston on the shores of Galveston Bay, commemorates the April 21, 1836, victory of Texans led by Sam Houston over Mexican forces led by Santa Ana. About 100 acres of the park are now under water due to subsidence, and part of the remaining area must now be protected from the Bay by dikes that trap local rain water, which must then be removed by pumps. Groundwater pumpage and subsidence at many localities in the Houston area have generated fault movement, leading to visible fracturing, surface offsets, and associated property damage.

Because of a growing awareness of subsidence-related problems on the part of the community and business leaders, in 1975 the Texas legislature was prompted to create the Harris–Galveston Coastal Subsidence District "… for the purpose of ending subsidence which contributes to, or precipitates, flooding, inundation, and overflow of any area within the District." This unique District was authorized to issue (or refuse) well permits, promote water conservation and education, and promote conversion from groundwater to surface-water supplies. It has largely succeeded in its primary objective of arresting subsidence in the coastal plain east of Houston; however, subsidence accelerated in fast-growing inland areas north and west of Houston which still rely on groundwater, contributing to creation of the Fort Bend Subsidence District by the legislature in 1989.

Land subsidence and sea-level rise (flooding) are problems in the Houston area, but the combination of topography and local climate in the Houston–Galveston Bay area also contributes to the flooding problem. Tidal exchange occurs between the Gulf and the bay system through the barrier-island and peninsula complex. The Houston climate is subtropical; temperatures range from 45° to 93°F, and on average about 47 inches of rain fall each year. The humid coastal plain slopes gently toward the Gulf at a rate of about 1 foot per mile. Two major rivers, the Trinity and San Jacinto, and many smaller ones traverse the plain before discharging into estuarine areas of the bay system. The Texas coast is subject to a hurricane or tropical storm about once every 2 years (McGowen et al., 1977). Storm tides associated with hurricanes have reached nearly 15 feet in Galveston. The flat-lying region is particularly prone to flooding from both riverine and coastal sources, and the rivers, their reservoirs, and an extensive system of bayous and manmade canals are managed as part of an extensive flood-control system.

Flooding events in the Houston area have increased in severity and frequency due to land subsidence. Near the coast, the net result of land subsidence is an apparent increase in sea level, or a relative sea-level rise: the net effect of global sea-level rise and regional land subsidence in the coastal zone. The sea level is in fact rising due to regional and global processes, both natural and human induced. The combined effects of the actual sea-level rise and natural consolidation of the sediments along the Texas Gulf Coast yield a relative sea-level rise from natural causes that locally may exceed 0.08 inches per year (Paine, 1993). Global warming is contributing to the current sea-level rise and is expected to result in a sea-level increase of nearly 4 inches by the year 2050 (Titus and Narayanan, 1995). However, human-induced subsidence has been by far the dominant cause of relative sea-level rise along the Texas Gulf Coast, exceeding 1 inch per year throughout much of the affected area. Extraction of groundwater and to a lesser extent oil and gas has contributed to this subsidence. Subsidence caused by oil and gas production is largely restricted to the field of production, as contrasted to the regional-scale subsidence typically caused by groundwater pumpage.

The bottom line is that, because land subsidence in the Houston area became evident, in the late 1960s groups of citizens began to work for a reduction in ground-water use. State legislators became educated about the problem, and in 1975 the Texas legislature passed a law that created the Harris–Galveston Coastal Subsidence District. The Subsidence District was authorized as a regulatory agency, with the power to restrict groundwater withdrawal by annually issuing or denying permits for large-diameter wells but was forbidden to own property such as water-supply and conveyance facilities. Preventing and mitigating the effects of subsidence in the Houston area are an ongoing process.

IN "THE MEADOWS," GAMBLING IS MORE THAN A GAME*

WEDNESDAY OCT. 11TH 1848   … Camped about midnight at a spring branch called Cayataus. Fair grass. This is what is called the "Vegas".

THURSDAY OCT. 12TH 1848   Staid in the camp we made last night all day to recruit the animals. They done finely. There is the finest stream of water here, for its size, I ever saw. The valley is extensive and I doubt not would by the aid of irrigation be highly productive. There is water enough in this rapid little stream to propel a grist mill with a dragger run of stones! And, oh! *such* water. It comes, too, like an oasis in the desert, just at the termination of a 50 m. stretch without a drop of water or a spear of grass.

—The Journal of Orville C. Pratt **(1848)**

More than 24 inches of precipitation fall annually in the Spring Mountains bounding Las Vegas Valley to the west, but less than 4 inches of rain fall annually on the valley floor; measurable amounts (greater than 0.01 inch) seldom occur more than 30 days each year. Temperatures range from below freezing in the mountains to more than 120°F on the valley floor. There are typically more than 125 days of 90°F or warmer temperatures each year in Las Vegas, Spanish for "the meadows."

At one time, the Las Vegas Valley was the fastest growing metropolitan area in the United States (U.S. Census, 1997). More people meant an increased demand for water. In this desert region, the continuing demand for water for municipal and industrial uses is being met with imported Colorado River System supplies and local

---

* Adapted from Pavelko, M.T. et al., in *Land Subsidence in the United States*, Galloway, D., Jones, D.R., and Ingebritsen, S.E., Eds., U.S. Geological Survey Circular 1182, U.S. Department of Interior, Reston, VA, 1999, pp. 49–64.

groundwater. The depletion of once-plentiful groundwater supplies is contributing to ground failures and land subsidence. Since 1935, compaction of the aquifer system has caused nearly 6 feet of subsidence and led to the formation of numerous earth fissures and the activation of several surface faults, creating hazards and potentially harmful impacts to the environment.

A major concern is that current water supplies are not expected to satisfy the anticipated water demand. The federally mandated limit placed on imported water supplied from nearby Lake Mead, a Hoover Dam reservoir on the Colorado River, will likely force a continued reliance on groundwater to supplement the limited imported water supplies. Unless some balanced use of the groundwater resource can be achieved, water supply-and-demand dynamics in this desert community will likely maintain problems of land subsidence and related ground failures in the Las Vegas Valley.

Land subsidence was not a problem prior to development in Las Vegas Valley, as there was a natural, although dynamic, balance between aquifer-system recharge and discharge. Over the short term, yearly and decadal climatic variations (e.g., drought, effects of El Niño) caused large variations in the amount of water available to replenish the aquifer system. But, over the long term, the average amount of water recharging the aquifer system was in balance with the amount discharging, chiefly from springs and by evapotranspiration. Estimates of the average annual, natural recharge of the aquifer system have varied from 25,000 to 35,000 acre-ft (Dettinger, 1989; Harrill, 1976; Malmberg, 1965; Maxey and Jameson, 1948).

In 1907, the first flowing well was drilled by settlers to support the settlement of Las Vegas, and there began to be more groundwater discharge than recharge (Domenico et al., 1964). Uncapped artesian wells were at first permitted to flow freely onto the desert floor, wasting large quantities of water. Intensive groundwater use led to steady declines in spring flows and groundwater levels throughout Las Vegas Valley. Spring flows began to wane as early as 1908 (Maxey and Jameson, 1948). By 1912, nearly 125 wells in Las Vegas Valley (60% of which were flowing artesian wells) were discharging nearly 15,000 acre-ft per year. With the construction of Hoover Dam came development of the military and industrial sectors and a rapidly increasing demand for water. In 1942, a water pipeline was constructed to bring water from Lake Mead to the Basic Magnesium Project (now called Basic Management, Inc.) in the City of Henderson. This pipeline marked the first supplementation of Las Vegas Valley groundwater and the beginning of surface water imports to the valley. In 1955, the Las Vegas Valley Water District (LVVWD) began to use this pipeline to supplement the growing water demands. By this time, the amount of groundwater pumped annually from wells had reached nearly 40,000 acre-ft, surpassing the estimated natural recharge to the valley aquifer system (Mindling, 1971).

In 1971, the capacity to import surface water into the valley was greatly expanded when a second, larger pipeline was constructed between Lake Mead and Las Vegas (Harrill, 1976). However, despite the steady increases in imported surface water deliveries, rising demand for water and federally stipulated limits on Lake Mead imports encouraged a continued dependence on the local groundwater resource.

As Las Vegas expanded, groundwater levels declined. Between 1912 and 1944, groundwater levels declined at an average rate of about 1 foot per year (Domenico et al., 1964). Between 1944 and 1963, some areas of the valley experienced water-level declines of more than 90 feet (Bell, 1981a). The City of North Las Vegas was the first area to experience large water-level declines but, as Las Vegas expanded, new wells were drilled, pumping patterns changed, and groundwater level declines spread to areas south and west of the City of North Las Vegas. Between 1946 and 1960, the area of the valley that could sustain flowing artesian wells shrank from more than 80 square miles (Maxey and Jameson, 1948) to less than 25 square miles (Domenico et al., 1964). By 1962, the springs that had supported the Native Americans, and those who followed, were completely dry (Bell, 1981a).

Since the 1970s, annual groundwater pumpage in the valley has remained between 60,000 and 90,000 acre-ft; most of that has been pumped from the northwestern part of the valley. By 1990, areas in the northeast experienced more than 300 feet of decline, and areas in the central (including downtown and The Strip) and southeastern (Henderson) sections experienced declines between 100 and 200 feet (Burbey, 1995).

In 1996, imports from Lake Mead provided Las Vegas Valley with approximately 356,000 acre-ft of water (Coache, 1996) and represented the valley's principal source of water. This amount included 56,000 acre-ft of return-flow credits for annual streamflow discharging into Lake Mead from Las Vegas Wash. Fed by urban runoff, shallow groundwater, reclaimed water, and stormwater, the Las Vegas Wash is a 12-mile-long channel; it is sometimes called an urban river, and it exists in its current capacity because of an urban population.

Using data from the USGS and the U.S Coast and Geodetic Survey between 1915 and 1941 to make comparisons, Maxey and Jameson in 1948 were the first to recognize land subsidence and related ground failures in Las Vegas Valley. Since, then, repeated surveys of various regional networks have shown continuous land subsidence throughout large regions within the valley.

The repeated surveys have shown that subsidence continued at a steady rate into the mid-1960s, after which rates began increasing through 1987 (Bell, 1981a; Bell and Price, 1993). Surveys made in the 1980s delineated three distinct, localized subsidence bowls, or zones, superimposed on a larger, valley-wide subsidence bowl. One of these smaller subsidence bowls, located in the northwestern part of the valley, subsided more than 5 feet between 1963 and 1987. Two other localized subsidence bowls, in the central (downtown) and southern (Las Vegas Strip) parts of the valley, subsided more than 2.5 feet between 1963 and 1987. The areas of maximum subsidence do not necessarily coincide with areas of maximum water-level declines. One likely explanation is that those areas with maximum subsidence are underlain by a larger aggregate thickness of fine-grained, compressible sediments (Bell and Price, 1993).

Although the passage of time and continuing study and observation of a limited number of benchmarks have made subsidence obvious to those monitoring the Las Vegas Valley, not all of the impacts of subsidence in the valley have been fully realized. Two important impacts that have been documented are (1) ground failures (localized ruptures of the land surface), and (2) permanent reduction of the storage capacity of the aquifer system. Other potential impacts, all of which potentially create legal issues related to mitigation, restoration, compensation, and accountability, that have not been studied extensively include the following:

DID YOU KNOW?

Determination of subsidence trends in time and in space is limited in part by the inherently sparse distribution of available benchmarks from which comparisons can be made. Subsidence is determined by comparing two elevations made at a vertical reference point—a benchmark—at two different times. The destruction and loss of historical benchmarks inevitably accompanied the march of time and cultural developments such as building and road construction. The loss of comparable reference points reduces the spatial detail of subsidence determinations and disrupts the continuity of subsidence monitoring unless care is taken to preserve benchmarks. These factors have limited the spatial detail of subsidence maps in Las Vegas and will continue to pose serious challenges to subsidence monitoring in the years to come. In 1990, the Nevada Bureau of Mines and Geology established more than 100 new benchmarks in Las Vegas Valley.

- Creation of flood-prone areas by altering natural and engineered drainage ways
- Creation of earth fissures connecting nonpotable or contaminated surface and near-surface water to the principal aquifers
- Replacement costs associated with protruding wells and collapsed well casings and well screens

Ground Failures

Earth fissures are the dominant and most spectacular type of ground failure associated with groundwater withdrawal in Las Vegas Valley. Earth fissures are tensile failures in subsurface materials that result when differential compaction of sediments pulls apart the earth materials. Buried, incipient earth fissures become obvious only when they reach the surface and begin to erode, often following extreme rains or surface flooding conditions. Earth fissures have been observed in Las Vegas Valley as early as 1925 (Bell and Price, 1993) but were not linked directly to subsidence until the late 1950s (Bell, 1981a). Most of the earth fissures are areally and temporally correlated with groundwater level declines. Subsidence still continues; a compilation using 1986–87 data (Bell and Price, 1993) showed that the location and rates of subsidence have remained relatively constant at least since 1963 (see Figure 6.3). Movement of preexisting surface faults has also been correlated to groundwater level changes and differential land subsidence in numerous alluvial basins (Bell, 1981a; Holzer, 1979, 1984). In Las Vegas Valley, earth fissures often occur preferentially along preexisting surface faults in the unconsolidated alluvium. They tend to form as a result of the warping of the land surface that occurs when the land subsides more on one side of the surface fault than the other. This differential land subsidence creates tensional stresses that ultimately result in fissuring near zones of maximum warping. The association of most earth fissures with surface faults suggests a causal relationship. The surface faults may act as partial barriers to groundwater flow, creating a contrast in groundwater levels across the fault, or many offset sediments of differing compressibility.

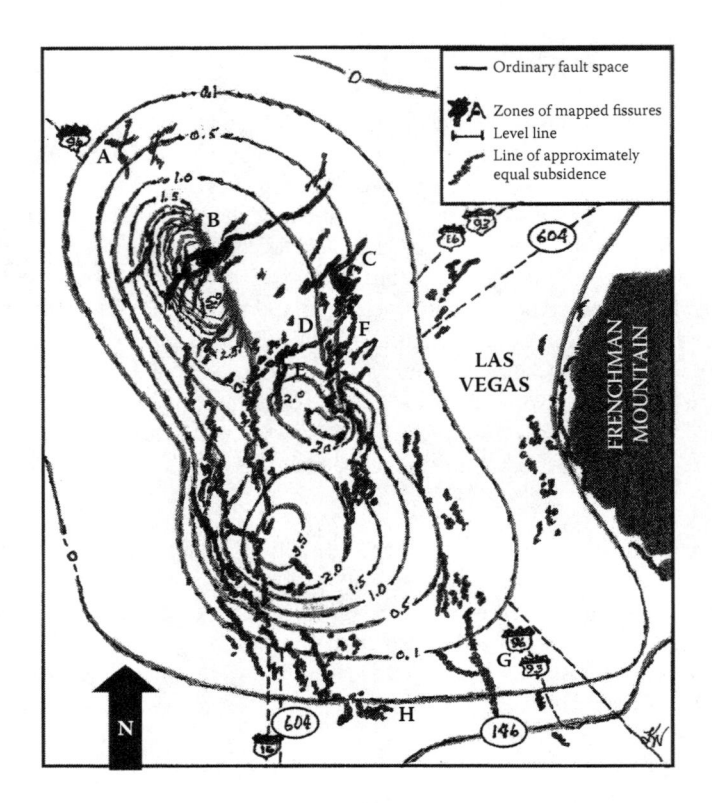

FIGURE 6.3 Quaternary faults and fissure zones in the Las Vegas area. Contours show subsidence measured only from 1963 to 1986–87. (Illustration by F.R. Spellman and Kathern Welsh, adapted from Bell, J.W. et al., *Las Vegas Valley: Land Subsidence and Fissuring Due to Ground-Water Withdrawal*, U.S. Geological Survey, Washington, DC, 2016.)

Rigid and precisely leveled structures are capable of being damaged due to land-surface displacements and tilts. Other damage related to fissuring includes cracking and displacement of roads, curbs, sidewalks, playgrounds, and swimming pools; warped sewage lines; ruptured water and gas lines; well failures resulting from shifted, sheared, and/or protruded well casings; differential settlement of railroad tracks; and a buckled drainage canal (Bell, 1981b). Earth fissures are also susceptible to erosion and can form wide, steep-walled gullies capable of redirecting surface drainage and creating floods and other hazards. Adverse impacts of ground failures may worsen as the valley continues to urbanize and more developed areas become affected.

REDUCED STORAGE CAPACITY

Reduction of storage capacity in the Las Vegas Valley aquifer system is another important consequence of aquifer system compaction. The volume of groundwater derived from the irreversible compaction of the aquifer system—*water of compaction*—is approximately equal to the reduced storage capacity of the aquifer system and represents a one-time quantity of water mined from the aquifer system.

Loss of aquifer system storage capacity is cause for concern, especially for a fast-growing desert metropolis that must rely in part on local groundwater resources. A study conducted by the Desert Research Institute (Mindling, 1971) estimated that, at times, up to 10% of the groundwater pumped from the Las Vegas Valley aquitard system has been derived from water of compaction. Assuming conservatively that only 5% of the total groundwater pumped between 1907 and 1996 was derived from water of compaction, the storage capacity of the aquifer system has been reduced by about 187,000 acre-ft. This may or may not be considered "lost" storage capacity: arguably, if this water is derived from an irreversible process, then this storage capacity has been used in the only way that it could have been. In any case, producing water of compaction represents mining groundwater from the aquifer system. Further, the reduced storage implies that, even if water levels recover completely, any future drawdowns will progress more rapidly.

The bottom line is that Las Vegas is dealing with a limited water supply. Moreover, managing land subsidence in the Las Vegas Valley is linked directly to the effective use of this limited water supply. With regard to the groundwater supply, more is appropriated by law and is being pumped in Las Vegas Valley than is available to be safely withdrawn from the groundwater basin (Coache, 1996; Nevada DCNR, 1992). Historic and recent rates of aquifer system depletion caused by overuse of the groundwater supply cannot be sustained without contributing further to land subsidence, earth fissures, and the reactivation of surface faults.

To arrest subsidence in the valley, groundwater levels must be stabilized or maintained above historic low levels. Stabilization or recovery of groundwater levels throughout the valley will require that the amount of groundwater pumped from the aquifers be less than or equal to the amount of water recharging the system. Eliminating any further decline will reduce the stresses contributing to the compaction of the aquifer system. Even so, a significant amount of land subsidence (residual compaction) will continue to occur until the aquifer system equilibrates fully with the stresses imposed by lowered groundwater level in the aquifers (Riley, 1969).

EARTH FISSURES AND SUBSIDENCE
IN SOUTH-CENTRAL ARIZONA[*]

There is a direct relationship between groundwater overdrafts in the deep alluvial basin of southern Arizona and widespread land subsidence within the region. Since 1900, groundwater has been pumped for irrigation, mining, and municipal use, and in some areas more than 500 times the amount of water that naturally replenishes the aquifer systems has been withdrawn (Schumann and Cripe, 1986). The resulting declines in groundwater levels—more than 600 feet in some places—have led to increased pumping costs, degraded the quality of groundwater in many locations, and led to the extensive and uneven permanent compaction of compressible

---

[*] Adapted from Carpenter, M.C., in *Land Subsidence in the United States*, Galloway, D., Jones, D.R., and Ingebritsen, S.E., Eds., U.S. Geological Survey Circular 1182, U.S. Department of Interior, Reston, VA, 1999, pp. 65–78.

fine-grained silt- and clay-rich aquitards. A total area of more than 3000 square miles has been affected by subsidence, including the expanding metropolitan areas of Phoenix and Tucson and some important agricultural regions nearby.

In southern Arizona, buildings, roads and highways, railroads, flood-control structures, and water and wastewater lines have been damaged by ground fissures from ground failure in areas or uneven or differential compaction. The presence and ongoing threat of subsidence and fissures forced a change in the planned route of the massive, federally financed Central Arizona Project (CAP) aqueduct that has delivered imported surface water from the Colorado River to central Arizona since 1985. In the CAP, Arizona now has a supplemental water supply that has lessened the demand and overdraft of groundwater supplies. Some CAP deliveries have been used in pilot projects to artificially recharge depleted aquifer systems. When fully implemented, recharge of this imported water will help to maintain water levels and forestall further subsidence and fissure hazards in some areas.

In Arizona, agriculture is synonymous with and dependent on irrigation. Irrigation is needed to grow crops because of the low annual rainfall and the high rate of potential evapotranspiration—more than 60 inches per year. Precipitation in south-central Arizona ranges from as low as 3 inches per year over some of the broad flat alluvial basins to more than 20 inches per year in the rugged mountain ranges. Large volumes of water can be stored in the intermontane basins, which contain up to 12,000 feet or more of sediments eroded from the various metamorphic, plutonic, volcanic, and consolidated sedimentary rocks that form the adjacent mountains. Groundwater is generally produced from the upper 1000 to 2000 feet of the basin deposits, which constitute the aquifer systems. Groundwater pumped from the aquifer systems became a reliable and heavily tapped source of irrigation water that fueled the development of agriculture during the early and mid-20th century. In many areas, the aquifer systems include a large fraction of fine-grained deposits containing silt and clay, which are susceptible to compaction when the supporting fluid pressures are reduced by pumping.

Pumping for irrigation has been increasing from the 1900s. By the mid-1960s, expected growth in the metropolitan Phoenix and Tucson areas, coupled with the already large groundwater-level declines and worsening subsidence problems, prompted Arizona water officials to push for and receive congressional approval for the CAP. Since then, growth in the metropolitan areas has exceeded expectations, and municipal, industrial, and domestic water use currently accounts for nearly 20% of Arizona's water demand.

Population growth, increased agricultural production, and industrial expansion have all combined to exact a toll on groundwater supplies and have increased subsidence. Subsidence first became apparent during the 1940s in several alluvial basins in southern Arizona where large quantities of groundwater were being pumped to irrigate crops. By 1950, earth fissures began forming around the margins of some of the subsiding basins. The areas affected then and subsequently included metropolitan Phoenix in Maricopa County and Tucson in Pima County, as well as important agricultural regions in Pinal and Maricopa counties near Apache Junction, Chandler Heights, Stanfield, and in the Picacho Basin; in Cochise County near Willcox and Bowie; and in La Paz County in the Harquahala Plain. By 1980, groundwater levels had declined at least 100 feet in each of these areas and between 300 and 500 feet in most of the areas.

Land subsidence was first verified in south-central Arizona in 1948 through the use of repeat surveys of benchmarks near Eloy (Robinson and Peterson, 1962). By the late 1960s, installation and monitoring of borehole extensometers at Eloy, Higher Road south of Mesa, and at Lake Air Force Base, as well as analysis of additional repeat surveys, indicated that land subsidence was occurring in several areas. The areas of greatest subsidence corresponded with the areas of greatest water level declines (Schuman and Poland, 1970).

By 1977, nearly 625 square miles had subsided around Eloy, where as much as 12.5 feet of subsidence was measured; another 425 square miles had subsided around Stanfield, with a maximum subsidence of 11.8 feet (Laney et al., 1978). Near Queen Creek, an area of almost 230 square miles had subsided more than 3 feet. In northeast Phoenix, as much as 5 feet of subsidence was measured between 1962 and 1982. By contrast, in the Harquahala Plain, only about 0.6 foot of subsidence occurred in response to about 300 feet of water-level decline, whereas near Willcox more than 5 feet of subsidence occurred in response to 200 feet of water level decline (Holzer, 1980; Schumann and Cripe, 1986; Strange, 1983). The relation between water level decline and subsidence varies between and within basins because of differences in the aggregate thickness and compressibility of susceptible sediments.

Groundwater decline continued, and by 1992 declines of more than 300 feet had caused aquifer system compaction and land subsidence of as much as 18 feet on and near Luke Air Force Base, about 20 miles west of Phoenix. Associated earth fissures occur in three zones of differential subsidence at Luke, which led to a flow reversal in a portion of the Dysart Drain, an engineered flood conveyance. In 1992, surface runoff from a rainstorm of 4 inches closed the base for 3 days. The sluggish Dysart Drain spilled over, flooding the base runways and more than 100 houses, and resulted in about $3 million in damage (Schumann, 1995).

Subsidence-related earth fissures, cracks, seams, or separations in the ground are common in many basins; some of the most spectacular examples occur in south-central Arizona, where they have been part of the landscape for at least 70 years. Earth fissures are the dominant mode of ground failure related to subsidence in alluvial-valley sediments in Arizona and are typically long linear cracks at the land surface with little or no vertical offset. The temporal and spatial correlation of earth fissures with declines in groundwater levels indicates that many of the earth fissures are induced and are related to groundwater pumpage. More than 50 fissure areas had been mapped in Arizona Prior to 1980 (Laney et al., 1978).

Most fissures occur near the margins of alluvial basins or near exposed or shallow buried bedrock in regions where differential land subsidence has occurred. They tend to be concentrated where the thickness of the alluvium changes markedly. In a very early stage, fissures can appear as hairline cracks less than 0.02 inch wide, interspersed with lines of sink-like depressions resembling rodent holes. When they first open, fissures are usually narrow vertical cracks less than about 1 inch wide and up to several hundred feet long. They can progressively lengthen to thousands of feet. Apparent depths of fissures range from a few feet to more than 30 feet; the greatest recorded depth is 82 feet for a fissure on the northwest flank of Picacho Peak (Johnson, 1980). Fissure depths of more than 300 feet have been speculated based on various indirect measurements, including horizontal movement, volume-balance calculations based on the volume of air space at the surface, and the amount of sediment transported into the fissures. Widening of fissures by collapse and erosion results in fissure gullies (Laney et al., 1978) that may be 30 feet wide and 20 feet deep. No horizontal shear (strike-slip movement) has been detected at earth fissures, and very few fissures show any obvious vertical offset. However, fissures monitored by repeated leveling surveys commonly exhibit a vertical offset of a few inches. Two notable exceptions are the Picacho earth fissure, which has more than 2 feet of vertical offset at many places along its 10-mile length, and a fissure near Chandler Heights that has about 1 foot of vertical offset.

Fissures can undercut and damage infrastructure and present a hazard to the public. Hazards associated with earth fissures are generally more local and include damage to homes and buildings, roads, dams, canals, and sewer and utility lines, in addition to providing a conduit for contaminated surface water to rapidly enter groundwater aquifers. Below are some of the hazards directly associated with earth fissures (ALSG, 2007):

- Cracked or collapsing roads
- Broken pipes and utility lines
- Damaged or breached canals
- Cracked foundation/separated walls
- Loss of agricultural land
- Livestock and wildlife injury or death
- Severed or deformed railroad track
- Damaged well casing or wellhead
- Disrupted drainage
- Contaminated groundwater aquifer
- Sudden discharge of ponded water
- Human injury or death

The bottom line is that, in an effort to arrest, control, and mitigate the effects of land subsidence in Arizona, importation of water for consumptive use and groundwater recharge, retirement of some farmlands, and water conservation measures have been put in place. These measures have resulted in a cessation of water level declines in many areas and the recovery of water levels in some areas. However, some basins are still experiencing subsidence, because much of the aquifer system compaction has occurred in relatively thick aquitards. It can take decades or longer for fluid pressures to equilibrate between the aquifers and the full thickness of many of these thick aquitards. For this reason, both subsidence and its abatement have

DID YOU KNOW?

In agricultural areas, fissures and fissure gullies are often obscured by cultivation. Reactivation of fissures can recur, only to be obscured again by cultivation. In some cases, farmers periodically fill fissures with soil and other materials because the gully formation processes are persistent. Such fissures are commonly known only to the farmers who cultivate the fields.

lagged pumping and recharge. A glimmer of hope is offered by data at the borehole extensometer near Eloy, where water levels have recovered more than 150 feet and compaction has decreased markedly.

REFERENCES AND RECOMMENDED READING

Alley, W.M., Reilly, T.E., and Franke, O.L. (1999). *Sustainability of Ground-Water Resources*, U.S. Geological Survey Circular 1186. Denver, CO: U.S. Geological Survey.

ALSG. (2007). *Land Subsidence and Earth Fissures in Arizona*. Arizona Land Subsidence Group, http://www.azgs.az.gov/Resources/CR-07-C_Dec07.pdf.

Bell, J.W. (1981a). *Subsidence in Las Vegas Valley*, Bulletin 95. Reno: Nevada Bureau of Mines and Geology.

Bell, J.W. (1981b). *Results of Leveling Across Fault Scarps in Las Vegas Valley, Nevada, April 1978–June 1981*, Open-File Report 81-5. Reno: Nevada Bureau of Mines and Geology.

Bell, J.W. and Price, J.G. (1993). *Subsidence in Las Vegas Valley, 1980–91: Final Project Report*, Open-File Report 93-4. Reno: Nevada Bureau of Mines and Geology.

Bell, J.W., Price, J.G., and Mifflin, M.D. (2016). *Las Vegas Valley: Land Subsidence and Fissuring Due to Ground-Water Withdrawal*. Washington, DC: U.S. Geological Survey (https://geochange.er.usgs.gov/sw/impacts/hydrology/vegas_gw/).

Burbey, T.J. (1995). *Pumpage and Water-Level Change in the Principal Aquifer of Las Vegas Valley, 1980–90*, Information Report 34. Carson City: Nevada Division of Water Resources.

Carpenter, M.C. (2013). South-central Arizona. In: *Land Subsidence in the United States*, U.S. Geological Survey Circular 1182 (Galloway, D., Jones, D.R., and Ingebritsen, S.E., Eds.), pp. 66–78. Reston, VA: U.S. Department of Interior.

Coache, R. (1996). *Las Vegas Valley Water Usage Report, Clark County, Nevada, 1996*. Carson City: Nevada Division of Water Resources.

Cone, T. (1997). The vanishing valley. *San Jose Mercury News West Magazine*, June 29, p. 9-15.

Coplin, LS. and Galloway, D. (2013). Houston–Galveston, Texas: managing coastal subsidence. In: *Land Subsidence in the United States*, U.S. Geological Survey Circular 1182 (Galloway, D., Jones, D.R., and Ingebritsen, S.E., Eds.), pp. 35–48. Reston, VA: U.S. Department of Interior.

Dawidowski, J.B. and Koolen, J.J. (1994). Computerized determination of the preconsolidation stress in compaction texting of field core samples. *Soil and Tillage Research*, 31(2): 277–282.

Dettinger, M.D. (1989). Reconnaissance estimates of natural recharge to desert basins in Nevada, U.S.A., by using chloride balance calculations. *Journal of Hydrology*, 106: 55–78.

Domenico, P.A., Stephenson, D.A., and Maxey, G.G. (1964). *Ground Water in Las Vegas Valley*, Water Resources Bulletin No. 29. Carson City: Nevada Division of Water Resources.

Galloway, D., Jones, D.R., and Ingebritsen, S.E., Eds. (1999). *Land Subsidence in the United States*, U.S. Geological Survey Circular 1182. Reston, VA: U.S. Department of Interior.

Green, J.H. (1964). *Compaction of the Aquifer System and Land Subsidence in the Santa Clara Valley, California*. U.S. Geological Survey Water-Supply Paper 1779-T. Reston, VA: U.S. Geological Survey.

Harrill, J.R. (1976). *Pumping and Ground-Water Storage Depletion in Las Vegas Valley, Nevada, 1955–74*. Bulletin No. 44. Carson City: Nevada Department of Conservation and Natural Resources.

Helm, D.C. (1975). One-dimensional simulation of aquifer system compaction near Pixley, California. 1. Constant parameters. *Water Resources Research*, 11: 465–478.

Holzer, T.L. (1979). *Leveling Data—Eglington Fault Scarp, Las Vegas Valley, Nevada*. Open-File Report 79-950. Reston, VA: U.S. Geological Survey.

Holzer, T.L. (1980). *Reconnaissance Maps of Earth Fissures and Land Subsidence, Bower and Willcox Areas, Arizona*, Miscellaneous Field Studies Map 1156. Reston, VA: U.S. Geological Survey.

Holzer, T.L. (1984). Ground failure induced by ground-water withdrawal from unconsolidated sediment. *Reviews in Engineering Geology*, 6: 67–106.

Ingebritsen, S.E. and Jones, D.R. (2013). Santa Clara Valley, California: a case of arrested subsidence. In: *Land Subsidence in the United States*, U.S. Geological Survey Circular 1182 (Galloway, D., Jones, D.R., and Ingebritsen, S.E., Eds.), pp. 15–22. Reston, VA: U.S. Department of Interior.

Ireland, R.L., Poland, J.F., and Riley, F.S. (1984). *Land Subsidence in the San Joaquin Valley, California, as of 1980*, U.S. Geological Survey Professional Paper 437-I. Reston, VA: U.S. Geological Survey.

Johnson, N.M. (1980). The Relation Between Ephemeral Stream Regime and Earth Fissuring in South-Central Arizona, master's thesis, University of Arizona, Tucson.

Laney, R.L., Raymond, R.H., and Winikka, C.C. (1978). *Maps Showing Water-Level Declines, Land Subsidence, and Earth Fissures in South-Central Arizona*, Water-Resources Investigations Report 78-83. Reston, VA: U.S. Geological Survey.

Leatherman, S.P., Chalfont, R., Pendleton, E.C., McCandless, T.L., and Funderburk, S. (1995). *Vanishing Lands: Sea Level, Society and Chesapeake Bay*. Annapolis, MD: U.S. Fish and Wildlife Service.

Malmberg, G.T. (1965). *Available Water Supply of the Las Vegas Ground-Water Basin, Nevada*, U.S. Geological Survey Water-Supply Paper 1780. Reston, VA: U.S. Geological Survey.

Manning, J.C. (1967). Report on the ground-water hydrology in the southern San Joaquin Valley. *American Water Works Association Journal*, 59: 1513–1526.

Maxey, G.B. and Jameson, C.H. (1948). *Geology and Water Resources of Las Vegas, Pahrump, and Indian Springs Valleys, Clark and Nye Counties, Nevada*, Water Resources Bulletin 5. Carson City: Nevada State Engineer.

McGowen, J.M., Garner, L.E., and Wilkinson, B.M. (1977). *The Gulf Shoreline of Texas: Process, Characteristics, and Factors in Use*, Circular 75-6. Austin: University of Texas.

Mendenhall, W.C., Dole, R.B., and Stabler, H. (1916). *Ground Water in San Joaquin Valley, California*, U.S Geological Survey Water-Supply Paper 398. Reston, VA: U.S. Geological Survey.

Mindling, A.L. (1971). *A Summary of Data Relating to Land Subsidence in Las Vegas Valley*. Reno: Center for Water Resources Research, Desert Research Institute, University of Nevada.

Nady, P. and Larragueta, I.L. (1983). *Development of Irrigation in the Central Valley of California*, Hydrologic Investigations Atlas HA-649. Reston, VA: U.S. Geological Survey.

Nevada DCNR. (1992). *Hydrographic Basin Summaries, 1990–1992*. Carson City: Nevada Department of Conservation and Natural Resources, Division of Water Resources and Water Planning.

Paine, J.G. (1993). Subsidence of the Texas coast: interferences from historical and later Pleistocene sea levels. *Tectonophysics*, 222: 445–458.

Pew, T.L. (1990). Land subsidence and earth-fissure formation caused by groundwater withdrawal in Arizona: a review. In: *Groundwater Geomorphology: The Role of Subsurface Water in Earth-Surface Process and Landforms* (Higgins, C.G. and Coates, D.R., Eds.), pp. 218–233. Boulder, CO: Geological Society of America.

Poland, J.F. and Green, J.H. (1962). *Subsidence in the Santa Clara Valley, California: A Progress Report*, U.S. Geological Survey Water-Supply Paper 1619-C. Reston, VA: U.S. Geological Survey.

Poland, J.F. and Ireland, R.L. (1988). *Land Subsidence in the Santa Clara Valley, California, as of 1982*, U.S. Geological Survey Professional Paper 497-F. Reston, VA: U.S. Geological Survey.

Poland, J.R., Lofgren, B.E., Ireland, R.L., and Pugh, R.G., (1975), *Land Subsidence in the San Joaquin Valley, California, as of 1972*, U.S. Geological Survey Professional Paper 437-H. Reston, VA: U.S. Geological Survey.

Riley, F.S. (1969). Analysis of borehole extensometer data from Central California. In: *Land Subsidence* (Tison, L.J., Ed.), Vol.. 2, pp. 423–431. Rennes, France: International Association of Scientific Hydrology..

Robinson, G.M. and Peterson, D.E. (1962). *Notes on Earth Fissures in Southern Arizona*, U.S. Geological Survey Circular 466. Reston, VA: U.S. Geological Survey.

Roethke, T. (1948). *The Lost Son and Other Poems by Theodore Roethke*. Garden City, NY: Doubleday & Co.

Schumann, H.H. (1995). Land subsidence and earth fissure hazards near Luke Air Force Base, Arizona. In: *Proceedings of U.S. Geological Survey Subsidence Interest Group Conference, Edwards Air Force Base, Antelope Valley, California, November 18–19, 1992: Abstracts and Summary*, U.S. Geological Survey Open-File Report 94-532 (Prince, K.R., Galloway, D.L., and Leake, S.A., Eds.), pp. 18–21. Sacramento, CA: U.S. Geological Survey.

Schumann, H.H. and Cripe, L.S. (1986). Land subsidence and earth fissures caused by groundwater depletion in southern Arizona, U.S.A. In: *Proceedings of the Third International Symposium on Land Subsidence, Venice, Italy, 19–25 March* (Johnson, A.I., Carbognin, L., and Ubertini, L., Eds.), pp. 841–851. Rennes, France: International Association of Scientific Hydrology.

Schumann, H.H. and Genauldi, R.B. (1986). *Land Subsidence, Earth Fissures, and Water-Level Change in Southern Arizona*. Tucson: Arizona Bureau of Geology and Mineral Technology, Geological Survey Branch.

Schumann, H.H. and Poland, J.F. (1969). *Land Subsidence, Earth Fissures, and Groundwater Withdrawal in South-Central Arizona, U.S.A.* Rennes, France: International Association of Scientific Hydrology.

Seiler, R.L., Skorupa, J.P., and Peltz, L. (1999). *Areas Susceptible to Irrigation-Induced Selenium Contamination of Water and Biota in the Western United States*, U.S. Geological Survey Circular 1180. Carson City, NV: U.S. Department of the Interior.

Siade, A.J., Nishikawa, T., Rewis, D.I., Martin, P., and Phillips, S.P. (2014). *Groundwater-Flow and Land-Subsidence Model of Antelope Valley, California*, Report 2014-5166. Reston, VA: U.S. Geological Survey.

Slaff, S. (1993). *Land Subsidence and Earth Fissures in Arizona*. Tucson: Arizona Geological Survey.

Strange, W.E. (1983). *Subsidence Monitoring for the State of Arizona*. Rockville, MD: National Oceanic and Atmospheric Administration.

Swanson, A.A. (1998). Land subsidence in the San Joaquin Valley, updated to 1995. In: *Land Subsidence Case Studies and Current Research: Proceedings of the Dr. Joseph F. Poland Symposium on Land Subsidence*, Special Publication No. 8 (Borchers, J.W., Ed.), pp. 75–79. Zanesville, OH: Association of Environmental & Engineering Geologists.

Terzaghi, K. (1925). Principles of soil mechanics. IV. Settlement and consolidation of clay. *Engineering News-Record*, 95(3): 874–878.

Titus, J.G. and Natayanan, V.K. (1995). *The Probability of Sea Level Rise*, EPA 230-R-95-008. Washington, DC: U.S. Environmental Protection Agency.

Tolman, C.F. and Poland, J.F. (1940). Ground-water infiltration, and ground-surface recession in Santa Clara Valley, Santa Clara County, California. *Transactions American Geophysical Union*, 21: 23–24.

U.S. Census. (1997). Las Vegas Metro Area Leads Nation in Population Growth [press release]. Washington, DC: U.S. Bureau of the Census.

Williamson, A.K., Prudic, D.E., and Swain, L.A. (1989). *Ground-Water Flow in the Central Valley, California*, U.S. Geological Survey Professional Paper 1401-D. Reston, VA: U.S. Geological Survey.

7 The Vanishing of Hampton Roads

Earth's energy budget is directly coupled to its water budget.

Healy et al. (2007)

Rain is grace; rain is the sky condescending to the earth; without rain, there would be no life.

Updike (1989)

INTRODUCTION

Hampton Roads is unfamiliar to many people, but those who are somewhat familiar with the name might ask, "Do you mean the body of water, or do you mean the location?" This question comes up because Hampton Roads is the name of *both* a body of water in Virginia and the surrounding metropolitan region in southeastern Virginia and northeastern North Carolina. The land area is also known as Tidewater. In this chapter, Hampton Roads refers to both the water body and the land region, because the topics of land subsidence and relative sea-level rise are relevant to the entire area. The total area is comprised of 527 square miles (1364 km²) and is made up of nine major cities: Norfolk, Virginia Beach, Chesapeake, Newport News, Hampton, Portsmouth, Suffolk, Poquoson, and Williamsburg. As a combined statistical area, it also includes Kitty Hawk, Elizabeth City, and North Carolina (see Figure 7.1). The entire area has a population of over 1.7 million. The body of water known as Hampton Roads is one of the world's largest natural harbors. It incorporates the mouths of the Elizabeth River, Nansemond River, and James River, along with several smaller rivers, and empties into Chesapeake Bay—a treasured estuary—near its mouth leading to the Atlantic Ocean.

> *Note:* Not only is the Chesapeake Bay region a treasured estuary but it also is a bastion of early American history. Moreover, it is not a bad place to live. Because I have been a resident of the area for more than 50 years, I think I am qualified to state this as fact. Most of those years have been spent studying the Bay area and specifically its water pollution problems (see Case Study 7.1 and Case Study 7.2); in other words, it has been a lifetime project for me. My focus of study has shifted dramatically, though, from pollution of the Bay to a more pressing problem— namely, the literal vanishing of the area. Review the following two case studies to see why.

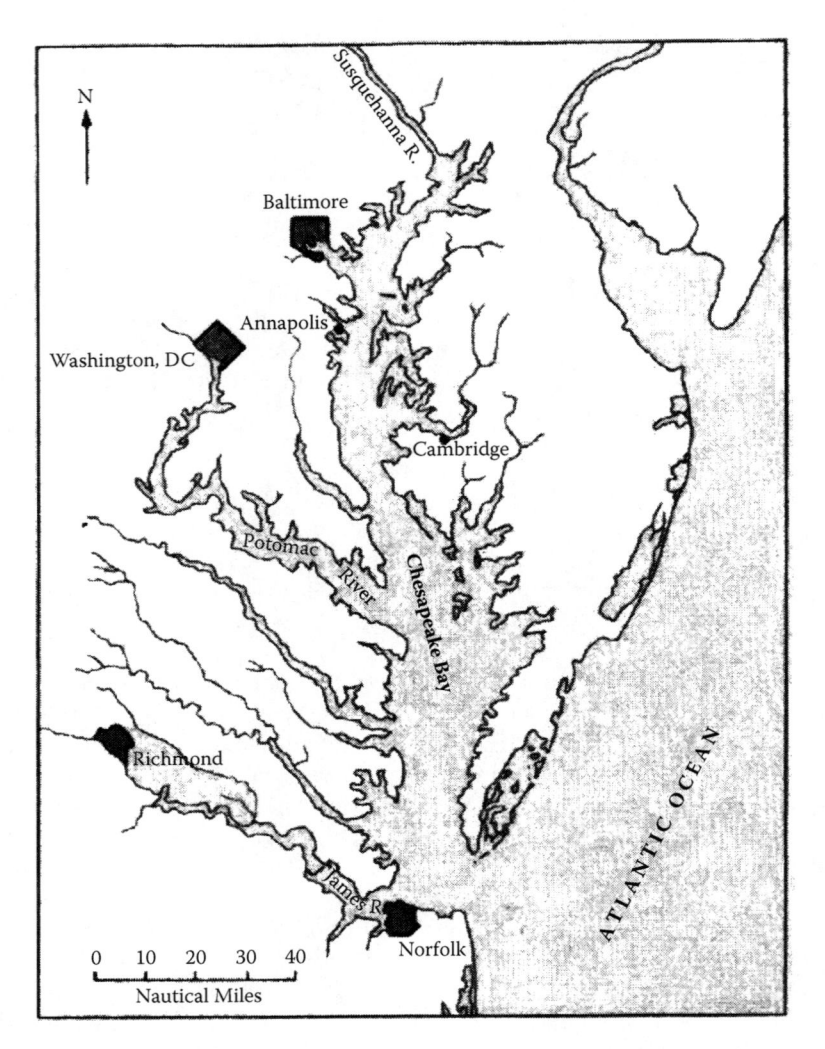

FIGURE 7.1   Hampton Roads region. (From Leatherman, S.P. et al., *Vanishing Lands: Sea Level, Society and Chesapeake Bay*, Chesapeake Bay Field Office, U.S. Fish and Wildlife Service, Annapolis, MD, 1995.)

Case Study 7.1. Chesapeake Bay Cleanup

The following newspaper article, written by the author, appeared in the January 2, 2005, issue of *The Virginian-Pilot*. It is an Op-Ed rebuttal to the article referenced in the text below. It should be pointed out that this piece was well received by many, but a few stated that it was nothing more than a rhetorical straw man. Of course, in contrast, I felt that the organizational critics were using the rhetorical Tin Man approach; that is, when you need to justify your cause and your organization's existence and you need more grease, you squawk.

The grease that many of these organizations require, however, is grease that is the consistency of paper-cloth and is colored green; thus, they squawk quite often. You be the judge.

CHESAPEAKE BAY CLEANUP: GOOD SCIENCE VS. "FEEL GOOD" SCIENCE

In your article, "Fee to help Bay faces anti-tax mood" (*The Virginian-Pilot*, 1/2/05), you pointed out that environmentalists call it the "Virginia Clean Streams Law." Others call it a "flush tax." I call the environmentalists' (and others') view on this topic a rush to judgment, based on "feel good" science vs. good science. The environmentalists should know better. Consider the following:

Environmental policymakers in the Commonwealth of Virginia came up with what is called the Lower James River Tributary Strategy on the subject of nitrogen (a nutrient) from the Lower James River and other tributaries contaminating the Lower Chesapeake Bay Region. When in excess, nitrogen is a pollutant. Some "theorists" jumped on nitrogen as being the cause of a decrease in the oyster population in the Lower Chesapeake Bay Region. Oysters are important to the local region. They are important for economical and other reasons. From an environmental point of view, oysters are important to the Lower Chesapeake Bay Region because they have worked to maintain relatively clean Bay water in the past. Oysters are filter-feeders. They suck in water and its accompanying nutrients and other substances. The oyster sorts out the ingredients in the water and uses those nutrients it needs to sustain its life. Impurities (pollutants) are aggregated into a sort of ball that is excreted by the oyster back into the James River.

You must understand that there was a time, not all that long ago (maybe 50 years ago), when oysters thrived in the Lower Chesapeake Bay. Because they were so abundant, these filter-feeders were able to take in turbid Bay water and turn it almost clear in a matter of three days. (How could anyone dredge up, clean, and then eat such a wonderful natural vacuum cleaner?)

Of course, this is not the case today. The oysters are almost all gone. Where did they go? Who knows?

The point is that they are no longer thriving, no longer colonizing the Lower Chesapeake Bay Region in numbers they did in the past. Thus, they are no longer providing economic stability to watermen; moreover, they are no longer cleaning the Bay.

Ah! But don't panic! The culprit is at hand; it has been identified. The "environmentalists" know the answer. They say it has to be nutrient contamination— namely, nitrogen is the culprit. Right?

Not so fast.

A local sanitation district and a university in the Lower Chesapeake Bay region formed a study group to formally, professionally, and scientifically study this problem. Over a five-year period, using biological nutrient removal (BNR) techniques at a local wastewater treatment facility, it was determined that the effluent leaving the treatment plant and entering the Lower James River consistently contained below 8 mg/L nitrogen (a relatively small amount) for five consecutive years.

The first question is: Has the water in the Chesapeake Bay become cleaner, clearer because of the reduced nitrogen levels leaving the treatment plant?

The second question is: Have the oysters returned?

Answers to both questions, respectively: no; not really.

Wait a minute. The environmentalists, the regulators, and other well-meaning interlopers stated that the problem was nitrogen. If nitrogen levels have been reduced in the Lower James River, shouldn't the oysters start thriving, colonizing, and cleaning the Lower Chesapeake Bay again?

You might think so, but they are not. It is true that the nitrogen level in the wastewater effluent was significantly lowered through treatment. It is also true that a major point source contributor of nitrogen was reduced with a corresponding decrease in the nitrogen level in the Lower Chesapeake Bay.

If the nitrogen level has decreased, then where are the oysters?

A more important question is: What is the real problem?

The truth is that no one at this point in time can give a definitive answer to this question.

Back to the original question: Why has the oyster population decreased?

One theory states that because the tributaries feeding the Lower Chesapeake Bay (including the James River) carry megatons of sediments into the bay (stormwater runoff, etc.), they are adding to the Bay's turbidity problem. When waters are highly turbid, oysters do the best they can to filter out the sediments but eventually they decrease in numbers and then fade into the abyss.

Is this the answer? That is, is the problem with the Lower Chesapeake Bay and its oyster population related to turbidity?

Only solid, legitimate, careful scientific analysis may provide the answer.

One thing is certain; before we leap into decisions that are ill-advised, that are based on anything but sound science, and that "feel" good, we need to step back and size up the situation. This sizing-up procedure can be correctly accomplished only through the use of scientific methods.

Don't we already have too many dysfunctional managers making too many dysfunctional decisions that result in harebrained, dysfunctional analysis—and results? Obviously, there is no question that we need to stop the pollution of Chesapeake Bay. However, shouldn't we replace the timeworn and frustrating position that "we must start somewhere" with good common sense and legitimate science?

The bottom line: We shouldn't do anything to our environment until science supports the investment. Shouldn't we do it right?

Case Study 7.2. Chesapeake Bay and Nutrients: A Modest Proposal

In Case Study 7.1, the technical problems and political controversy regarding nutrient contamination of the Chesapeake Bay were discussed. Nutrient pollution in Chesapeake Bay and other water bodies is real and ongoing, and the controversy over what the proper mitigation procedures might be is intense and never-ending (and very political). Nutrients are substances that all living organisms need for growth and reproduction. Two major nutrients, nitrogen and phosphorus, occur naturally in water, soil, and air. Nutrients are present in animal and human waste and in chemical fertilizers. All organic material such as leaves and grass clippings contains nutrients. These nutrients cause algal growth and depletion of oxygen in the Bay which leads to the formation of dead zones lacking in oxygen and aquatic life.

Nutrients can find their way into the Bay from anywhere within the 64,000-square-mile Chesapeake Bay watershed (USFWS, 2011), and that is the problem. All streams, rivers, and storm drains in this huge area eventually lead to the Chesapeake. The activities of over 13.6 million people in the watershed have overwhelmed the Bay with excess nutrients. Nutrients come from a wide range of sources, including sewage treatment plants (20 to 22%), industry, agricultural fields, lawns, and even the atmosphere. Nutrient inputs are divided into two general categories: point sources and nonpoint sources.

Sewage treatment plants, industries, and factories are the major point sources. These facilities discharge wastewater containing nutrients directly into a waterway. Although each facility is regulated for the amount of nutrients that can be legally discharged, at times violations still occur. In this text, it is the wastewater treatment plant, a point source discharger, that is of concern to us. It should be pointed out that wastewater treatment plants (approximately 350 units outfalling effluent to nine major rivers and other locations all flowing into the Chesapeake Bay region) discharge somewhere between 20 and 22% of total nutrients into the Bay.

Many target these point source dischargers as being the principal causes of oxygen depletion and creators of dead zones within the Bay. If this is true, one needs to ask the question: Why is wastewater outfalled into the Chesapeake Bay in the first place? Water is the new oil, and if we accept this as fact we should therefore preserve and use our treated wastewater with great care and even greater utility. Thus, it makes good sense to take the wastewater that is currently being discharged into Chesapeake Bay and reuse it. This recycling of water saves raw water supplies in reservoirs and aquifers and limits the amount of wastewater that is discharged from wastewater treatment plants into public waterways, such as Chesapeake Bay. Properly treated wastewater could be used for many other purposes that raw water now serves, such as irrigating lawns, parks, gardens, golf courses, and farms; fighting fires; washing cars; controlling dust; cooling industrial machinery, towers, and nuclear reactors; making concrete; and cleaning streets. It is my contention, of course, that when we get thirsty enough we will find another use for properly treated and filtered wastewater, and when this occurs we certainly will not use this treated wastewater for any purpose other than quenching our thirst. In reality, we are doing this already, but this is the subject for another forthcoming book: *Water Is the New Oil: Sustaining Freshwater Supplies.*

Let's get back to the Chesapeake Bay problem. The largest sources of nutrients dumped into the Bay are nonpoint sources. These nonpoint sources pose a greater threat to the Chesapeake ecosystem, as they are much more difficult to control and regulate. It is my view that, because of the difficulty of controlling runoff from agricultural fields, the lack of political will, and the technical difficulty of preventing such flows, wastewater treatment plants and other end-of-pipe dischargers have become the targets of convenience for regulators. The problem is that regulators are requiring the expenditure of hundreds of millions to billions of dollars to upgrade wastewater treatment to biological nutrient removal (BNR), tertiary treatment, or the combining of microfiltration membranes with a biological process to produce superior quality effluent—these requirements are commendable, interesting, and achievable but not at all necessary.

What is the alternative, the answer to the dead zone problem—the lack of oxygen in various locations in Chesapeake Bay? Putting it simply, take a portion of the hundreds of millions of dollars earmarked for upgrading wastewater treatment plants (which is a total waste, in my opinion) and build mobile, floating platforms containing electromechanical aerators or mixers. These platforms should be outfitted with diesel generators and accessories to provide power to the mixers. The mixer propellers should be adjustable so they are able to mix at a water depth of as little as 10 feet or can be extended to mix at a depth of 35 feet. Again, these platforms are mobile. When a dead zone appears in the bay, the mobile platforms with their mixers are moved to a center portion of the dead zone area and energized at the appropriate depth. These mobile platforms are anchored to the Bay bottom and so arranged to accommodate shipping to ensure that maritime traffic is not disrupted. The idea is to churn the dead zone water and sediment near the benthic zone and force a geyser-like effect above the surface to aerate the Bay water in the dead zone regions. Nothing adds more oxygen to water than natural or artificial aeration. Of course, while aerating and forcing oxygen back into the water, bottom sediments containing contaminants will also be stirred up and sent to the surface, and temporary air pollution problems will occur around the mobile platforms. Some will view this churning up of contaminated sediments as a bad thing, not a good thing. I, in contrast, suggest that removing contaminants from the Bay by volatizing them is a very good thing.

How many of these mobile mixer platforms will be required? It depends on the number of dead zones. Enough platforms should be constructed to handle the warm season's average number of dead zones that appear in the Bay. Will this modest proposal—using aerators to eliminate dead zones—actually work? I do not have a clue. This proposal makes more sense to me, though, than spending billions of dollars on upgrading wastewater treatment plants and effluent quality when they account for only 20 to 22% of the actual problem. The regulators and others do not have the political will or insight to go after the runoff that is the real culprit in contaminating Chesapeake Bay with nutrient pollution.

Recall that it was that great mythical hero Hercules, the world's first environmental engineer, who said that "dilution is the solution to pollution." I agree with this; however, in this text, the solution to preventing dead zones in Chesapeake Bay is to prohibit discharge of wastewater from point sources (i.e., reuse to prevent abuse) and to aerate the dead zones caused by agricultural runoff.

Reuse wastewater or inject it to prevent abuse? Yes, that is what this book is all about. It is about injecting treated wastewater (that meets drinking water quality standards) into the Potomac Aquifer to prevent abuse, which will be discussed further later in the chapter. For now, it is important to understand that land subsidence in the Hampton Roads region is real and ongoing and this event is exacerbated by relative sea-level rise. And, if you live in or depend upon the region, then land subsidence and sea-level rise constitute abuse.

Over the years the problems associated with global sea-level rise have received a lot of attention. Projections of the consequences of future rises in sea levels are stark for those coastal communities in locations where land areas are barely above sea level. This problem is especially acute in Hampton Roads, where land areas in most locations are barely above sea level and are therefore subject to inundation from rising seawater, as well as from storm surges and unusually high tidal events. The U.S. Fish and Wildlife Service's Chesapeake Bay Field Office, based in Annapolis, Maryland, published an excellent account of sea-level rise in the Chesapeake Bay region in 1995 titled *Vanishing Lands* (Leatherman et al., 1995). This handbook describes the formation of the Chesapeake ecosystem, its characteristics, how it is being changed by sea-level rise, and how modern civilization may alter the rate of sea-level rise.

HAMPTON ROADS AND SEA-LEVEL RISE

Of all the potential impacts of climate change, whether natural or human induced, a global rise in sea level appears to be the most certain and the most dramatic. As shown in Figure 7.2, for the last 5000 years the rate of sea level rise was only 3 feet per 1000 years. In the Chesapeake Bay region, the relative rise in sea level has been about 1 foot during the last 100 years. Although scientists view this rapid rate as possibly a temporary acceleration, many scientists believe that it signals a new trend in response to global warming. If the rate of rise accelerates in the near future, as

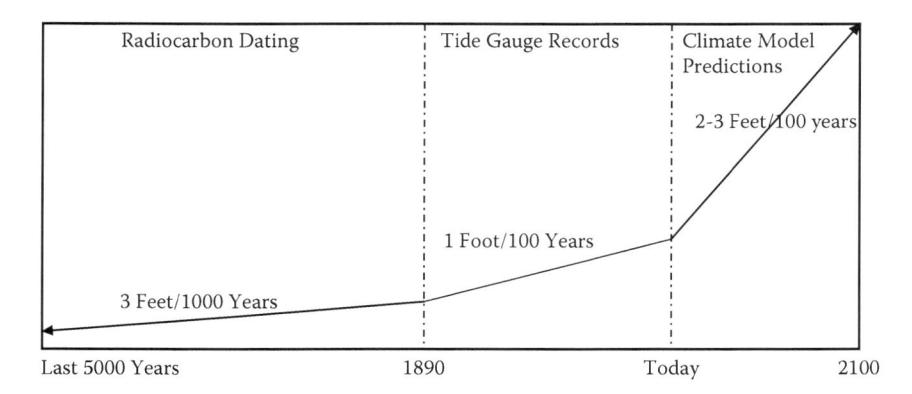

FIGURE 7.2   Sea-level rise in the Hampton Roads region past and present. (From Leatherman, S.P. et al., *Vanishing Lands: Sea Level, Society and Chesapeake Bay*, Chesapeake Bay Field Office, U.S. Fish and Wildlife Service, Annapolis, MD, 1995.)

projected, it could have serious repercussions for Chesapeake Bay. Because water levels are measured relative to the land, relative sea-level rise in the Chesapeake Bay region has two components: global water level and land subsidence. A worldwide, or eustatic, sea-level rise is caused by water released from melting glaciers and thermal expansion of seawater. Both are related to global warming and have amounted to a 6-inch rise in the last century. Let's take a closer look at the global warming problem.

GLOBAL WARMING

At present, many feel that humanity is conducting an unintended, uncontrolled, globally pervasive experiment whose ultimate consequences could be second only to nuclear war. The Earth's atmosphere is being changed at an unprecedented rate by pollutants resulting from human activities, inefficient and wasteful fossil fuel use and the effects of rapid population growth in many regions. These changes are already having harmful consequences over many parts of the globe.

—World Conference on the Changing Atmosphere closing statement, June 1988

The preceding quotation clearly states the issue. But what is global warming? Global warming has been defined as a long-term rise in the average temperature of Earth. This would appear to be what is happening today, even though the geological record shows that abrupt climate changes occur from time to time (Crowley and North, 1988). Here is another question, one asked by many people who question the validity of the concept of global warming as an environmental hazard: Is global warming actually occurring? The answer to this question is of enormous importance to all life on Earth—and is the subject of intense debate throughout the world. All of the debate regarding the existence of global warming cannot dispute the historical record, which points out that measurements made in central England, Geneva, and Paris from about 1700 until the present suggest a general *downward* trend in surface temperature (Thompson, 1995).

For the sake of discussion, let's assume that global warming is occurring (and with rising sea-levels this point is difficult to argue against). With this assumption in place, we must ask other questions: (1) Why is global warming occurring? (2) How can we be absolutely sure that it is occurring? (3) What will be the ultimate effects? (4) What can and are we going to do about it? These questions are difficult to answer. The real danger is that we may not be able to definitively answer these

DID YOU KNOW?

Climate change can have broad effects on *biodiversity*—the number and variety of plant and animal species in a particular location. Although species have adapted to environmental change for millions of years, a quickly changing climate could require adaption on larger and faster scales than in the past. Those species that cannot adapt are at risk of extinction. Even the loss of a single species can have cascading effects because organisms are connected through food webs and other inactions (USEPA, 2016a).

DID YOU KNOW?

Oceans and the atmosphere are constantly interacting—exchanging heat, water, gases, and particles. As the atmosphere warms, the ocean absorbs some of this heat. The amount of heat stored by the ocean affects the temperature of the ocean both at the surface and at great depths. Warming of the Earth's oceans can affect and change the habitat and food supplies for many kinds of marine life—from plankton to polar bears. The oceans also absorb carbon dioxide from the atmosphere. When it dissolves in the ocean, carbon dioxide reacts with seawater to form carbonic acid. As people put more carbon dioxide into the atmosphere, the oceans absorb some of this extra carbon dioxide, which leads to more carbonic acid. An increasingly acidic ocean can have negative effects on animal life, such as coral reefs (USEPA, 2016a).

questions before it is too late—when we have reached the point that the process has progressed beyond the power of humans to effect prevention or mitigation. This situation raises a red flag—a huge red flag—and additional questions. Are we to stand by and do nothing? Are we to simply ignore the potential impact of this problem? Are we to take the consequences of global warming lightly? Are we not to take precautionary actions now instead of later—much later, when it is too late? Indeed, a red flag has been raised (i.e., the cause-and-effect relationship with the greenhouse effect), but there is still time before it begins to wave in the climate change that is inevitable—when mitigation becomes more difficult, more expensive, and impossible to do. Exactly what is the nature of the problem of global warming? We may not provide all the answers here, but we are about to launch into a discussion of the entire phenomenon and its potential impact on Earth.

GREENHOUSE EFFECT

Earth's greenhouse effect, of course, took its name because of its similarity to the effects of a greenhouse. Because the glass walls and ceilings of a greenhouse are largely transparent to short-wave radiation from the sun, surfaces and objects inside the greenhouse absorb the radiation. The radiation, once absorbed, transforms into long-wave (infrared) radiation (heat), which radiates out from the greenhouse interior, but the glass prevents the long-wave radiation from escaping again and the warm rays are absorbed. The interior of the greenhouse becomes much warmer than the air outside, because of the heat trapped inside.

For the Earth and its atmosphere, a very similar greenhouse effect takes place. The short-wave and visible radiation that reaches Earth is absorbed by the surface as heat. The long heat waves are then radiated back out toward space, but the atmosphere instead absorbs many of them. This is a natural and normally balanced process, and indeed is essential to our life on Earth. The problem comes when changes in the atmosphere radically change the amount of absorption and therefore the amount of heat retained. Scientists speculate that this may have begun happening as various air pollutants caused the atmosphere to absorb more heat.

DID YOU KNOW

The problem with Earth's greenhouse effect is that human activities are now rapidly intensifying this natural phenomenon, which may lead to global warming. There is much debate, confusion, and speculation about this potential consequence. Scientists are not entirely sure of or agree about whether the recently perceived worldwide warming trend is due to greenhouse gases or some other cause, or whether it is simply a wider variation in the normal heating and cooling trends they have been studying. But, if it continues unchecked, the process may lead to significant global warming, with profound effects. Human impacts on the greenhouse effect are quite real. In fact, the rate at which the greenhouse effect is intensifying is now more than five times what it was over the last 100 years (Hansen et al., 1989).

That this phenomenon takes place at a local level when air pollution causes heat islands in and around urban centers is not questioned. The main contributors to this problem are the greenhouse gases: water vapor, carbon dioxide, carbon monoxide, methane, volatile organic compounds (VOCs), nitrogen oxides, chlorofluorocarbons (CFCs), and surface ozone. These gases delay the escape of infrared radiation from the Earth into space, thus causing a general climatic warming. Note that scientists stress that this a natural process—indeed, the earth would be 33°C cooler than it is today if the "normal" greenhouse effect did not exist (Hansen et al., 1986).

GREENHOUSE EFFECT AND GLOBAL WARMING COMBINED

Supporters of the global warming theory assume that human activities are significantly altering the Earth's normal and necessary greenhouse effect. The human activities they blame for this increase of greenhouse gases include burning fossil fuels, deforestation, and the use of certain aerosols and refrigerants. From information based on recent or short-term observations, many scientists have observed that the last decade has been the warmest since temperature recordings began in the late 19th century. They can see that the general rise in temperature coincides with the Industrial Revolution and its accompanying increase in fossil fuel use. Other evidence supports the global warming theory. In places that are synonymous with ice and snow—the Arctic and Antarctica, for example—we are seeing evidence of receding ice and snow cover.

Taking a long-term view, scientists look at temperature variations over thousands or even millions of years. Such an approach produces findings suggesting that global warming is nothing more than a short-term variation in Earth's climate, based on historical records showing that the Earth's temperature does vary widely, growing colder with ice ages and then warming again. On the other side of the argument, some people point out that the 1980s saw 9 of the 12 warmest temperatures ever recorded, and the Earth's average surface temperature rose approximately 0.6°C (1°F) in the last century (Titus and Narayanan, 1995). But, at the same time, still others offer as evidence that the same decade also saw three of the coldest years: 1984, 1985, and 1986. So what is really going on? We are not certain.

Assuming that we are indeed seeing long-term global warming, we must determine what is causing it. But, again, we face the problem that scientists cannot be sure of the precise causes of the greenhouse effect. Our current, possible trend in global warming may simply be part of a much longer trend of warming since the last ice age. We have learned much in the past two centuries of science, but little is actually known about the causes of the worldwide global cooling and warming that sent the Earth through major and smaller ice ages. The data we need would have to reach back over millennia, and we simply do not possess enough long-term data to support our theories.

Factors Involved with Global Warming and Cooling

Currently, scientists can point to six factors they think could be involved in long-term global warming and cooling:

1. Long-term global warming and cooling could result if changes in the Earth's position relative to the sun occur (i.e., changes in the Earth's orbit around the sun), with higher temperatures resulting when the two are closer together and lower ones when they are farther apart.
2. Long-term global warming and cooling could result from major catastrophes (e.g., meteor impacts, massive volcanic eruptions) throwing pollutants into the atmosphere that then block out solar radiation.
3. Long-term global warming and cooling could result if changes in albedo (reflectivity of Earth's surface) occur. If the Earth's surface were more reflective, for example, the amount of solar radiation radiated back toward space instead of absorbed would increase, lowering temperatures on Earth.
4. Long-term global warming and cooling could result if the amount of radiation emitted by the sun changes.
5. Long-term global warming and cooling could result if the shape and relationship of the land and oceans change.
6. Long-term global warming and cooling could result if the composition of the atmosphere changes.

"If the composition of the atmosphere changes"—this final factor, of course, defines our current concern: Have human activities had a cumulative impact large enough to affect the total temperature and climate of Earth? Right now, we cannot be sure. The problem concerns us, and we are alert to it, but we are not certain, because, again, we do not know what we do not know about global warming or climate change.

How Is Climate Change Measured?

Worldwide, scientists are trying to establish ways to determine whether or not greenhouse-induced global warming is occurring. Scientists are currently looking for signs of what is collectively referred to as a greenhouse *signature* or *footprint*. If global warming is occurring, eventually it will be obvious to everyone—but what we really want is clear, advanced warning. Thus, scientists are currently attempting to collect and then decipher a mass of scientific evidence to find those signs to give us clear advance warning. These signs are currently believed to include changes in the following (Franck and Brownstone, 1992):

1. *Global temperature patterns*, with continents being warmer than oceans, lands near the Arctic warming more than the tropics, and lower atmosphere warming while the higher stratosphere becomes cooler
2. *Atmospheric water vapor*, with increasing amounts of water evaporating into the air as a result of the warming (more in the tropics than in the higher latitudes), thus intensifying the warming process because water vapor is considered a greenhouse gas
3. *Sea surface temperature*, with a fairly uniform rise in the temperature of oceans at their surface and an increase in the temperature differences among oceans around the globe
4. *Seasonality*, with changes in the relative intensity of the seasons and warming effects being especially noticeable during the winter and in higher latitudes

In a measured, scientific way, these signs provide a general overview of some of the changes that would be expected to occur with global warming. Note, however, that from the point of view of life on Earth, changes resulting from long-term global warming would be drastic and profoundly serious. Probably the most dramatic of these effects—and the one with the most far-reaching results—would be sea-level rise.

GLOBAL WARMING AND SEA-LEVEL RISE

In the past few decades, human activities (e.g., burning fossil fuels, leveling forests, producing synthetic chemicals such as CFCs) have released into the atmosphere huge quantities of carbon dioxide and other greenhouse gases. These gases are warming the planet at an unprecedented rate. If current trends continue, they are expected to raise Earth's average surface temperature by at least 1.5°C to 4.5°C (or more), in the next century—with warming at the poles perhaps two to three times as high as warming at the middle latitudes (Wigley et al., 1986).

If we assume that global warming is inevitable and is already underway, what, then, can we do? Obviously, we cannot jump off the planet and head for greener pastures. We live on Earth and are stuck here for the time being. So, it makes good sense to try to understand the dynamics of change that are evolving around us and take whatever prudent actions we can to mitigate the situation. Such an approach should be applied to mitigating the effect that global warming is having on the rise in sea level. This rise is already underway and with it comes increased storm damage, pollution, and subsidence of coastal lands.

The rise in sea level is already underway? Absolutely. Consider the following information published by the U.S. Environmental Protection Agency (Titus and Narayanan, 1995):

1. *Global warming is most likely to raise sea levels 15 cm by the year 2050 and 34 cm by the year 2100.* In addition, there is also a 10% chance that climate change will contribute 30 cm by 2050 and 65 cm by 2100. These estimates do not include sea-level rise caused by factors other than greenhouse warming.

2. *There is a 1% chance that global warming will raise sea level 1 meter in the next 100 years and 4 meters in the next 200 years.* By the year 2200, there is also a 10% chance of a 2-meter contribution. Such a large rise in sea level could occur if the Antarctic ocean temperatures warm 5°C and Antarctic ice streams respond more rapidly than most glaciologists expect, or if Greenland temperatures warm by more than 10°C. Neither of these scenarios is likely.

3. *By the year 2100, climate change is likely to increase the rate of sea-level rise by 4.2 mm/yr.* There is also a 1-in-10 chance that the contribution will be greater than 10 mm/yr, as well as a 1-in-10 chance that it will be less than 1 mm/yr.

4. *Stabilizing global emissions in the year 2050 would likely reduce the rate of sea-level rise by 28% by the year 2100, compared with what it would be otherwise.* These calculations assume that we are uncertain about the future trajectory of greenhouse gas emissions.

5. *Stabilizing emissions by the year 2025 could cut the rate of sea-level rise in half.* If a high global rate of emissions growth occurs in the next century, sea level is likely to rise 6.2 mm/yr by 2100; freezing emissions in 2025 would prevent the rate from exceeding 3.2 mm/yr. If less emissions growth were expected, freezing emissions in 2025 would cut the eventual rate of sea-level rise by one-third.

6. *Along most coasts, factors other than anthropogenic climate change will cause the sea to rise more than the rise resulting from climate change alone.* These factors include compaction and subsidence of land, groundwater depletion, and natural climate variations. If these factors do not change, global sea level is likely to rise 45 cm by the year 2100, with a 1% chance of a 112-cm rise. Along the coast of New York, which typifies the United States, sea level is likely to rise 26 cm by 2050 and 55 cm by 2100. There is also a 1% chance of a 55-cm rise by 2050 and a 120-cm rise by 2100.

Along with these USEPA findings, additional lines of evidence corroborate that the global mean sea level has been rising for at least the last 100 years. Such evidence can be found in tide gauge records, the erosion of 70% of the world's sandy coasts and 90% of America's sandy beaches, and the melting and retreat of mountain glaciers (Broecker, 1987). The correspondence between the two curves of rising global temperatures and rising sea levels over the last 100 years appears to be more than coincidental (Edgerton, 1991).

Major uncertainties are present in any estimates of future sea-level rise. The problem is further complicated by our lack of understanding of the mechanisms contributing to relatively recent rises in sea level. In addition, different outlooks for climatic warming dramatically affect estimates. In all this uncertainty, one thing is sure—estimates of sea-level rise will undergo continual revision and refinement as time passes and more data are collected.

MAJOR PHYSICAL EFFECTS OF SEA-LEVEL RISE

With increased global temperatures, global sea-level rise will occur at a rate unprecedented in human history (Edgerton, 1991). Changes in temperature and sea level will be accompanied by changes in salinity levels; for example, a coastal freshwater aquifer is influenced by two factors: pumping and mean sea level. If pumping withdrawals exceed recharge, the water table is drawn down and saltwater penetrates inland. Another problem occurs when the sea level rises and the coastline moves inland, reducing aquifer area. Additional problems brought about by changes in temperature and sea level are seen in tidal flooding, oceanic currents, biological processes of marine creatures, runoff and landmass erosion patterns, and saltwater intrusion.

The most important direct physical effects of sea-level rise can be seen on coastal beach systems. At current rates of sea-level rise of 1 to 2 mm/year, significant coastal erosion is already occurring. Two major factors contribute to beach erosion. First, deeper coastal waters enhance wave generation, thus increasing their potential for overtopping barrier islands. Second, shorelines and beaches will attempt to establish new equilibrium positions according to what is known as the *Bruun rule*; these adjustments will include a recession of shoreline and a decrease in shore slope (Bruun, 1962, 1986).

MAJOR DIRECT HUMAN EFFECTS OF SEA-LEVEL RISE

Along with the physical effects of sea-level rise, in one way or another, directly or indirectly, accompanying effects have a direct human side, especially concerning human settlements and the infrastructure that accompanies them: highways, airports, waterways, water supply and wastewater treatment facilities, landfills, hazardous waste storage areas, bridges, and associated maintenance systems. Sea-level rise could also cause intrusion of saltwater into groundwater supplies (Edgerton, 1991).

To point out that this infrastructure would be placed under tremendous strain by a rising sea level coupled with other climatic change is to understate the possible consequences. Indeed, the impact on infrastructure is only part of the direct human impact. There is widespread agreement among scientists, for example, that any significant change in world climate resulting from warming or cooling would (1) disrupt world food production for many years, (2) lead to a sharp increase in food prices, and (3) cause considerable economic damage. Just how much of a rise in sea level are we talking about? According to Titus and Narayanan (1995, p. 123), "If the experts on whom we relied fairly represent the breadth of scientific opinion, the odds are fifty-fifty that greenhouse gases will raise sea level at least 15 cm by the year 2050, 35 cm by 2100, and 80 cm by 2200."

GLOBAL CLIMATE CHANGE

Is global warming a hoax? Is Earth's climate changing? Are warmer times or colder times on the way? Is the greenhouse effect going to affect our climate, and, if so, do we need to worry about it? Will the tides rise and flood New York? Does the ozone hole portend disaster right around the corner? These and many other questions related to climate change have come to the attention of us all, thanks to a constant barrage of newspaper headlines, magazine articles, and television news reports. We have seen reports on El Niños and their devastation of the west coast of the United States, as well as Peru and Ecuador. We have also seen a reduction in the number and magnitude of hurricanes that annually blast the eastern coast of the United States.

To illustrate the constant barrage of newspaper headlines, listed below is just a sampling of the climate change news stories and global warming headlines published in many locations throughout the globe in the month of April 2008 (Bloch, 2009):

April 1—Global Warming Awareness and Apathy
April 2—Oceans Under Stress from Global Warming
April 7—Australian Drought Affected Areas Grow
April 13—Fossil Fuel Carbon Emissions Over 8 Gigatons
April 22—UK Migrating Birds Numbers Drop
April 28—March Warmest on Record Globally

Scientists have been warning us of the catastrophic harm that can be done to the world by atmospheric warming. One view states that the effect could bring record droughts, record heat waves, record smog levels, and an increasing number of forest fires. Another caution put forward warns that the increasing atmospheric heat could melt the world's icecaps and glaciers, causing ocean levels to rise to the point where some low-lying island countries would disappear, while the coastlines of other nations would be drastically altered for ages—or perhaps for all time.

What's going on? We hear plenty of theories put forward by doomsayers, but are they correct? If they are correct, what does it all mean? Does anyone really know the answers? Should we be concerned? Should we invest in waterfront property in Antarctica? Should we panic? No. Although no one really knows the answers and although we certainly should be concerned, no real cause for panic exists. Still, though, should we take some type of decisive action, should we come up with quick answers and put together a plan to fix these problems? What really needs to be done? What can we do? Is there anything we can do?

The key question here is "What really needs to be done?" We can study the facts, the issues, and the possible consequences, but the key to successfully combating these issues is to stop and seriously evaluate the problems. We need to let scientific fact, common sense, and cool-headedness prevail—shooting from the hip is not called for, makes little sense, and could have colossal consequences for us all.

Another question that has merit here is, "If there truly is a problem with global climate change, will we take the correct actions before it is too late?" The key words here are "correct actions." Eventually, we may have to take action, but we do not yet know what those actions should be.

From the author's perspective, one thing is certain: Sooner or later, college-level environmental health courses address global warming and global climate change. Through time and experience, the teachers of such courses have learned that whether we call it global warming, global climate change (humankind-induced global warming, under a broader label), or an inconvenient truth, the topic is a conundrum, a riddle the answer of which is a pun. As such, before diving into the many emotionally charged, heated classroom discussion about this "hot" topic (pun intended), we are reminded by two celebrated statements of just how complicated a conundrum can be:

1. What is black and white and read all over? *A newspaper.*
2. Why is a man in jail like a man out of jail? *The answer is that there is no answer, there is no way out.*

Consider this: Any damage we do to our atmosphere affects the other two media of water and soil—and biota (including us). Thus, the endangered atmosphere is a major concern to all of us.

The Past

Before beginning our discussion of the past, we need to define the era we refer to when we say "the past." Table 7.1 summarizes the entire expanse of time from Earth's beginning to present. Table 7.2 provides the sequence of geological epochs over the past 65 million years, as dated by modern methods. The Paleocene through Pliocene together make up the Tertiary period; the Pleistocene and the Holocene compose the Quaternary period.

When we think about general and climatic conditions in the prehistoric past, two things generally come to mind—ice ages and dinosaurs. Of course, in the immense span of time that prehistory covers, those two eras represent only a brief moment in time, so let's look at what we know about the past and about Earth's climate and

TABLE 7.1
Geologic Eras and Periods

Era	Period	Millions of Years before Present
Cenozoic	Quaternary	2.5–present
	Tertiary	65–2.5
Mesozoic	Cretaceous	135–65
	Jurassic	190–135
	Triassic	225–190
Paleozoic	Permian	280–225
	Pennsylvanian	320–280
	Mississippian	345–320
	Devonian	400–345
	Silurian	440–400
	Ordovician	500–440
	Cambrian	570–500
Precambrian		4600–570

TABLE 7.2
Geologic Epochs

Epoch	Million Years Ago
Holocene	0.01–0
Pleistocene	1.6–0.01
Pliocene	5–1.6
Miocene	24–5
Oligocene	35–24
Eocene	58–35
Paleocene	65–58

conditions. One thing to consider is that geological history shows us that the normal climate of the Earth was so warm that subtropical weather reached to 60° north and south latitude, and polar ice was entirely absent.

Only during less than about 1% of the Earth's history did glaciers advance and reach as far south as what is now the temperate zone of the northern hemisphere. The latest such advance, which began about 1,000,000 years ago, was marked by geological upheaval and (perhaps) the advent of human life on Earth. During this time, vast ice sheets advanced and retreated and coarsely ground their way over the continents, reducing mountains to fine dirt.

A Time of Ice

Nearly 2 billion years ago, the oldest known glacial epoch occurred. A series of deposits of glacial origin in southern Canada, extending east to west about 1000 miles, shows us that within the last billion years or so apparently at least six major phases of massive, significant climatic cooling and consequent glaciation occurred at intervals of about 150 million years. Each lasted perhaps as long as 50 million years.

Examination of land and oceanic sediment core samples clearly indicates that in more recent times (the Pleistocene epoch to the present), many alternating episodes of warmer and colder conditions occurred over the last 2 million years (during the middle and early Pleistocene epochs). In the last million years, at least eight such cycles have occurred, with the warm part of the cycle lasting a relatively short interval.

During the Great Ice Age (the Pleistocene epoch), a series of ice advances began that at times covered over one quarter of Earth's land surface. Great sheets of ice thousands of feet thick, these glaciers moved across North America many times, reaching as far south as the Great Lakes. An ice sheet thousands of feet thick spread over Northern Europe, sculpting the land and leaving behind lakes, swamps, and terminal moraines as far south as Switzerland. Each succeeding glacial advance was apparently more severe than the previous one. Evidence indicates that the most severe began about 50,000 years ago and ended about 10,000 years ago. Several interglacial stages separated the glacial advances, during which average temperatures were higher than ours today, and, consequently, the ice melted.

Wait a minute! Temperatures were higher than today? Yes, they were. Think about that as we proceed.

Because one-tenth of the globe's surface is still covered by glacial ice, scientists consider the Earth still to be in a glacial stage. The ice sheet has been retreating since the climax of the last glacial advance, and world climates, although fluctuating, are slowly warming.

From our observations and from well-kept records, we know that the ice sheet is in a retreating stage. The records clearly show that a marked worldwide retreat of ice has occurred over the last 100 years. World famous for its 50 glaciers and 200 lakes, Glacier National Park in Montana does not present the same visual experiences it did 100 years ago. In 1937, a 10-foot pole was put into place at the terminal edge of one of the main glaciers. The sign is still in place, but the terminal end of the glacier has retreated several hundred feet back up the slope of the mountain. Swiss resorts built during the early 1900s to offer scenic glacial views now have no ice in sight. Theoretically, if glacial retreat continues, melting all of the world's ice supply, sea levels would rise more than 200 feet, flooding many of the world's major cities. New York and Boston would become aquariums.

The question of what causes ice ages is one scientists still grapple with. Theories range from changing ocean currents to sunspot cycles. Of one fact we are absolutely certain, however; an ice age event occurs because of a change in Earth's climate. But what could cause such a drastic change?

Climate results from uneven heat distribution over Earth's surface. It is caused by the Earth's tilt—the angle between the Earth's orbital plane around the sun and its rotational axis. This angle is currently 23.5 degrees, but it has not always been that. The angle, of course, affects the amount of solar energy that reaches the Earth and where it falls. The heat balance of the Earth, which is driven mostly by the concentration of carbon dioxide (CO_2) in the atmosphere, also affects long-term climate. If the pattern of solar radiation changes or if the amount of CO_2 changes, climate change can result. Abundant evidence that the Earth does undergo climatic change exists, and we know that climatic change can be a limiting factor for the evolution of many species.

Evidence (primarily from soil core samples and topographical formations) tells us that changes in climate include events such as periodic ice ages characterized by glacial and interglacial periods. Long glacial periods lasted up 100,000 years; temperatures decreased about 9°F, and ice covered most of the planet. Short periods lasted up to 12,000 years, with temperatures decreasing by 5°F and ice covering 40° north latitude and above. Smaller periods (e.g., the "Little Ice Age," which occurred from about 1000 to 1850 AD) experienced about a 3°F drop in temperature. *Note*: Despite its name, the Little Ice Age was a time of severe winters and violent storms, not a true glacial period. These ages may be or not be significant, but consider that we are currently in an interglacial period and that we may be reaching its apogee. What does that mean? No one knows with any certainty.

Let's look at the effects of ice ages (i.e., effects we think we know about). Changes in sea levels could occur. Sea level could drop by about 100 meters during a full-blown ice age, exposing the continental shelves. Increased deposition during melt would change the composition of the exposed continental shelves. Less evaporation would change the hydrological cycle. Significant landscape changes could occur—on the scale of the Great Lakes formation. Drainage patterns throughout most of the world and topsoil characteristics would change. Flooding on a massive scale could occur. How these changes would affect us depends on whether you live in Northern Europe, Canada, Seattle, Washington, around the Great Lakes, or near a seashore.

We are not sure what causes ice ages, but we have some theories. To generate a full-blown ice age (massive ice sheet covering most of the globe), as scientists point out, certain periodic or cyclic events or happenings must occur. Periodic fluctuations would have to affect the solar cycle, for instance; however, we have no definitive proof that this has ever occurred.

Another theory speculates that periods of volcanic activity could generate masses of volcanic dust that would block or filter heat from the sun, thus cooling down the Earth. Some speculate that the carbon dioxide cycle would have to be periodic or cyclic to bring about periods of climate change. There is reference to a so-called Factor 2 reduction, causing a 7°F temperature drop worldwide. Others speculate that another global ice age could be brought about by increased precipitation at the poles due to changing orientation of continental land masses. Others theorize that a global ice age would result if the mean temperatures of ocean currents decreased. But the question is how? By what mechanism? Are these plausible theories? No one is sure—this is speculation.

Speculation aside, what are the most probable causes of ice ages on Earth? According to the *Milankovitch hypothesis*, ice age occurrences are governed by a combination of factors: (1) the Earth's change of altitude in relation to the Sun (the way it tilts in a 41,000-year cycle and at the same time wobbles on its axis in a 22,000-year cycle), making the time of its closest approach to the Sun come at different seasons; and (2) the 92,000-year cycle of eccentricity in its orbit around the sun, changing it from an elliptical to a near circular orbit, the most severe period of an ice age coinciding with the approach to circularity.

We have a lot of speculation about ice ages and their causes and their effects. We know that ice ages occurred—we know that they caused certain things to occur (e.g., formation of the Great Lakes), and although there is a lot we do not know, we recognize the possibility of recurrent ice ages. Right now, no single theory is sound, and doubtless many factors are involved. Keep in mind that the possibility does exist that we are still in the Pleistocene ice age. It may reach another maximum in another 60,000 plus years or so.

Warm Winter

The headlines we see in the paper sound authoritative: "1997 Was the Warmest Year on Record" ... "Scientists Discover Ozone Hole Is Larger Than Ever" ... "Record Quantities of Carbon Dioxide Detected in Atmosphere." Or maybe you saw the one that read "January 1998 Was the Third Warmest January on Record." Other reports indicate that we are undergoing a warming trend, but conflicting reports abound. This section discusses what we think we know about climate change.

Two environmentally significant events took place late in 1997: El Niño's return and the Kyoto Conference on Global Warming and Climate Change. News reports blamed El Niño for just about anything that had to do with weather conditions throughout the world. Some incidents were indeed El Niño related or generated: the out-of-control fires, droughts, floods, the stretches of dead coral with no sign of fish in the water, and few birds around certain Pacific atolls. The devastating storms that struck the west coasts of South America, Mexico, and California were also probably related to El Niño. El Niño's effect on the 1997 hurricane season, one of the mildest on record, is not in question, either.

Does a connection exist between El Niño and global warming or global climate change? On December 7, 1997, the Associated Press reported that, while delegates at the global climate conference in Kyoto haggled over greenhouse gases and emission limits, a compelling question had emerged: "Is global warming fueling El Niño?" (Anon., 1997). Nobody knows for sure because we need more information than we have today. The data we do have, however, suggests that El Niños are getting stronger and more frequent. Some scientists fear that the increasing frequency and intensity of El Niños (records show that two of the last century's three worst El Niños came in 1982 and 1997) may be linked to global warming. At the Kyoto conference, experts said the hotter atmosphere is heating up the world's oceans, setting the stage for more frequent and extreme El Niños. Weather-related phenomena seem to be intensifying throughout the globe. Can we be sure that all of this is related to global warming? No. Without more data, more time, more science, we cannot be sure.

According to the Associated Press coverage of the Kyoto conference, scientist Richard Fairbanks reported that he found startling evidence supporting our need for concern. During 2 months of scientific experiments conducted in autumn 1997 on Christmas Island, the world's largest atoll in the Pacific Ocean, he witnessed a frightening scene. The water surrounding the atoll was 7°F higher than average for that time of year, which upset the balance of the environmental system. According to Fairbanks, 40% of the coral was dead, the warmer water had killed off or driven away fish, and the atoll's plentiful bird population was almost completely gone.

No doubt, El Niños are having an acute impact on the globe; however, we do not know if these events are caused by or intensified by global warming. What do we know about global warming and climate change? *USA Today* (Anon., 1997) discussed the results of a report issued by the Intergovernmental Panel on Climate Change. They interviewed Jerry Mahlman of the National Oceanic and Atmospheric Administration and Princeton University, and presented the following information about what most scientists agree on:

- There is a natural greenhouse effect and scientists know how it works; without it, Earth would freeze.
- The Earth undergoes normal cycles or warming and cooling on grand scales. Ice ages occur every 20,000 to 100,000 years.
- Globally, average temperatures have risen 1°F in the past 100 years, within the range that might occur normally.
- The level of manmade carbon dioxide in the atmosphere has risen 30% since the beginning of the Industrial Revolution in the 19th century and is still rising.
- Levels of manmade carbon dioxide will double in the atmosphere over the next 100 years, generating a rise in global average temperatures of about 3.5°F (larger than the natural swings in temperature that have occurred over the past 10,000 years).

- By 2050, temperatures will rise much higher in northern latitudes than the increase in global average temperatures. Substantial amounts of northern sea ice will melt, and snow and rain in the northern hemisphere will increase.
- As the climate warms, the rate of evaporation will rise, further increasing warming. Water vapor also reflects heat back to Earth.

Causes of Global Warming

What is global warming? To answer this question we need to discuss the greenhouse effect. Water vapor, carbon dioxide, and other atmospheric gases (greenhouse gases) help warm the Earth. Earth's average temperature would be closer to 0°F than its actual 60°F without the greenhouse effect. But, as gases are added to the atmosphere, the average temperature could increase, changing orbital climate.

How does the greenhouse effect actually work? As noted earlier, Earth's greenhouse effect took its name because of its similarity to the effects of a greenhouse. Because the glass walls and ceilings of a greenhouse are largely transparent to short-wave radiation from the sun, surfaces and objects inside the greenhouse absorb the radiation. The radiation, once absorbed, transforms into long-wave (infrared) radiation (heat), which radiates out from the greenhouse interior, but the glass prevents the long-wave radiation from escaping again and the warm rays are absorbed. The interior of the greenhouse becomes much warmer than the air outside, because of the heat trapped inside.

Earth and its atmosphere undergo a very similar process. Short-wave and visible radiation reaching Earth is absorbed by the surface as heat. The long heat waves radiate back out toward space, but the atmosphere absorbs many of them, trapping them. This natural and balanced process is essential to supporting life systems on Earth. Changes in the atmosphere can radically change the rate of absorption and therefore the amount of heat the atmosphere retains. In recent decades, scientists have speculated that various air pollutants have caused the atmosphere to absorb more heat. At the local level, with air pollution, the greenhouse effect causes heat islands in and around urban centers, a widely recognized phenomenon.

Again, as noted earlier, the main contributors to this effect are the greenhouse gases: water vapor, carbon dioxide, carbon monoxide, methane, volatile organic compounds, nitrogen oxides, chlorofluorocarbons, and surface ozone. These gases cause a general climatic warming by delaying the escape of infrared radiation from the Earth into space. Scientists stress that this is a natural process. Indeed, as observed earlier, if the normal greenhouse effect did not exist then the Earth would be far cooler than it currently is (Hansen et al., 1986).

Human activities, though, are rapidly intensifying this natural phenomenon which may lead to problems of warming on a global scale. The rate at which the greenhouse effect is intensifying is now more than five times what it was during the 1800s (Hansen et al., 1989). Much debate, confusion, and speculation about this potential consequence of global warming exist, because scientists cannot yet agree as to whether the recently perceived worldwide warming trend is because of greenhouse gases, is due to some other cause, or is simply a wider variation in the normal heating and cooling trends they have been studying. Without a doubt, the human impact on the greenhouse effect is real; it has been measured and detected. Unchecked, the greenhouse effect may lead to significant global warming, with profound effects on our lives and our environment.

Supporters of the global warming theory assume that human activities are significantly altering the Earth's normal and necessary greenhouse effect. The human activities they blame for this increase of greenhouse gases include burning fossil fuels, deforestation, and the use of certain aerosols and refrigerants. From information based on recent or short-term observations, many scientists have observed that the last decade has been the warmest since temperature recordings began in the late 19th century. They can see that the general rise in temperature coincides with the Industrial Revolution and its accompanying increase in fossil fuel use. Other evidence supports the global warming theory. In places that are synonymous with ice and snow—the Arctic and Antarctica, for example—we are seeing evidence of receding ice and snow cover.

Trying to pin down definitively whether or not changing our anthropogenic activities could have any significant effect on lessening global warming, though, is difficult. Scientists look at temperature variations over thousands or even millions of years, taking a long-term view of Earth's climate. The variations in Earth's climate are wide enough that they cannot definitively show that this global warming is anything more than another short-term variation. Historical records show that the Earth's temperature does vary widely, as it grows colder with ice ages and then warms again; because we cannot be certain of the causes of those climate changes, we cannot be certain of what is causing the current warming trend.

Still, debate abounds for the argument that our climate is warming and our activities are part of the equation. The 1980s saw 9 of the 12 warmest temperatures ever recorded, and the Earth's average surface temperature rose approximately 0.6°C (1°F) in the last century (USEPA, 2009a). An article in *Time* magazine (Anon., 1998) reported that scientists are increasingly convinced that the Earth is getting hotter because of the buildup in the atmosphere of carbon dioxide and other gases produced in large part by the burning of fossil fuels. Each month from January through July 1998, for example, set a new average global temperature record, and if that trend continues the surface temperature of the Earth could rise by about 1.8 to 6.3°F by 2100. At the same time, as noted earlier, others argue that the 1980s also saw three of the coldest years: 1984, 1985, and 1986. The debate does not end there, however. For example, NASA's Goddard Institute for Space Studies made a correction to data that seemed to show that nine of the ten hottest years in U.S. history occurred since 1995; as it turns out, it was only three. The more vocal global warming deniers have been using this error to prove that the whole idea of global warming-induced climate change is a hoax. Others argue that this one mistake does not undermine the entire global warming creditability issue.

If global warming is, indeed, occurring, then we can expect winters to be longer, and summers hotter. Over the next 100 years, sea level will rise as much as a foot or so. Is this bad? Depends on where you live. Keep in mind, however, that not only could sea level rise 1 foot over the next 100 years, but it could also continue to do so for many hundreds of years. Another point to consider is that we have routine global temperature measurements for only about 100 years. Even these are unreliable, because instruments and methods of observation have changed over that course of time.

The only conclusion we can safely draw about climate and climate change is that we do not know if drastic changes are occurring. We could be at the end of a geological ice age. Evidence indicates that during interglacials temperatures increase before they plunge. Are we ascending the peak temperature range? We have no way to tell. To what extent does our human activity impact climate? Have anthropogenic effects become so marked that we have affected the natural cycle of ice ages (which lasted for roughly the last 5 million years)? Maybe we just have a breathing spell of a few centuries before the next advance of the glaciers. If this is the case, if we are at the apogee of the current interglacial, then we have to ask ourselves a few questions: Is global warming the lesser of two evils when compared to the alternative, global cooling? If we are headed into another glacial freeze, in this era of expanding population and decreasing resources, where will we get the energy to keep all of us warm?

Obvious Questions

1. Is global warming or global climate change real?
2. Are humans responsible for global climate change?
3. Is global climate change cyclical or human driven?

The bottom line is that at this point it does not really matter whether global warming is cyclical or human driven. The evident and undeniable point is that sea levels are rising globally and in Hampton Roads. But, there is another problem, one that compounds the problem of rising sea levels: land subsidence in Hampton Roads.

LAND SUBSIDENCE IN HAMPTON ROADS[*]

This chapter of the book was prefaced by a discussion of the problems in the Hampton Roads area and in particular those concerning the Chesapeake Bay, such as the continuing appearance of dead zones, nutrient pollution, sediment contamination, and sea-life decline. Without a doubt these are serious problems that continue to garner the attention of officials responsible for monitoring and managing the health of Chesapeake Bay. As serious as these problems are, it is the ongoing rise in relative sea levels that is beginning to impact Hampton Roads and offers a future that can best be described as foreboding and quite wet unless certain mitigation practices are put into place, not tomorrow but today and continuing into the future. Part of the relative sea-level rise problem is due to land subsidence, and it is this problem that is addressed in the section.

---

[*] Adapted from Eggleston, J. and Pope, J., *Land Subsidence and Relative Sea-Level Rise in the Southern Chesapeake Bay Region*, Circular 1392, U.S. Geological Survey, Reston, VA, 2013.

As explained earlier, land subsidence is the sinking or lowering of the land surface. In the United States, most land subsidence is caused by human activities (Galloway et al., 1999). With regard to land subsidence problems in the western region of the country, after recognition of the problems associated with groundwater drawdown and resulting land subsidence the areas affected set up monitoring networks and ultimately adopted new water-management practices to prevent or arrest land subsidence. In the Hampton Roads area, data indicate that land subsidence has been responsible for more than half of the relative sea-level rise measured there, suggesting that the problem will be ongoing in the future. This is bad news for those residing in the area. Land subsidence is a serious issue because the increased flooding has and will continue to have important economic, environmental, and human health consequences for the heavily populated and ecologically important southern Chesapeake Bay region.

Local or isostatic factors contribute to a *relative* sea-level rise in Hampton Roads through subsidence or sinking of the land. In the Chesapeake Bay area, subsidence of land is due to both geologic factors and excessive withdrawal of groundwater which has amounted to 6 inches in the last 100 years at rates of 0.039 to 0.189 inches per year (1.1 to 4.8 mm per year). Consequently, there has been a relative increase of sea level in the Chesapeake Bay area of 1 foot in the last century (see Figure 7.1). More specifically, land subsidence in the region is the result of flexing of the Earth's crust from glacial isostatic adjustment in response to glacier formation and melting. In addition, more than half of the observed subsidence is the result of the aquifer system in Hampton Roads that has been compacted by extensive groundwater extraction at rates of 1.5 to 3.7 mm/yr. This helps explain why the southern Chesapeake Bay region has the highest rate of sea-level rise on the Atlantic Coast of the United States (Zervas, 2009). Because the communities in the region must grapple with flooding problems that contribute to the disappearance of existing land, which will continue to worsen in the future, it is important to understand and potentially manage land subsidence.

Experience has shown that the rates and locations of land subsidence change over time, so accurate measurements and predictive tools are needed to improve our understanding of land subsidence. Although rates of land subsidence are not as high on the Atlantic Coast as they have been in the Houston–Galveston area or the Santa Clara Valley, land subsidence is important because of the low-lying topography and susceptibility to sea-level rise in the southern Chesapeake Bay region. Although the focus of this discussion is on land subsidence in the Hampton Roads and Chesapeake Bay Region, the lessons learned here can be applied to other affected areas.

WHY LAND SUBSIDENCE IS A CONCERN IN THE CHESAPEAKE BAY REGION

In the Chesapeake Bay region, increased flooding, wetland and coastal ecosystem alteration, and damage to infrastructure and historical sites are all the result of land subsidence. This problem is not a new one; it is well known to regional planners who have gained understanding of what land subsidence is—that is, why, where, and how fast it is occurring, now and in the future.

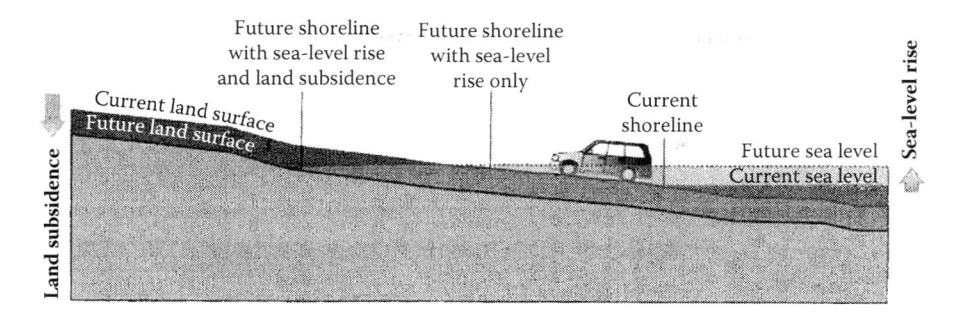

FIGURE 7.3   Shoreline retreat caused by a combination of sea-level rise and land subsidence. (From Eggleston, J. and Pope, J., *Land Subsidence and Relative Sea-Level Rise in the Southern Chesapeake Bay Region*, Circular 1392, U.S. Geological Survey, Reston, VA, 2013.)

LAND SUBSIDENCE CONTRIBUTES TO RELATIVE SEA-LEVEL RISE

Land subsidence contributes to the relative sea-level rise that has been measured in the Chesapeake Bay (Figure 7.3). Tidal-station measurements of sea levels, however, do not distinguish between water that is rising and land that is sinking—the combined elevation changes are termed *relative sea-level rise*. Global sea-level rise and land subsidence increase the risk of coastal flooding and contribute to shoreline retreat.

LAND SUBSIDENCE INCREASES FLOODING RISK

As relative sea levels rise, shorelines retreat and the magnitude and frequency of near-shore coastal flooding increase. This is particularly a problem in Norfolk, Virginia, where, during a coastal storm event and corresponding high tide, downtown Norfolk streets flood; at times, it can flood to several feet. Although land subsidence can be slow, its effects accumulate over time. As pointed out in Chapter 6, this has been an expensive problem in the Houston–Galveston area and the Santa Clara Valley (Galloway et al., 1999) and likely contributes to current flooding problems in the Norfolk and other parts of the southern Chesapeake Bay region. Between 59,000 and 176,000 residents living near the shores of the southern Chesapeake Bay could be either permanently inundated or regularly flooded by 2100 (McFarlane, 2012). This estimate is based on 2010 census data, using the spring high tide as a reference elevation and assuming a 1-m relative sea-level rise. Damage to personal property was estimated to be $9 billion to $26 billion, and 120,000 acres of economically valuable land could be inundated or regularly flooded, under these same assumptions. Historic and cultural resources are also vulnerable to increased flooding from relative sea-level rise in the southern Chesapeake Bay, particularly at shoreline sites near tidal water, such as the 17th-century historic Jamestown site.

Another historic site in southern Chesapeake Bay is First Landing State Park, where this book began with Mr. Grasshopper and Mr. Rabbit's conversation. If relative sea-level rise continues unabated, it would not be all that long before Mr. Rabbit would have to learn to swim, get a raft, or don a life jacket, and Mr. Grasshopper would have to jump and leap and remain in flight so as not to drown.

It should be pointed out that the shoreline area in southern Hampton Roads is not the only area prone to flooding. Land subsidence can also increase flooding in areas away from the coast in low-lying areas such as Franklin, Virginia. The city of Franklin is about 60 road miles west of Hampton Roads. The Blackwater River Basin, which encompasses Franklin and other local areas, can be subject to increased flooding as the land sinks. In fact, Franklin and the counties of Isle of Wight and Southampton have experienced large floods (Federal Emergency Management Agency, 2002). Land subsidence may be altering the topographic gradient that drives the flow of the river and possibly contributing to the flooding.

LAND SUBSIDENCE CAN DAMAGE WETLAND AND COASTAL MARSH ECOSYSTEMS

Wetland and marsh ecosystems in low-lying coastal areas are sensitive to small changes in elevation (Cahoon et al., 2009). Salt marshes, which are widespread in the southern Chesapeake Bay region, are dependent on tidal dynamics for their existence. Small changes in either land or sea elevations can alter sediment deposition, organic production and plant growth, and the balance between freshwater and seawater (Morris et al., 2002). The effects of sea-level rise on tidal wetlands are numerous and already apparent in local wetlands. These effects include the following:

- Shoreline erosion
- Habitat loss
- Changes in tidal amplitude
- Landward migration of tidal waters
- Landward migration of habitats
- More frequent inundation
- Changes in plant and animal species composition
- Changes in tidal flow patterns
- Migration of estuarine salinity gradients
- Changes in sediment transport

Sea-level rise has a direct effects on tidal wetlands, but shoreline environments also are affected by land subsidence. When land subsides, it subjects shorelines to increased wave action, which increases erosion and washover. This type damage is happening in the Chesapeake Bay because of relative sea-level rise (Erwin et al., 2011; Kirwan and Guntenspergen, 2012; Kirwan et al., 2013). Major changes in the coastal and marine ecosystem of the southern Chesapeake Bay are expected to be caused by relative sea-level rise; these changes will likely be more severe if land subsidence continues.

LAND SUBSIDENCE CAN DAMAGE INFRASTRUCTURE

Buildings, bridges, canals, water/wastewater treatment plants, electrical substations, communication towers, pipes and other components of a region's infrastructure can be damaged from relative groundwater rise or from differential settling in areas with high subsidence gradients (Galloway et al., 1999). As land sinks and sea level continues to rise, groundwater levels rise toward the land surface in coastal areas, which can cause problems for subterranean structures, septic fields, buried pipes and tanks, buried cables, and infrastructure not designed for elevated groundwater levels. Storm and wastewater interceptor lines in urban areas are vulnerable because land subsidence can alter the topographic gradient driving the flow through the sewers, causing increased flooding and more frequent sewage discharge from combined sewer overflows.

Aquifer Compaction

When groundwater is pumped from the Potomac aquifer system in southern Chesapeake Bay, pressure decreases. The pressure change is reflected by water levels in wells, with water levels decreasing as aquifer system pressure decreases. This is happening over most of the southern Chesapeake Bay region, with the greatest water-level decrease seen near the pumping centers of Franklin and West Point, Virginia (see Figure 7.4).

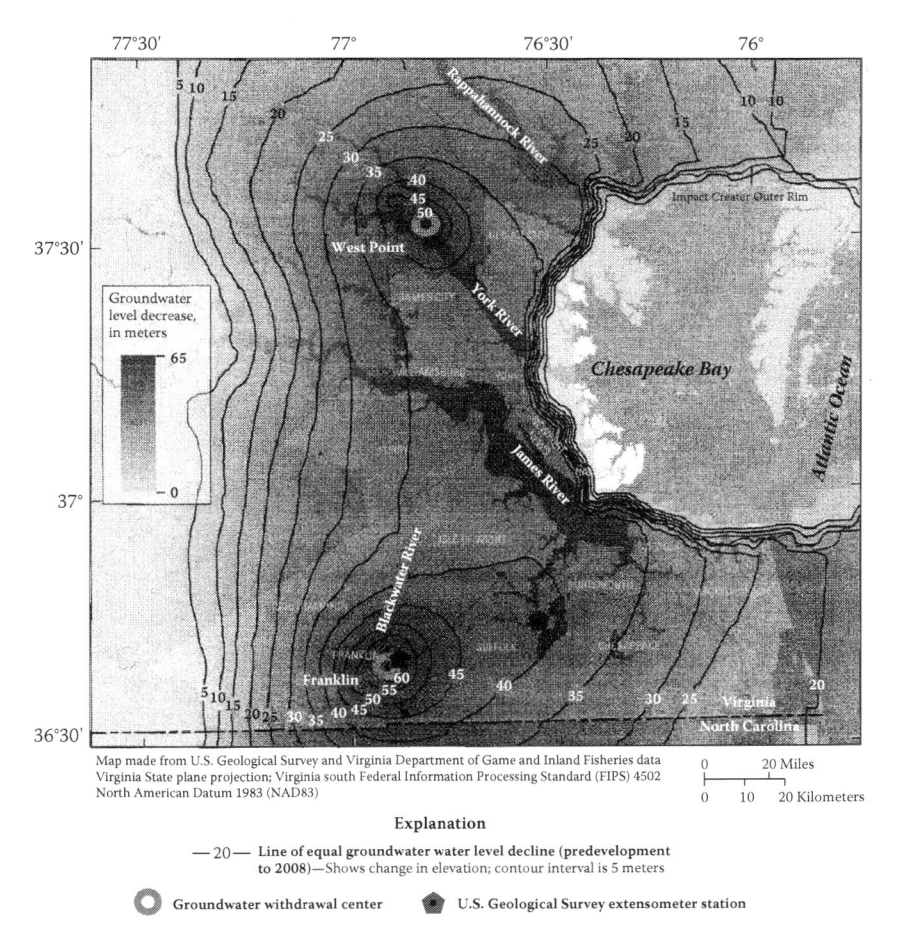

Map made from U.S. Geological Survey and Virginia Department of Game and Inland Fisheries data
Virginia State plane projection; Virginia south Federal Information Processing Standard (FIPS) 4502
North American Datum 1983 (NAD83)

Explanation

— 20 — Line of equal groundwater water level decline (predevelopment to 2008)—Shows change in elevation; contour interval is 5 meters

⬤ Groundwater withdrawal center ⬟ U.S. Geological Survey extensometer station

FIGURE 7.4 Decreases in groundwater levels from 1990 to 2008. (From Eggleston, J. and Pope, J., *Land Subsidence and Relative Sea-Level Rise in the Southern Chesapeake Bay Region*, Circular 1392, U.S. Geological Survey, Reston, VA, 2013.)

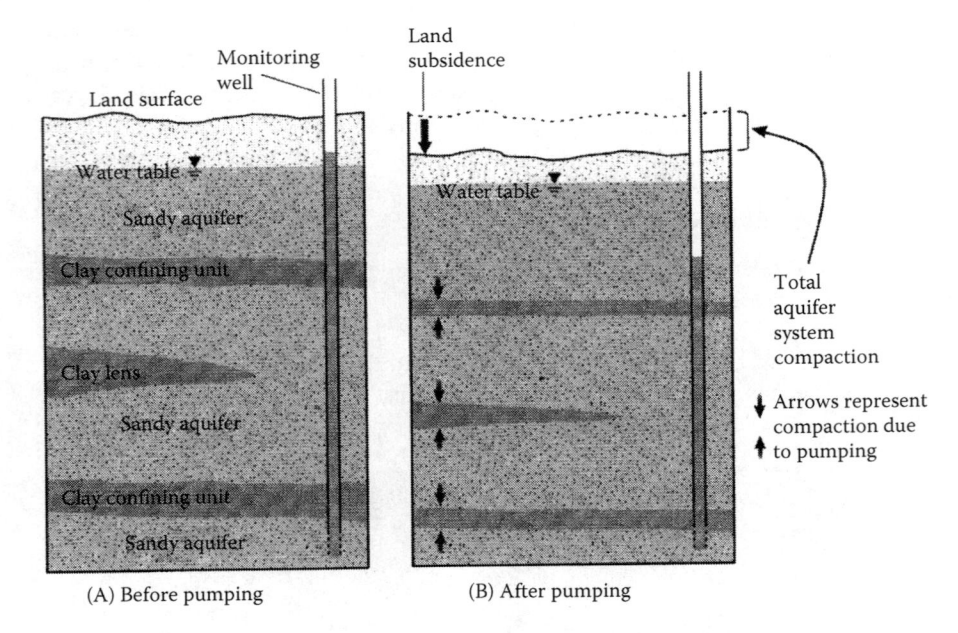

(A) Before pumping   (B) After pumping

FIGURE 7.5   Aquifer system compaction caused by groundwater withdrawals (A) before and (B) after pumping. (From Eggleston, J. and Pope, J., *Land Subsidence and Relative Sea-Level Rise in the Southern Chesapeake Bay Region*, Circular 1392, U.S. Geological Survey, Reston, VA, 2013.)

As water levels decrease, the aquifer system compacts, causing the land surface to subside (see Figure 7.5). Water levels have decreased over the entire Virginia Coastal Plain and in the Potomac aquifer, which is the deepest and thickest aquifer in the southern Chesapeake Bay region and supplies about 75% of groundwater withdrawn from the Virginia Coastal Plain aquifer system (Heywood and Pope, 2009).

Three factors determine the amount of aquifer system compaction: water-level decline, sediment compressibility, and sediment thickness. If any of these three factors increase in magnitude, then the amount of aquifer system compaction and land subsidence increases. Because all three of these factors vary spatially across the southern Chesapeake Bay region, rates of load subsidence caused by aquifer system compaction also vary spatially across the region.

Figure 7.6 shows the Virginia Coastal Plain aquifer system, which consists of many stacked layers of sand and clay. Although groundwater is withdrawn primarily from the aquifers (sandy layers), most compaction occurs in confining units and clay lenses, the relatively impermeable layers sandwiched between and within the aquifers (Pope and Burbey, 2004). The compression of the clay layers is mostly non-recoverable, meaning that, if groundwater levels later recover and increase, then the aquifer system does not expand to its previous volume and the land surface does not rise to its previous elevations (Pope, 2002). It has been estimated that 95% of the water removed from storage in the Virginia Coastal Plain aquifer system between 1891 and 1980 was derived from the confining layers (Konikow and Neuzil, 2007).

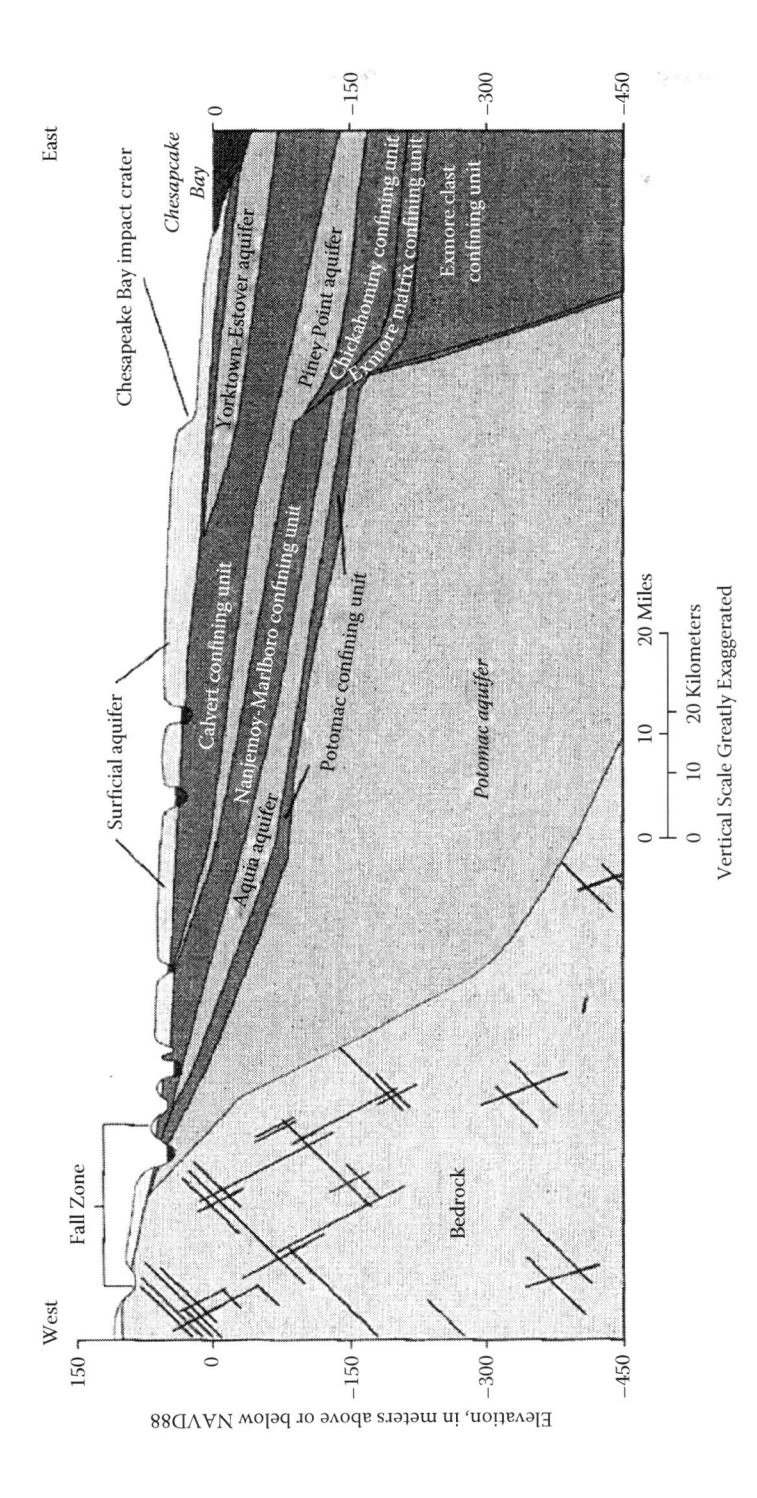

FIGURE 7.6 Section illustrating layering in the Virginia Coastal Plain aquifer system from west to east. Elevation is relative to North American Vertical Datum of 1988 (NAVD88). (From Eggleston, J. and Pope, J., *Land Subsidence and Relative Sea-Level Rise in the Southern Chesapeake Bay Region*, Circular 1392, U.S. Geological Survey, Reston, VA, 2013.)

The timing of aquifer system compaction is also important. After groundwater levels drop, compaction can continue for many years or decades. When groundwater is pumped from an aquifer, pressure decreases in the aquifer. The pressure decrease then slowly propagates into clay layers that are adjacent to or within the aquifer. As long as pressure continues to decrease in the clay layers, compaction continues.

The layered sediments of the Virginia Coastal Plain aquifer system range in grain size from very fine (silts and clays) to coarse (sand and shell fragments) (McFarland and Bruce, 2006). Based on the hydrogeologic framework of McFarland and Bruce (2006) and Heywood and Pope (2009), confining layers outside of the bolide impact crater occupy about 16% of the total aquifer system thickness, an average of 100 m out of the total average thickness of 619 m. These continuing layers have high specific storage (compressibility) estimated to be 0.00015 per meter (Pope and Burbey, 2004). Clay layers overlying and within the Potomac Aquifer are compressing as aquifer pressure decrease migrate vertically and laterally from pumping wells (Eggleston and Pope, 2013).

GLACIAL ISOSTATIC ADJUSTMENT

The last ice age occurred about 16,000 years ago, when great sheets of ice covered much of the Northern Hemisphere. Although the ice melted long ago, the land once under and around the ice is still rising and falling in reaction to its ice burden. This ongoing movement of land is referred to as glacial isostatic adjustment or postglacial rebound (see Figure 7.7). Here is how it works: Imagine lying down on a soft mattress for awhile and then getting up. You will likely see an indentation in the mattress

FIGURE 7.7   Glacier on retreat in Alaska allowing for isostatic land adjustment. (Photograph by F.R. Spellman.)

where your body had been and a puffed-up area around the indentation. It takes the mattress awhile to return back to its original shape. Similarly, Earth is constantly on the move. We cannot always see the movement unless we witness an earthquake or its aftermath or a volcanic eruption such as the Mount St. Helens event in 1980 or land fissures and sinkholes opening in the ground around us (NOAA, 2015).

The Virginia Coastal Plain aquifer consists of layered sediments overlaying crystalline bedrock. Bedrock is not solid and unyielding but actually flexes and moves in response to stress. Bedrock in the mid-Atlantic region is moving slowly downward in response to melting of the Laurentide ice sheet that covered Canada and the northern United States during the last ice age (Boon et al., 2010; Sella et al., 2007). When the ice sheet still existed, the weight of the ice pushed the underlying Earth's crust downward and, in response, areas away from the ice sheet were forced upward (called *glacial forebulge*; see Figure 7.7). The southern Chesapeake Bay region is in the glacial forebulge area and was forced upward by the Laurentide ice sheet. The ice sheet started melting about 18,000 years ago and took many thousands of years to disappear entirely. As the ice melted and its weight was removed, glacial forebulge areas, which previously had been forced upward, began sinking and continue to sink today. Again, this movement of the Earth's crust in response to ice loading or melting is glacial isostatic adjustment. Data from GPS and carbon dating of marsh sediments indicate that regional land subsidence in response to glacial isostatic adjustment in the southern Chesapeake Bay region may have a current rate of about 1 mm/yr (Engelhart, 2010; Engelhart et al., 2009). This downward velocity rate is uncertain and probably not uniform across the region.

REFERENCES AND RECOMMENDED READING

Anon. (1997). Global warming: politics and economics further complicate the issue. *USA Today*, December 7, pp. A-1–A-2.

Anon. (1998). Global warming: it's here … and almost certain to get worse. *Time*, August 24.

Associated Press. (1997). Does warming feed El Niño? *Virginian-Pilot* (Norfolk, VA), December 7, p. A-15.

Associated Press. (1998a). Tougher air pollution standards too costly, Midwestern states say. *Lancaster New Era* (Lancaster, PA), September 25.

Associated Press. (1998b). Ozone hole over Antarctica at record size. *Lancaster New Era* (Lancaster, PA), September 28.

Bloch, M. (2009). *Global Warming—A Hoax?* Carbonify.com, http://www.carbonify.com/articles/global-warming-hoax.htm.

Boon, J.D., Brubaker, J.M., and Forrest, D.M. (2010). *Chesapeake Bay Land Subsidence and Sea Level Change: An Evaluation of Past and Present Trends and Future Outlook*, Special Report 425. Gloucester Point: Virginia Institute of Marine Science (http://www.vims.edu/GreyLit/VIMS/sramsoe425.pdf).

Broecker, W. (1987). Unpleasant surprises in the greenhouse? *Nature*, 328: 123–126.

Bruun, P. (1962). Sea level rise as a cause of shore erosion. *Journal of Waterways Harbors Division, ASCE*, 88: 117–130.

Bruun, P. (1986). Worldwide impact of sea level rise on shorelines. In: *Effects of Changes in Stratospheric Ozone and Global Climate*. Vol. 4. *Sea Level Rise* (Titus, J.G., Ed.), pp. 99–128. Washington, DC: U.S. Environmental Protection Agency.

Cahoon, D.R., Reed, D.J., Kolker, A.S. et al. (2009). Coastal wetland sustainability. In: *Coastal Sensitivity to Sea-Level Rise: A Focus on the Mid-Atlantic Region* (Titus, J.G., Anderson, K.E., Cahoon, D.R. et al., Eds.), pp. 57–72. Washington, DC: U.S. Environmental Protection Agency.

Crowley, T.J. and North, G.R. (1996). Abrupt climate change and extinction events in Earth's history. *Science*, 240: 996.

Davis, M. L. and Cornwell, D. A. (1991). *Introduction to Environmental Engineering*. New York: McGraw-Hill.

Edgerton, L. (1991). *The Rising Tide: Global Warming and World Sea Levels*. Washington, DC: Island Press.

Eggleston, J. and Pope, J. (2013). *Land Subsidence and Relative Sea-Level Rise in the Southern Chesapeake Bay Region*, Circular 1392. Reston, VA: U.S. Geological Survey.

Engelhart, S.E. (2010). Sea-Level Changes Along the U.S. Atlantic Coast: Implications for Glacial Isostatic Adjustment Models and Current Rates of Sea-Level Change, PhD dissertation. *Publicly Accessible Penn Dissertations*, 407 (http://repository.upenn.edu/cgi/viewcontent.cgi?article=1136&context=edissertations).

Engelhart, S.E., Horton, B.P., Douglas, B.C., Peltier, W.R., and Tornqvist, T.E. (2009). Spatial variability of late Holocene and 20th century sea-level rise along the Atlantic coast of the United States. *Biology*, 37: 1115–1118.

Erwin, R.M., Brinker, D.F., Watts, B.D., Costanzo, G.R., and Morton, D.D. (2011). Islands at bay: rising seas, eroding islands, and waterbird habitat loss in Chesapeake Bay, USA. *Journal of Coastal Conservation*, 15: 51–60.

FEMA. (2002). *Flood Insurance Study of Franklin, Virginia, Community 510060*, revised. Washington, DC: Federal Emergency Management Agency.

Franck, I. and Brownstone, D. (1992). *The Green Encyclopedia*. New York: Prentice-Hall.

Galloway, D., Jones, D.R., and Ingebritsen, S.E., Eds. (1999). *Land Subsidence in the United States*, U.S. Geological Survey Circular 1182. Reston, VA: U.S. Department of Interior.

Graedel, T.E. and Crutzen, P.J. (1989). The changing atmosphere. *Scientific American*, 261: 58–68.

Hansen, J.E., Lacis, A., Rind, D., and Russell, G. (1986). Climate sensitivity to increasing greenhouse gases. In: *Greenhouse Effect and Sea Level Rise: A Challenge for This Generation* (Barth, M.C. and Titus, J.G., Eds.), Chapter 2. New York: Van Nostrand Reinhold.

Hansen, J.E., Lacis, A., and Prather, M. (1989). Greenhouse effect of chlorofluorocarbons and other trace gases. *Journal of Geophysical Research*, 94(D13): 16,417–16,421.

Healy, R.W., Winter, T.C., LaBaugh, J.W., and Franke, O.L. (2007). *Water Budgets Foundations for Effective Water-Resources and Environmental Management*, Circular 1308. Reston, VA: U.S. Geological Survey.

Heywood, C.E. and Pope, J.P. (2009). *Simulation of Groundwater Flow in the Coastal Plain Aquifer System of Virginia*, Scientific Investigations Report 2009-5039. Reston, VA: U.S. Geological Survey (http://pubs.usgs.gov/sir/2009/5039/).

Kirwan, M.L. and Guntenspergen, G.R. (2012). Feedbacks between inundation, root production, and shoot growth in a rapidly submerging brackish marsh. *Journal of Ecology*, 100(3): 760–770.

Kirwan, M.L., Langley, J.A., Guntenspergen, G.R., and Megonigal, J.P. (2013). The impact of sea-level rise on organic matter decay rates in Chesapeake Bay brackish tidal marshes. *Biogeosciences*, 10: 14689–14708.

Konikow, L.F. and Neuzil, C.E. (2007). A method to estimate groundwater depletion from confining layers. *Water Resources Research*, 43(7): 931–936.

Ladurie, E.L. (1971). *Times of Feast, Times of Famine: A History of Climate Since the Year 1000*. New York: Doubleday.

Leatherman, S.P., Chalfont, R., Pendleton, E.C., McCandless, T.L., and Funderburk, S. (1995). *Vanishing Lands: Sea Level, Society and Chesapeake Bay.* Annapolis, MD: Chesapeake Bay Field Office, U.S. Fish and Wildlife Service.

Masters, G.M. (1991). *Introduction to Environmental Engineering and Science.* Englewood Cliffs, NJ: Prentice-Hall.

McFarland, E.R. and Bruce, T.S. (2006). *The Virginia Coastal Plain Hydrogeologic Framework*, U.S. Geological Survey Professional Paper 1731. Reston, VA: U.S. Geological Survey (http:pubs.water.usgs.gov/pp1731/).

McFarlane, B.J. (2012). *Climate Change in Hampton Roads. Phase III. Sea Level Rise in Hampton Roads, Virginia.* Chesapeake, VA: Hampton Roads Planning District Commission.

Morris, J.T., Sundareshwar, P.V., Nietch, C.T. et al. (2002). Responses of coastal wetlands to rising sea level. *Ecology*, 83(10): 2869–2877.

NOAA. (2015). *What Is Glacial Isostatic Adjustment?* Silver Spring, MD: National Oceanic and Atmospheric Administration (http://oceanservice.noaa.gov/facts/glacial-adjustment.html).

Pope, J.P. (2002). Characterization and Modeling of Land Subsidence Due to Groundwater Withdrawals from the Confined Aquifers of the Virginia Coastal Plain, master's thesis, Virginia Polytechnic Institute, Blacksburg.

Pope, J.P. and Burbey, T.J. (2004). Multiple-aquifer characteristics from single borehole extensometer records. *Ground Water*, 42(1): 45–58.

Ramanathan, V. (2006). Atmospheric brown clouds: health, climate and agriculture impacts. *Pontifical Academy of Sciences Scripta Varia*, 106: 47–60.

Randall, D.A., Wood, R.A., Bony, R. et al. (2007). Climate models and their evaluation. In: *Climate Change 2007: The Physical Science Basins* (Solomon, S., Qin, D., Manning, M. et al., Eds.), pp. 589–662. Cambridge, U.K.: Cambridge University Press.

Sella, G.F., Stein, S., Dixon, T.H, Craymer, M., James, T.S., Mazzotti, S., and Dokka, R.L. (2007). Observation of glacial isostatic adjustment in "stable" North America with GPS. *Geophysical Research Letters*, 34(2): L02306.

Spellman, F.R. and Whiting, N. (2006). *Environmental Science and Technology: Concepts and Applications.* Boca Raton, FL: CRC Press.

Stanhill, G. and Moreshet, S. (2004). Global radiation climate changes in Israel. *Climatic Change*, 22: 121–138.

Thompson, D.J. (1995). The seasons, global temperature, and precession. *Science*, 268(5207): 59–68.

Titus, J.G. and Narayanan, V.K. (1995). *The Probability of Sea Level Rise*, EPA 230-R-95-008. Washington, DC: U.S. Environmental Protection Agency.

Travis, D.J., Carleton, A.M., and Lauritsen, R.G. (2002). Contrails reduce daily temperature range. *Nature*, 418: 601.

USEPA. (2009). *National Ambient Air Quality Standards (NAAQS).* Washington, DC: U.S. Environmental Protection Agency.

USEPA. (2016a). *Climate Change Impacts: Climate Impacts on Ecosystems.* Washington, DC: U.S. Environmental Protection Agency (https://www.epa.gov/climate-impacts/climate-impacts-ecosystems).

USEPA. (2016b). *Technology Transfer Network Support Center for Regulatory Atmospheric Modeling (SCRAM).* Washington, DC: U.S. Environmental Protection Agency (https://www.epa.gov/scram).

USFWS. (2011). *Nutrient Pollution.* Washington, DC: Chesapeake Bay Field Office, U.S. Fish and Wildlife Service (www.fws.gov/chesapeakebay/nutrient.html).

Updike, J. (1989). *Self-Consciousness: Memoirs.* New York: Random House.

USGS. (2016). *Hydroelectric Power Water Use.* Reston, VA: U.S. Geological Survey (https://water.usgs.gov/edu/wuhy.html).

USFWS. (2011). *Nutrient Pollution*. Annapolis, MD: Chesapeake Bay Field Office, U.S. Fish and Wildlife Service.

Wigley, T.M., Jones, P.D., and Kelly P.M. (1986). Empirical climate studies: warm world scenarios and the detection of climatic change induced by radioactively active gases. In: *The Greenhouse Effect, Climatic Change, and Ecosystems* (Bolin, B. et al., Eds.), pp. 271–323. New York: Wiley.

Zervas, C. (2009). *Sea Level Variations of the United States, 1854–2006*, Technical Report NOS CO-OPS 053. Silver Spring, MD: National Oceanic and Atmospheric Administration (https://tidesandcurrents.noaa.gov/publications/Tech_rpt_53.pdf).

Zurer, P.S. (1988). Studies on ozone destruction expand beyond Antarctic. *Chemical and Engineering News*, 66(22): 16–25.

8 Measuring and Monitoring Land Subsidence

A river seems a magic thing. A magic, moving, living part of the very earth itself—for it is from the soil, both from its depth and from its surface, that a river has its beginning.

Gilpin (1949)

MEASURING SUBSIDENCE[*]

Land subsidence can be effectively and accurately measured using several reliable and proven techniques. Multiple measuring or monitoring techniques are often used together to understand different aspects of land subsidence (Figure 8.1 and Table 8.1). Because rates and locations of land subsidence change over time, repeat measurements at multiple locations are often needed to improve our understanding of this complex phenomenon and to guide computer models that forecast future subsidence. Extensometers measure changes in aquifer system thickness, whereas other methods measure land surface elevation, from which subsidence is calculated by subtracting measurements over time.

BOREHOLE EXTENSOMETERS

Extensometers are used to measure changes in the length of an object. In applications related to land subsidence measurement, a borehole extensometer measures compaction or expansion of an aquifer system independently of other vertical movements, such as crustal and tectonic motions (Galloway et al., 1999). An extensometer measures change in aquifer system thickness by recording changes in the distance between two points in a well (see Figure 8.2). Usually the two measurement points are established at the top and bottom of a well to measure total aquifer system compaction between the land surface and the bottom of the aquifer system. Alternatively, specific intervals within a well can be measured—for example, to measure compaction of just one aquifer within a layered aquifer system. Extensometer measurements are often combined with surface monitoring techniques to determine the portion of total land subsidence attributable to aquifer system compaction (Poland, 1984).

---

[*] Adapted from Eggleston, J. and Pope, J., *Land Subsidence and Relative Sea-Level Rise in the Southern Chesapeake Bay Region*, Circular 1392, U.S. Geological Survey, Reston, VA, 2013.

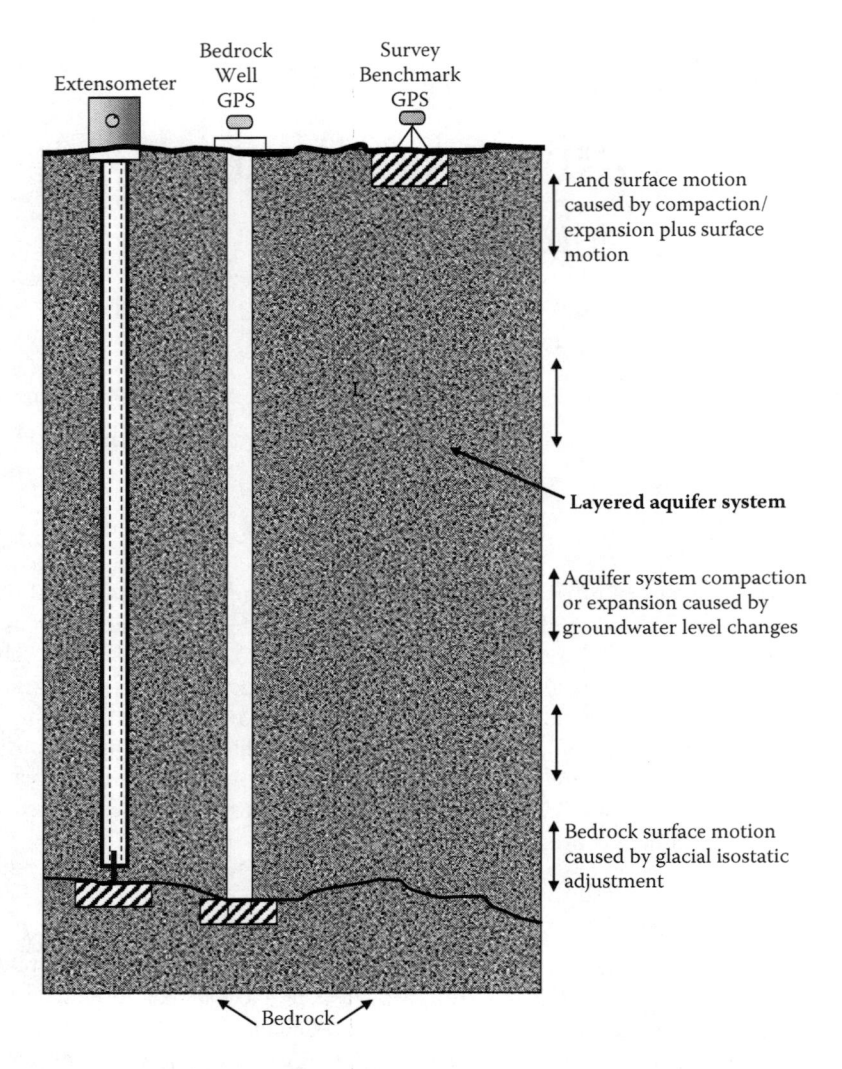

FIGURE 8.1 Subsidence monitoring methods. Survey benchmark global positioning system (GPS) to measure land surface motion, bedrock well GPS to measure bedrock surface motion, and extensometer to measure aquifer system compaction or expansion. (From Eggleston, J., *Land Subsidence Monitoring in Hampton Roads: Progress Report*, U.S. Geological Survey, Reston, VA, 2016.)

TIDAL STATIONS

A tidal station measures sea elevation at one location. To determine long-term trends, sea-level measures are averaged over time to remove the effect of waves, tides, and other short-term fluctuations. In the southern Chesapeake Bay, tidal stations have been in operation for many decades (Table 8.2). Tidal stations are significant and valuable because they indicate relative sea-level rise.

TABLE 8.1
Land Subsidence Monitoring Methods

Method	Type of Data	Measures Aquifer System Compaction Independently	Spatial Coverage	Temporal Detail
Borehole extensometer	Aquifer system thickness at one location; continuous record	Yes	Low	High
Tidal station	Sea elevation at one location; continuous record	No	Low	High
Geodetic surveying	Land elevation at one or several locations; multiple times or continuous record	No	Low to moderate	Low to high
Remote sensing (InSAR)	Land elevations over a wide area; multiple times	No	High	Moderate

Source: Eggleston, J. and Pope, J., *Land Subsidence and Relative Sea-Level Rise in the Southern Chesapeake Bay Region*, Circular 1392, U.S. Geological Survey, Reston, VA, 2013.

Note: InSAR, interferometric synthetic aperture radar.

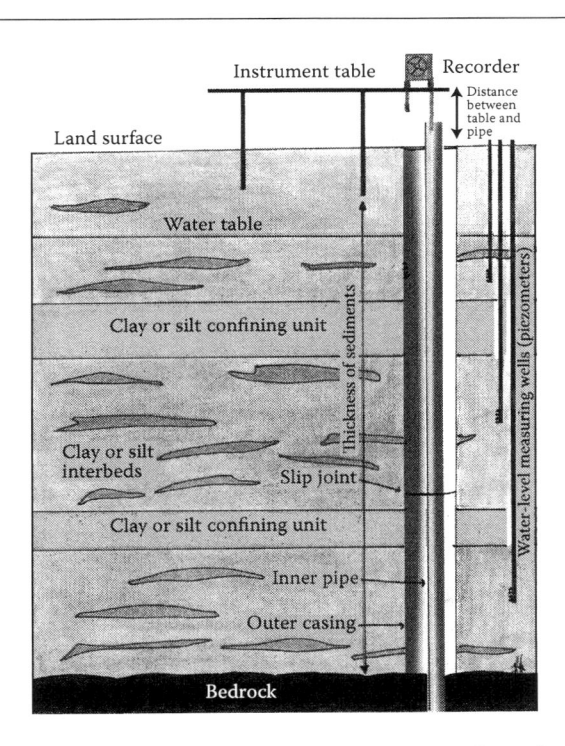

FIGURE 8.2 Borehole extensometer with recording equipment. (Illustration by F.R. Spellman and Kathern Welsh, adapted from Leake, S.A., *Land Subsidence from Ground-Water Pumping*, U.S. Geological Survey, Reston, VA, 2016.)

TABLE 8.2

Relative Sea-Level Rise At Selected National Oceanic and Atmospheric Administration Tidal Stations in the Southern Chesapeake Bay Region

			Rate of Relative Sea-Level Rise	
ID	Site Name	Period	Measured (mm/yr)	95% CI
8632200	Kiptopeake, Virginia	1951–2006	3.5	±0.42
8637624	Gloucester Point, Virginia	1950–2006	3.8	±0.47
8638610	Sewells Point, Virginia	1927–2006	4.4	±0.27
8638660	Portsmouth, Virginia	1935–2006	3.8	±0.45
		Average	3.9	±0.40

Source: Eggleston, J. and Pope, J., *Land Subsidence and Relative Sea-Level Rise in the Southern Chesapeake Bay Region*, Circular 1392, U.S. Geological Survey, Reston, VA, 2013.

Note: CI, confidence interval; mm/yr, millimeters per year.

GEODETIC SURVEY

Geodetic surveying is the measurement of land surface coordinates. Geodetic surveying is most commonly performed with either global positioning system (GPS) technology that reads signals from satellites to obtain very detailed location and time information or traditional optical leveling equipment. Because historical geodetic survey records are available for the southern Chesapeake Bay region, geodetic surveying can be used to determine cumulative land subsidence over many decades. Benchmark stations are established and, for as long as they remain undisturbed, can be surveyed multiple times to determine elevation changes between surveys. The Continuously Operating Reference Stations (CORS) network, managed by the National Geodetic Survey, is a network of long-term GPS stations throughout the United States that includes stations in the Southern Chesapeake Bay region. Each CORS station continuously records three-dimensional position data (north–south, east–west, and up–down), allowing rates of change to be calculated over time (Snay and Soler, 2008). A CORS station records ground position at one site and is designed to operate for many years. Besides stationary GPS sites such as the CORS stations, portable GPS receivers can be used to expand spatial coverage. In the Houston–Galveston area, GPS receivers mounted on trailers have been used to collect data at up to four different sites each month (Bawden et al., 2012; Galloway et al., 1999). The portable GPS approach has acceptable subcentimeter accuracy, provides greater spatial coverage than stationary GPS, and has a lower cost than interferometric synthetic aperture radar (InSAR) technology.

InSAR

Interferometric synthetic aperture radar (InSAR) is a radar technique used in geodesy and remote sensing. InSAR has been used to investigate surface deformation resulting from land subsidence (Galloway and Hoffman, 2007). With InSAR, an elevation change as little as 5 mm can be measured over hundreds or thousands of square kilometers with a horizontal spatial resolution down to 20 m (Pritchard, 2006). Interferograms (maps) show land-surface elevation changes that are produced by combining two synthetic aperture rate (SAR) images acquired by multiple satellite or airborne passes over the same area at different times. InSAR analysis has the advantage of measuring subsidence over a large area, whereas traditional geodetic leveling and GPS surveying are performed at only one or a handful of locations during a survey (Sneed et al., 2002; Stork and Sneed, 2002). Using InSAR in the Chesapeake Bay region has potential limitations. Subsidence rates determined by InSAR might have errors that are larger than the subsidence rates observed in the region (1.1 to 4.8 mm/yr). The region's high humidity and dense vegetation would create spurious radar signals, require the use of persistent scatter techniques, and result in lower measurement resolution than is found in more arid regions (Raucoules et al., 2009). Also, available satellite data cover only a relatively short time span. The best available synthetic aperture radar satellite data for the southern Chesapeake Bay region are for 1992 to 2000, so the time of accumulated subsidence determined from these data would be no more than 8 years. Despite these limitations, InSAR could be used to identify hotspot areas of subsidence. Such mapping could be useful for identifying unexpected areas of subsidence, focusing attention on important areas, and picking locations for other ground-based subsidence monitoring techniques (Eggleston and Pope, 2013).

IMPORTANCE OF LAND SUBSIDENCE MONITORING

Land subsidence, or sinking of the land surface, has occurred and is still occurring in many locations throughout the United States. Earlier, we pointed out a few of the most impacted areas in the United States, but the focus of this book is on Hampton Roads, Virginia (the area of southern Chesapeake Bay, the largest estuary in the United States), where land subsidence is an ongoing and serious issue, but also where innovation and technology and far thinking might be able to arrest or correct the problem. In Hampton Roads, the boundary between land and water is low lying. Land subsidence is important in the region because it causes increased flooding, alters wetland and coastal ecosystems, and damages infrastructure and historical sites such as Jamestown, Virginia.

As population densities continue to increase in the Hampton Roads region, the flood hazard unfortunately is increasing, as well, due to locally high rates of land subsidence with global sea-level rise. The combination of global sea-level rise and coastal subsidence is doubly significant even in locations where the occurrence of major hurricanes is relatively rare compared to other regions of the country. In the Hampton Roads region, in particular, extratropical cyclones or "nor'easters" that have not caused significant flooding in the past will begin to do so—and with greater frequency—as sea level continues to rise *relative* to the land. Flood hazard mitigation complicated by land subsidence and relative sea-level rise at a minimum requires an understanding of the land and by–ocean processes involved that contribute in complex and often unpredictable ways. The fact is that the rates and locations of land subsidence are not well known throughout the Hampton Roads area because monitoring has been insufficient in recent decades. Monitoring data are needed to better understand rates and locations of land subsidence and to plan for preventing or mitigating its potentially damaging effects.

Recurrent flooding problems have prompted concern about land subsidence in Hampton Roads (Sweet et al., 2014). In addition, these concerns are compounded by evidence that groundwater pumping and associated aquifer depressurization have caused past land subsidence (Holdahl and Morrison, 1974; Pope and Burbey, 2004) and by measurements showing that relative sea-level rise is faster in Hampton Roads than elsewhere on the Atlantic Coast (Sallenger et al., 2012). Because the rates and locations of land subsidence are not well known in the Hampton Roads area because of insufficient monitoring, risks commonly associated with coastal land subsidence—increased flooding, alteration of wetland and coastal ecosystems, and damage to infrastructure and historical sites—cannot be accurately assessed. More frequent monitoring at multiple locations using multiple complementary methods is needed to build an understanding of subsidence and to plan how to avoid or mitigate the effects of subsidence.

Before land subsidence can be understood it must be monitored. Monitoring data provide the foundation for understanding why, where, and how fast land subsidence is occurring, both now and in the future. Because rates of land subsidence change over time and vary from one location to another, monitoring should be done at multiple locations for multiple years. Monitoring data are used for the following purposes (Eggleston, 2016)

- To avoid or mitigate problems caused by land subsidence (e.g., urban planners, resource managers, and politicians can use monitoring data to guide their decisions)
- To answer questions such as why is subsidence occurring?
- To predict future land subsidence (predictive models that can test mitigation strategies require monitoring data for accuracy and reliability)
- To make maps showing critical areas for mitigating land subsidence

Land subsidence monitoring measures the following:

- Land surface motion
- Bedrock surface motion
- Changes in aquifer system thickness

MONITORING METHODS

Land subsidence is detected by measuring land surface positions over time and calculating rates of change by subtraction. Reliable and accurate techniques for measuring land subsidence in Hampton Roads include borehole extensometers, tidal stations, geodetic surveying, and remote sensing (InSAR) (Eggleston, 2016).

ONGOING MONITORING BY BOREHOLE EXTENSOMETERS

As shown in Figure 8.1, borehole extensometers are wells designed for measuring compaction or expansion of an aquifer system (Galloway et al., 1999). Extensometers typically are paired with monitoring wells so that correlation between groundwater-level changes and aquifer compaction can be determined. In Hampton Roads from 1996 to 2016, no borehole extensometers were active; however, historic extensometer data are available for the periods from 1979 to 1995 for an extensometer located at Franklin Virginia, and from 1982 to 1995 for an extensometer located at Suffolk, Virginia (Pope and Burbey, 2004). These older existing extensometers at Franklin and Suffolk have recently been equipped by the U.S. Geological Survey (USGS) with digital potentiometers, dial gauges, and satellite telemetry to provide aquifer compaction measurements with submillimeter (0.01 mm) accuracy. Data are being collected to test if the extensometer stations can be reactivated to detect aquifer compaction and expansion. The extensometers will be monitored for several months and, if monitoring results are successful, the extensometers may be reactivated on a long-term basis. The possibility of installing GPS antennas on the extensometers, to determine contributions to subsidence from glacial isostatic rebound, will also be investigated. Michelle Sneed, a USGS expert on subsidence and extensometers, was brought in to consult on land subsidence monitoring options in Hampton Roads. She described how, in California, extensometers provide the basis for understanding how land subsidence is related to groundwater withdrawals, for calibrating InSAR estimates of land subsidence, and for calibrating predictive models of land subsidence. Extensometers there provide data used for water-resource planning and subsidence-mitigation planning (Eggleston, 2016).

Ongoing Monitoring by Geodetic Surveying

Geodetic surveying is the measurement of land surface position. GPS technology is now widely used to perform geodetic surveying. Permanent GPS stations, such as the network of Continuously Operating Reference Stations (CORS) operated by the National Geodetic Survey (NGS), provide continuous information about land surface motion at single locations. CORS stations typically achieve centimeter-scale accuracy for absolute vertical position measurement and millimeter-scale accuracy for differential vertical position measurement. Permanent geodetic stations, such as CORS, also provide valuable information for calibrating remote sensing measurements of subsidence. Survey networks, consisting of multiple high-integrity monuments (benchmarks) that are installed on land and periodically occupied with GPS antennas to measure land surface position, can also provide valuable regional estimates of land subsidence.

A separate type of geodetic surveying that would be valuable for understanding land subsidence in Hampton Roads is the use of GPS antennas on bedrock wells to measure bedrock surface motion (Figure 8.1). This can be done at any new extensometer that is constructed. Existing bedrock wells, such as those at Franklin and Suffolk, may also be available as platforms for this type of monitoring.

The NGS, the lead U.S. federal agency for surveying and geodetic science, operates the CORS network of benchmark stations that continuously record land surface positions in fine detail in three dimensions. The CORS network includes five benchmark sites in Hampton Roads. Various other organizations have established continuous monitoring GPS antennas at benchmark stations in Hampton Roads that are not part of the CORS network. For example, the NASA Langley Research Center in

Hampton, Virginia, established four benchmark sites with GPS antennas in 2015. In some cases, data from these non-CORS stations are available and, if a site has been constructed and operated following NGS guidelines (Floyd, 1978; NGS, 2013), the resulting data can be of high quality and useful for subsidence calculation. The NGS is currently analyzing historic surveys of first-order benchmark sites on the Atlantic Coast, including in Hampton Roads, to determine rates of subsidence over the past century. This study will produce maps of subsidence rates over multiple time periods (Eggleston, 2016).

ONGOING MONITORING BY TIDAL STATIONS

For many decades, the National Oceanic and Atmospheric Administration (NOAA) has operated tidal stations to provide continuous water-level data at four sites in Hampton Roads. Data are publically available at no cost from NOAA's website, https://tidesandcurrents.noaa.gov/sltrends/sltrends.html (Eggleston, 2016).

ONGOING MONITORING BY REMOTE SENSING

Interferometric synthetic aperture radar (InSAR) is a remote sensing technique used to measure land surface elevation changes over wide areas, such as over the entire Hampton Roads area. InSAR can be used to determine and map critical areas of land subsidence, select locations for detailed geodetic surveying, and plan strategies for preventing and mitigating land subsidence (Bawden et al., 2003). Accuracy of InSAR subsidence estimates will be important in Hampton Roads, because subsidence rates in the area have been measured at 1.1 to 4.8 millimeters, as compared to typical error for InSAR of 5 to 10 mm. The high atmospheric humidity and dense vegetation found in Hampton Roads can reduce InSAR accuracy. Problems with error can be overcome by analyzing a large number of satellite scenes, applying persistent scatter analysis techniques, using InSAR data collected over multiple years, and by using L-band or X-band rather than C-band InSAR data (Eggleston, 2016). Probably the most valuable aspect of InSAR remote sensing is its capacity to input valuable data for detailed mapping of regional subsidence over time. The type of remote sensing data used to map subsidence, InSAR, has been collected for Hampton Roads by various satellites since 1992, and such data are currently being collected by several international satellites. In 2020, a new U.S. satellite, NISAR, is scheduled to begin collecting InSAR data over Hampton Roads.

THE BOTTOM LINE

The U.S. Geological Survey is cooperating with federal, state, and local government agencies to study and better understand the problem of land subsidence in the southern Chesapeake Bay region. In order to make informed decisions, local resource managers, planners, politicians, and regulators need in-depth knowledge of the particulars involved with relative sea-level rise and land subsidence in the Hampton Roads area. This knowledge is necessary for addressing the increased flood risks and preventing land subsidence. The real bottom line is that the intended purpose of this book is to satisfy the elements described above—to explain what relative sea-level rise and land subsidence are; to provide nuts-and-bolts explanations (in the chapters to follow) for planners, managers, and others; and to provide a methodology and technology that can potentially prevent, provide rebound from, or mitigate the impact of land subsidence in the Hampton Roads region.

REFERENCES AND RECOMMENDED READING

Bawden, G.W., Sneed, M., Stork, S.V., and Galloway, D.L. (2003). *Measuring Human-Induced Land Subsidence from Space*. U.S Geological Survey Fact Sheet 069-03. Reston, VA: U.S Geological Survey (https://pubs.er.usgs.gov/publication/fs06903).

Bawden, G.W., Johnson, M.R., Kasmarek, M.C., Brandt, J., and Middleton, C.S. (2012). *Investigation of Land Subsidence in the Houston–Galveston Region of Texas by Using the Global Positioning System and Interferometric Synthetic Aperture Radar, 1993–2000*, Scientific Investigations Report 2012-5211. Reston, VA: U.S. Geological Survey (https://pubs.usgs.gov/sir/2012/5211/).

Boon, J.D., Brubaker, J.M., and Forrest, D.R. (2010). *Chesapeake Bay Land Subsidence and Sea Level Change: An Evaluation of Past and Present Trends and Future Outlook*, Special Report No. 425. Norfolk: Virginia Institute of Marine Science.

Eggleston, J. (2016). *Land Subsidence Monitoring in Hampton Roads: Progress Report*. Reston, VA: U.S. Geological Survey.

Eggleston, J. and Pope, J. (2013). *Land Subsidence and Relative Sea-Level Rise in the Southern Chesapeake Bay Region*, Circular 1392. Reston, VA: U.S. Geological Survey.

Floyd, R.P. (1978). *Geodetic Bench Marks*, NOAA Manual NOS NGS 1. Rockville, MD: National Oceanic and Atmospheric Administration (http://www.ngs.noaa.gov/PUBS_LIB/GeodeticBMs.pdf).

Galloway, D.L. and Hoffmann, J. (2007). The application of satellite differential SAR interferometry-derived ground displacements in hydrogeology. *Hydrogeology Journal*, 15(1): 133–154.

Galloway, D., Jones, D.R., and Ingebritsen, S.E., Eds. (1999). *Land Subsidence in the United States*, U.S. Geological Survey Circular 1182. Reston, VA: U.S. Department of Interior.

Gilpin, L. (1949). *The Rio Grande: River of Destiny*. New York: Duell, Sloan and Pearce.

Holdahl, S.R. and Morrison, N. (1974). Regional investigations of vertical crustal movements in the U.S. using precise relevelings and mareograph data. *Tectonophysics*, 23(4): 373–390.

Leake, S.A. (2016). *Land Subsidence from Ground-Water Pumping*. Reston, VA: U.S. Geological Survey.

NGS. (2013). *Guidelines for New and Existing Continuously Operating Reference Stations (CORS)*. Silver Spring, MD: National Geodetic Survey National Ocean Survey (http://www.ngs.noaa.gov/CORS/Establish_Operate_CORS.html).

Poland, J.F., Ed. (1984). *Guidebook to Studies of Land Subsidence Due to Ground-Water Withdrawal*. Geneva: United Nations Educational, Scientific and Cultural Organization (UNESCO) (https://wwwrcamnl.wr.usgs.gov/rgws/Unesco/PDF-Chapters/Guidebook.pdf).

Pope, J.P. and Burbey, T.J. (2004). Multiple-aquifer characteristics from single borehole extensometer records. *Ground Water*, 42(1): 45–58.

Pritchard, M.E. (2006). InSAR, a tool for measuring Earth's surface deformation: *Physics Today*, 59(7): 68–69.

Raucoules, D., Bourgine, B., de Michele, M. et al. (2009). Validation and intercomparison of persistent scatterers interferometry—PSIC4 project results. *Journal of Applied Geophysics*, 68(3): 335–347.

Sallenger, A.H., Doran, K.S., and Howd, P.A. (2012). Hotspot of accelerated sea-level rise on the Atlantic Coast of North America. *Nature Climate Change*, 2(12): 884–888.

Snay, R.A. and Soler, T. (2008). Continuously operating reference station (CORS): history, applications, and future enhancements. *Journal of Surveying Engineering*, 134(4): 95–104.

Sneed, M., Stork, S.V., and Ikehara, M.E. (2002). *Detection and Measurement of Land Subsidence Using Global Position System and Interferometric Synthetic Aperture Radar, Coachella Valley, California, 1998–2000*, Water-Resources Investigations Report 02-4239. Reston, VA: U.S. Geological Survey.

Stork, S.V. and Sneed, M. (2002). *Houston–Galveston Bay Area, Texas, from Space—A New Tool for Mapping Land Subsidence*, U.S. Geological Survey Fact Sheet 2002-110. Reston, VA: U.S. Geological Survey (https://pubs.usgs.gov/fs/fs-110-02/pdf/FS_110-02.pdf).

Sweet, W., Park, J., Marra, J., Zervas, C., and Gill, S. (2014). *Sea Level Rise and Nuisance Flood Frequency Changes Around the United States*, NOAA Technical Report NOS CO-OPS 073. Silver Spring, MD: National Oceanic and Atmospheric Administration (https://tidesandcurrents.noaa.gov/publications/NOAA_Technical_Report_NOS_COOPS_073.pdf).

Zervas, C. (2009). *Sea Level Variations of the United States, 1854–2006*, Technical Report NOS CO-OPS 053. Silver Spring, MD: National Oceanic and Atmospheric Administration (https://tidesandcurrents.noaa.gov/publications/Tech_rpt_53.pdf).

9 Every Problem Has a Solution

When we examine sea level records on the east coasts of continents, we see surprisingly large variations at periods on the order of 100 months or longer.

Hong et al. (2000)

Among those who crunch the numbers for city and county budgets, a big wild card in the decade ahead is the Chesapeake Bay cleanup.

Mayfield (2016)

Future generations will inherit clean waterways and be able to keep them clean.

—Hampton Roads Sanitation District

When we use up one resource and then another resource and then another, there is one critical resource mankind never runs out of—that one item that sustains us—in a word it is innovation, innovation, innovation.

—Frank R. Spellman

The oysters lay as thick as stones.

—Captain John Smith, writing about the Chesapeake Bay area in 1608

HAMPTON ROADS SANITATION DISTRICT

Its genesis was driven by oysters. No, not the genesis of Chesapeake Bay; its genesis was driven by a heavy, unstoppable, all-knowing hand. Hampton Roads Sanitation District (HRSD), arguably the premier wastewater treatment district on the globe, became a viable governor-appointed, state-commission-monitored entity because of a significant decline in the oyster population in Chesapeake Bay. As a case in point, consider that in the Hampton Roads region of Chesapeake Bay in 1607, when Captain John Smith and his team settled in Jamestown, oysters up to 13 inches in size were plentiful—more than could ever be harvested and consumed by the handful of early settlers. This population of oysters and other aquatic lifeforms remained plentiful until the population gradually increased in the Bay region.

Overharvesting of oysters by the increased numbers of humans living in the Chesapeake Bay region has been a major factor in the decline of the oyster population; however, the real culprit is pollution. Before the Bay became polluted by sewage, sediment, and garbage, oysters could handle natural pollution from stormwater runoff and other sources. It has been estimated that a century ago, when there was a much larger oyster population than today, the oyster population could filter pollutants from the Bay and clean it up in as little as 4 days. By the 1930s, however, the declining oyster population was overwhelmed by the increasing pollution levels.

For years, the author has stated that pollution is a judgment call. Why a judgment call? Because people's opinions differ as to what they consider to be a pollutant based on their assessment of benefits and risks to their health and economic wellbeing. For example, visible and invisible chemicals spewed into the air or water by an industrial facility might be harmful to people and other forms of life living nearby, but if the facility is required to install expensive pollution controls it might have to shut down or move away. Workers who would lose their jobs and merchants who would lose their livelihoods might feel that the risks posed by polluted air and water are minor weighed against the benefits of profitable employment. The same level of pollution can also affect two people quite differently. Some forms of air pollution, for example, might cause a slight irritation for a healthy person but cause life-threatening problems for someone with chronic obstructive pulmonary disease, such as emphysema. Differing priorities lead to differing perceptions of pollution (e.g., concern about the level of pesticides in foodstuffs that leads to wholesale banning of insecticides is unlikely to help the starving). No one wants to hear that cleaning up the environment is going to have a negative impact on them. Public perception lags behind reality because the reality is sometimes unbearable. This perception lag is clearly demonstrated in Case Study 9.1 and Case Study 9.2.

Case Study 9.1. Cedar Creek Composting

The Cedar Creek Composting (CCC) facility was built in 1970. A 44-acre site designed to receive and process wastewater biosolids from six local wastewater treatment plants, CCC composted biosolids at the rate of 17.5 dry tons per day. CCC used the aerated static pile (ASP) method to produce pathogen-free, humus-like material that could be used beneficially as an organic soil amendment. The final compost product was successfully marketed under a registered trademark name.

Today, the Cedar Creek Composting facility is no longer in operation. The site was shut down in early 1997. From an economic point of view, CCC was highly successful. When a fresh pile of compost had completed the entire composting process (including curing), dump truck after dump truck would line the street outside the main gate, waiting in the hope to buy a load of the popular product. Economics was not the problem. In fact, CCC could not produce enough compost fast enough to satisfy the demand. No, economics was not the problem.

What *was* the problem, then? The answer to this is actually twofold: social and then eventually legal considerations. Social limitations were imposed by the community in which the compost site was located. In 1970, the 44 acres CCC occupied were located in an out-of-town, rural area. CCC's only neighbor was a regional, small airport on its eastern border. CCC was completely surrounded by woods on the other three sides. The nearest town was 2 miles away. But, by the mid-1970s, things started to change. Population growth and its accompanying urban sprawl quickly turned forested lands into housing complexes and shopping centers. CCC's western border soon became the site of a two-lane road that was upgraded to four and then to six lanes. CCC's northern fence separated it from a mega-shopping mall. On the southern end of the facility, houses, playgrounds, swimming pools, tennis courts, and a golf course were built. CCC became an island surrounded by urban growth. Further complicating the situation was the airport; it expanded to the point that, by 1985, three major airlines used the facility.

CCC's ASP composting process was not a problem before the neighbors moved in. We all know dust and odor control problems are not problems until the neighbors complain—and complain they did. CCC was attacked from all four sides. The first complaints came from the airport. The airport complained that dust from the static piles of compost was interfering with air traffic control.

The new, expanded highway brought several thousand new commuters right up alongside CCC's western fence line. Commuters started complaining any time the compost process was in operation; they complained primarily about the odor—the thick, earthy smell permeated everything.

After the enormous housing project was completed and people took up residence there, complaints were raised on a daily basis. The new homeowners complained about the earthy odor and the dust that blew from the compost piles onto their properties downwind from the site. The shoppers at the mall also complained about the odor.

City Hall received several thousand complaints over the first few months before they took any action. The city environmental engineer was told to approach CCC's management to see if some resolution of the problem could be effected. CCC management listened to the engineer's concerns but stated that there wasn't a whole lot that the site could do to rectify the problem.

As you might imagine, this was not the answer the city fathers were hoping to get. Feeling the increasing pressure from local inhabitants, commuters, shoppers, and airport management people, the city brought the local state representatives into the situation. The two state representatives for the area immediately began a campaign to close down the CCC facility.

CCC was not powerless in this struggle—after all, CCC was there first, right? The developers and those people in those new houses didn't have to buy land right next to the facility, right? Besides, CCC had the USEPA on their side. CCC was taking a waste product no one wanted, one that traditionally ended up in the local landfill (taking up valuable space), and turning it into a beneficial reuse product. CCC was helping to conserve and protect the local environment, a noble endeavor.

The city politicians didn't really care about noble endeavors, but they did care about the concerns of their constituents, the voters. They continued their assault through the press, electronic media, legislatively, and by any other means they could bring to bear.

CCC management understood the problem and felt the pressure. They had to do something, and they did. Their environmental engineering division was assigned the task of coming up with a plan to mitigate not only CCC's odor problem but also its dust problem. After several months of research and a pilot study, CCC's environmental engineering staff came up with a solution. The solution included enclosing the entire facility within a self-contained structure. The structure would be equipped with a state-of-the-art ventilation system and two-stage odor scrubbers. The engineers estimated that the odor problem could be reduced by 90% and the dust problem reduced by 98.99%. CCC management thought they had a viable solution to the problem and was willing to spend the $5.2 million to retrofit the plant.

After CCC presented their mitigation plan to the city council, the council members made no comment but said that they needed time to study the plan. Three weeks later, CCC received a letter from the mayor stating that CCC's efforts to come up with a plan to mitigate the odor and dust problems at CCC were commendable and to be applauded but were unacceptable.

From the mayor's letter, CCC could see that the focus of attack had now changed from a social to a legal issue. The mayor pointed out that he and the city fathers had a legal responsibility to ensure the good health and well-being of local inhabitants and that certain legal limitations would be imposed and placed on the CCC facility to protect their health and welfare.

Compounding the problem was the airport. Airport officials also rejected CCC's plan to retrofit the compost facility. Their complaint (written on official FAA paper) stated that the dust generated at the compost facility was hazarding flight operations, and even though the problem would be reduced substantially by engineering controls the chance of control failure was always possible, and then an aircraft could be endangered. From the airport's point of view, this was unacceptable.

Several years went by, with local officials and CCC management contesting each other on the plight of the compost facility. In the end, CCC management decided they had to shut down its operation and move to another location, so they closed the facility.

After shutdown, CCC management staff immediately started looking for another site to build a new wastewater biosolids-to-compost facility. They are still looking. To date, their search has located several pieces of property relatively close to the city (but far enough away to preclude any dust and odor problems), but they have had problems finalizing any deal. Buying the land is not the problem; getting the required permits from various county agencies to operate the facility is. CCC officials were turned down in each and every case. The standard excuse? Not in my backyard. Have you heard this phrase before? It is so common now that it is usually abbreviated as NIMBY. Whether back in the day or at present, NIMBY is alive and well. To this very day, CCC officials are still looking for a location for their compost facility; they are not all that optimistic about their chances of success in this matter.

Case Study 9.2. Eau de Paper Mill

With regard to certain unbearable facets of reality, consider, for example, the residents of Franklin, Virginia, and their reeking paper mill. For those of us who live close to Franklin—it is 50 miles from Norfolk/Virginia Beach—there is no need to read the road signs. The nose knows when it's close to Franklin. The uninitiated, after a stream of phew-eees courtesy of Eau de paper mill, ask that same old question: How can anyone stand to live in a town that smells like a cocktail mixture of swamp, marsh, sulfur mine, and rotten eggs? Among those who live inside the city limits and, in particular, the 1100 who work at the paper mill, few seem to appreciate the question. When the question is asked, smiles fade; attitudes get defensive. The eventual response is "What smell?" Then, waiting for that quizzical look to appear on the face of the questioner, the local's eyes will twinkle and with a chuckle he will say, "Oh, you must mean that smell of money."

So, again, what is pollution? Our best answer: Pollution is a judgment call. And preventing pollution demands continuous judgment.

So, now you might be asking, "What do composting biosolids and paper mill operations have to do with oyster depletion, pollution of Chesapeake Bay, and the HRSD?" Well, it does not take a leap of faith or an understanding of hieroglyphics to appreciate several key points being made here. First, along with the overharvesting of oysters (including crabs and other species of Bay life), pollution has degraded the waters of Chesapeake Bay and directly contributed to the decrease in the oyster population therein. Second, just as for the compost facility and paper mill, pollution did not become an issue in the Bay until the "neighbors" complained, and complain they did. In the 1930s, they complained not only about the reduced population of sea life in southern Chesapeake Bay (and other regions of the Bay) but also about the accumulation of floating sewage and the amplification of nasty odors emanating from the biodegrading sewage, which grabbed the attention of passersby, would-be swimmers, boaters, and fishermen. Third, again, pollution is a judgment call, and in 1940 the judgment was made by voters to authorize the governor to appoint a representative commission to oversee pollution mitigation of the southern Chesapeake Bay region. Thus, the Hampton Roads Sanitation District (HRSD) came into being. HRSD is a political subdivision of the Commonwealth of Virginia with a service area that includes 18 counties and cities encompassing 2800 square miles in its southeastern Virginia service area (see Figure 9.1). HRSD's collection system consists of more than 500 miles of piping, 6 to 66 inches in diameter. HRDS possesses more than 100 active pump stations that pump raw wastewater to 9 major treatment plants in Hampton Roads and 4 smaller plants in the Middle Peninsula. The combined capacity of HRSD facilities is 249 million gallons per day (MGD).

Probably another question rumbling through the reader's brain matter at this point is, "Has HRSD solved the pollution problem in Chesapeake Bay?" This question leads us to another question: "Has the oyster population rebounded?" The answer

HRSD Service Area

A Political Subdivision of the Commonwealth of Virginia

Major facilities include the following:

1. Atlantic, Virginia Beach
2. Chesapeake-Elizabeth, Va. Beach
3. Army Base, Norfolk
4. Virginia Initiative, Norfolk
5. Nansemond, Suffolk
6. Boat Harbor, Newport News
7. James River, Newport News

8. Williamsburg, James City County
9. York River, York County
10. West Point, King William County
11. Central Middlesex, Middlesex County
12. Urbanna, Middlesex County
13. King William, King William County

Serving the Cities of Chesapeake, Hampton, Newport News, Norfolk, Poquoson, Portsmouth, Suffolk, Virginia Beach, Williamsburg, and the Counties of Gloucester, Isle of Wight, James City King and Queen, King William, Mathews, Middlesex and York

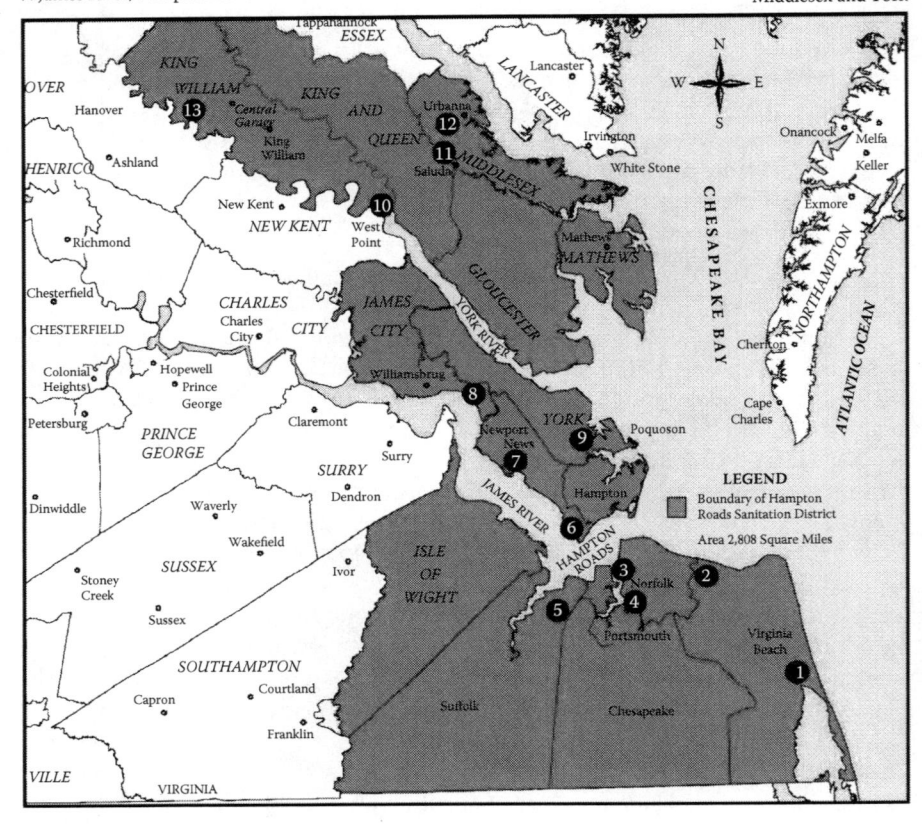

FIGURE 9.1 Hampton Roads Sanitation District (HRSD) service area. (Adapted from HRSD, http://www.hrsd.com/servicearea.shtml.)

to both these questions is yes—*to a point*. On an ongoing, 24/7 basis, HRSD treats wastewater to a quality better than can be found in the James River, Elizabeth River, York River, and other river systems in the region. Those who have no knowledge of wastewater treatment, of HRSD, or the conditions of the rivers in this region might wonder about this statement. But, it is true. It is all about the human-made water cycle. In this case, we are talking about urban water cycles. Water and wastewater professionals maintain a continuous urban water cycle (see Figure 9.2) on a daily basis, which can be summed up as follows (Jones, 1980):

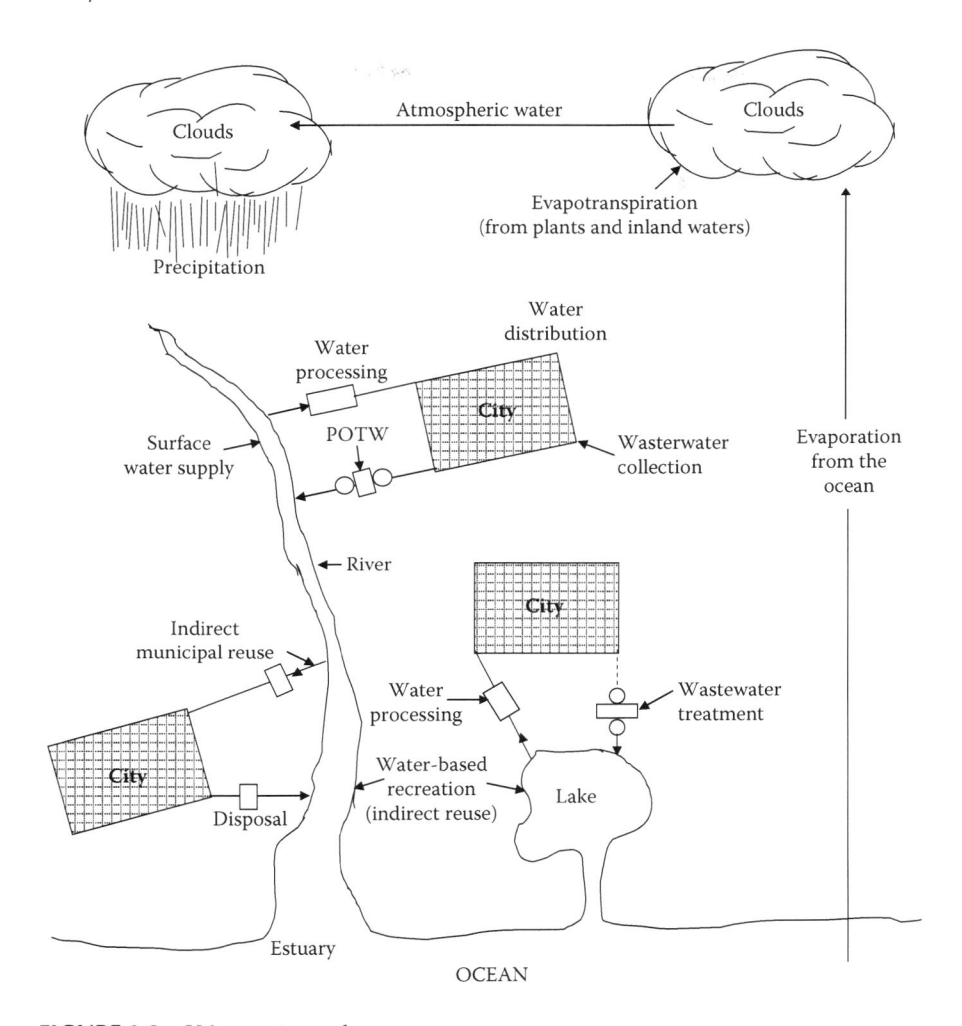

FIGURE 9.2 Urban water cycle.

Delivering services is the primary function of municipal government. It occupies the vast bulk of the time and effort of most city employees, is the source of most contacts that citizens have with local governments, occasionally becomes the subject of heated controversy, and is often surrounded by myth and misinformation. Yet, service delivery remains the "hidden function" of local government.

DID YOU KNOW?

An artificially generated water cycle, or the urban water cycle, consists of (1) a source, either surface or groundwater; (2) water treatment and distribution; (3) use and reuse; and (4) wastewater treatment and disposition, as well as the connection of the cycle to the surrounding hydrological basins.

So, let's get back to the answer that ended with "to a point." There is no doubt that Chesapeake Bay is cleaner and that the sea life, including oysters, is happier today because of the efforts of HRSD; however, the problem of making the Bay cleaner is compounded by two factors. First, there are more than 300 wastewater treatment plants that outfall treated water to the Chesapeake via its 9 major river systems and numerous tributaries. These treatment plants, separate and isolated from HRSD's 13 plants, do the best they can to treat wastewater to a cleaner product (effluent) than the influent they received from various sources. However, some of these more than 300 other plants treat only to primary treatment levels, and their effluent is not as clean as secondary and tertiary plant effluent. Second, HRSD treats wastewater to a top-notch water quality level, but treating wastewater to remove nutrients is a complicated and expensive undertaking. Biological nutrient removal and other nutrient removal technologies are available and in use in many locales, but the technology is expensive—to the point where the treatment technology needed and used might overtax the ratepayers.

Chesapeake Bay occasionally suffers from dead zones due to algae blooms, which occur when excessive nutrients within the Bay cause an explosion of plant life that results in the depletion of the oxygen in the water needed by fish and other aquatic life. Algae blooms are usually the result of urban runoff (of lawn fertilizers, etc.). The potential tragedy is a fish kill, when bay life dies in one mass execution. Algae blooms and dead zones and the resulting fish kill events are major issues, of course, and when they are combined with the problems of relative sea-level rise and land subsidence, it can readily be seen that maintaining the health of Chesapeake Bay and its inhabitants is a multifaceted proposition.

THE SOLUTION TO POLLUTION IN CHESAPEAKE BAY*

The title of this chapter states that every problem has a solution. With regard to the problems associated with Chesapeake Bay, as well as land subsidence, and relative sea-level rise in the Hampton Roads region, HRSD has developed the innovative Sustainable Water Initiative for Tomorrow (SWIFT) program. Do not confuse the acronym SWIFT with the adjectives fast, speedy, rapid, hurried, immediate, or quick. SWIFT is a long-term project whose timeline establishes installation of the technical equipment and operational procedures by 2030.

The SWIFT goal is to inject treated wastewater into the subsurface; specifically, it is designed to inject wastewater treated to drinking water quality into the Potomac Aquifer. Injection of water into the subsurface is expected to raise groundwater pressures, thereby potentially expanding the aquifer system, raising the land surface, and counteracting land subsidence occurring in the Virginia Coastal Plain. In 2016, construction began on a project at the HRSD Nansemond treatment plant in Suffolk, Virginia, to test injection into the aquifer system. HRSD asked the U.S. Geological Survey to prepare a proposal for installation of an extensometer monitoring station at the test site to monitor groundwater levels and aquifer compaction and expansion.

---

* Based on U.S. Geological Survey, *Proposal to Establish an Extensometer Station at the Nansemond Wastewater Treatment Plant in Suffolk, VA*, presented to Hampton Roads Sanitation District, May 20, 2016.

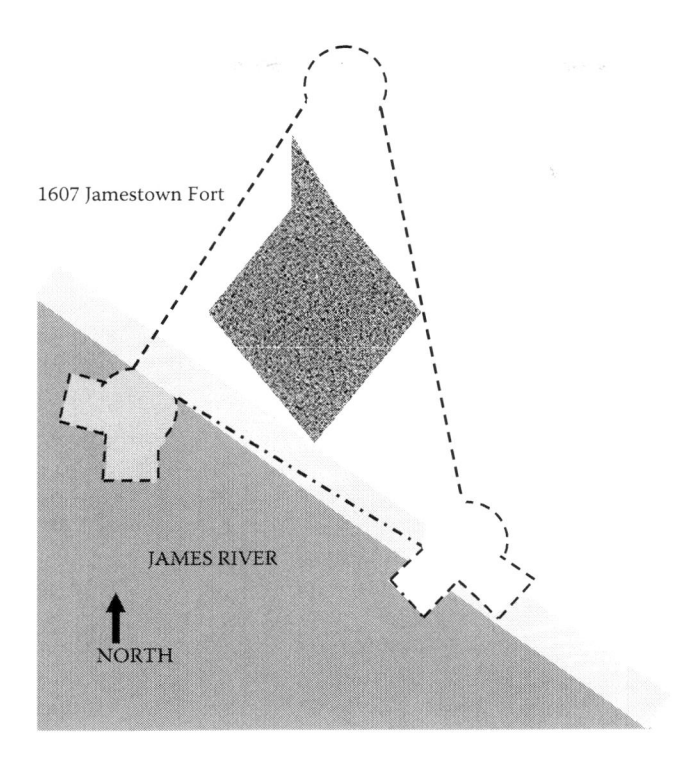

FIGURE 9.3 Outline of the original Jamestown Settlement in 1607; most of the western bulwarks and a portion of the eastern bulwarks are now inundated by the James River.

PROBLEM

SWIFT is designed to counter land subsidence at various locations in the Hampton Roads area of southern Chesapeake Bay, where land subsidence rates of 1.1 to 4.8 millimeters per year have been observed (Eggleston and Pope, 2013; Holdahl and Morrison, 1974). An obvious indicator of relative sea-level rise in the Hampton Roads region can be seen in Figure 9.3. The dashed lines outline the original 1607 fort palisades at the historic Jamestown Settlement on the James River. As shown, the western section of the original bulwark is completely inundated today; a section of the eastern bulwark is also covered by the James River. A major cause of land subsidence is extensive groundwater pumping, which causes regional aquifer system compaction (Pope and Burbey, 2004). Although aquifer system compaction was measured from the late 1970s to the mid 1990s at two stations in Virginia (Suffolk and Franklin), it has not been measured anywhere in Virginia since 1995.

Regular and highly accurate measurements of aquifer compaction are needed to provide information critical for understanding coastal flooding; to protect water resources, natural habitat, and historic sites such as the Jamestown Settlement; and to plan urban and coastal infrastructure in the region. Injection of treated wastewater (treated to drinking water quality) is expected to counteract land subsidence, or raise land surface elevations in the region. Careful monitoring of aquifer system compaction and groundwater levels can be used to optimize the injection process and to improve fundamental understanding of the relation between groundwater pressures and aquifer system compaction and expansion.

There is more to HRSD's treated wastewater injection project, SWIFT, than just arresting or mitigating land subsidence and relative sea-level rise in the Hampton Roads region. One of the additional goals of the project is to stop discharge of treated wastewater from seven of its plants, which would mean 18 million pounds a year less of nitrogen, phosphorus, and sediment outfalling into the bay. Assuming SWIFT works as designed this is a huge benefit to Chesapeake Bay in that it may help to prevent or reduce the formation of algae bloom dead zones. Not only would success with the project benefit the bay but it would also be a huge benefit for the ratepayers at HRSD. To meet regulatory guidelines to remove nutrients from discharged treated wastewater would cost millions of dollars and almost non-stop retrofitting at the treatment plants to keep up with advances in treatment technology and regulatory requirements. Another goal of HRSD's SWIFT project is to restore or restock potable groundwater supplies in the local aquifers. The drawdown of water from the groundwater supply has contributed to land subsidence and a reduction of water available for potable use.

The planned restocking of the Hampton Roads groundwater supply with injected wastewater treated to potable water quality is not without its critics, who state that HRSD's wastewater injection project would contaminate potable water aquifers. This is where the so-called "yuck factor" comes into play. The yuck factor, in this particular instance, has to do with the thought that groundwater for consumptive use will be contaminated, basically, with toilet water. This is the common view of many of the critics who feel that HRSD's SWIFT project is nothing more than direct reuse of wastewater; that is, a pipe-to-pipe connection of toilet water to their home water taps.

What the critics and others do not realize is that we are already using and drinking treated and recycled toilet water. Figure 9.4 shows an example of an urban water cycle where wastewater is indirectly used. The point is, whether we like it or not, we are using recycled wastewater for potable water use. With regard to the idea that HRSD's SWIFT project would contaminate existing aquifers with toilet water, it is important to point out that this water will be treated (and already is at the York River Treatment Plant) to drinking water quality, with *to drinking water quality* being key here. A sophisticated and extensive train of unit drinking water quality treatment processes produces treated wastewater that the HRSD general manager and several others have drunk right out of the process, which is discussed in detail later in the text. The bottom line is that the yuck factor involved in drinking treated toilet water is being grossly overstated, as pointed out in Sidebar 9.1.

SIDEBAR 9.1. WASTEWATER YUCK FACTOR OVERSTATED

When that great mythical hero, Hercules, arguably the world's first environmental engineer, was ordered by Eurystheus to perform his fifth labor—clean up King Augeas' stables—he was faced with a mountain of horse and cattle waste piled high in the stable area. His method to dispose of the waste was to divert a couple of river streams to the inside of the stable area so that all of the animal waste could simply be deposited into the river streams: out of

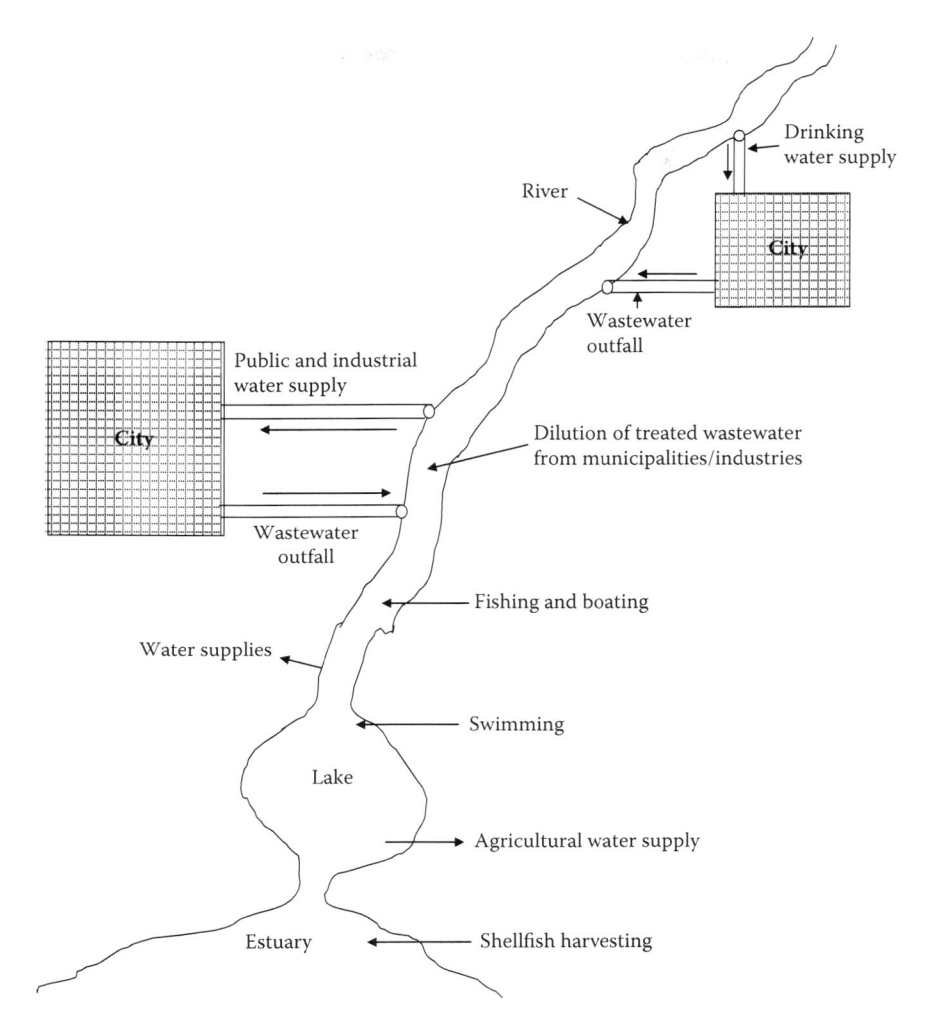

FIGURE 9.4 Indirect wastewater reuse.

sight, out of mind. The waste simply flowed downstream. Hercules understood the principal point in pollution control technology that is pertinent to this very day and to this discussion: *dilution is the solution to pollution.*

When people say they would never drink toilet water, they have no idea what they are talking about. The fact is that we drink recycled wastewater every day (Spellman, 2015). In Hampton Roads, for example, HRSD's wastewater treatment plants discharge treated water to the major rivers in the region. Many of the region's rivers are sources of local drinking water supplies. Even local groundwater supplies are routinely infiltrated with surface water inputs, which, again, are commonly supplied by treated wastewater (and sometimes infiltrated by raw sewage that is accidentally spilled).

My compliments to Ted Henifin, General Manager of HRSD, who stated in a recent local newspaper article that he would be the first to drink treated wastewater effluent from the unit treatment processes at York River Treatment Plant. My only argument with his statement is that, because of Mother Nature's water cycle, the one we all learned about in grade school, we have been drinking toilet water all along. I have yet to find anything yucky about it or its taste.

The SWIFT project includes construction of an extensometer monitoring station with the ability to accurately measure land-surface elevations, bedrock-surface elevations, and changes in aquifer system thickness. Monitoring of groundwater levels and aquifer system elastic response will benefit operation of the wastewater injection system at the Nansemond treatment plant in Suffolk and provide guidance for future wastewater injection facilities for the SWIFT project.

OBJECTIVES

The objectives of the SWIFT project are to

- Design and construct an extensometer station for collection of aquifer system thickness data with submillimeter accuracy for long-term (decadal) operation.
- Operate and maintain the station for at least 3 years to collect data describing groundwater levels, land-surface and bedrock vertical motion, and changes in aquifer system thickness.

NANSEMOND WASTEWATER TREATMENT PLANT EXTENSOMETER PLAN

The HRSD extensometer station at Nansemond Wastewater Treatment Plant will be designed and constructed to produce reliable and accurate data describing aquifer thickness, groundwater levels, land-surface elevation, and bedrock for at least 20 years and likely for 40 years or more.

Extensometer Station Design

The USGS, based on experience with its National Research Drilling Program (NDRP), has obtained unique and extensive experience in designing and installing extensometers in many locations around the United States, including California, Texas, and Louisiana. Based on this past experience, the USGS recommended including the following features (see Figure 9.5) in the extensometer instrument:

- Casing slip joints to accommodate the vertical stress and strain that accompany aquifer compaction/expansion and that can cause casing failure
- Casing centralizers and a counterweighted fulcrum at the surface to reduce friction and striking between the extensometer pipe and casing
- Combined mechanical strain gauges and digital potentiometers to achieve both accuracy and long-term record continuity

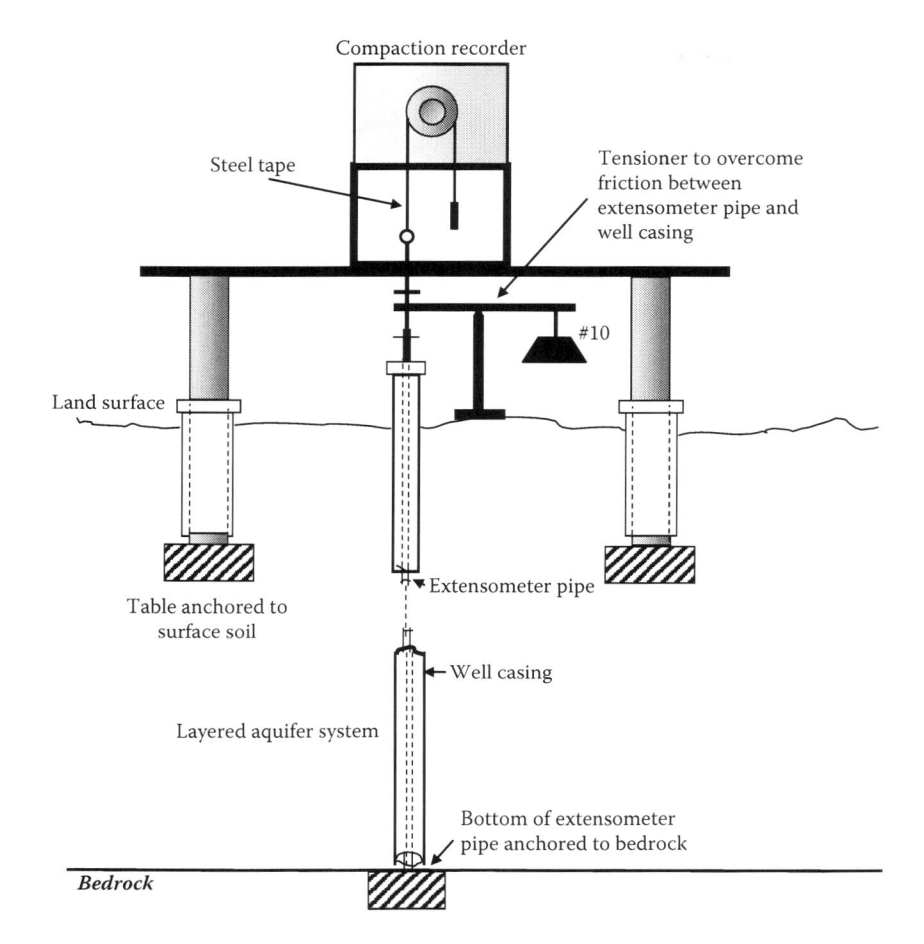

FIGURE 9.5 Type of extensometer instrument to be used at the HRSD Nansemond Wastewater Treatment Plant in Suffolk, VA.

- Deep surface-mount support piers for the reference table to exclude surface soil compaction and heaving from the aquifer compaction/expansion signals
- Heavy steel beam frames to support surface equipment

Extensometer Station Monitoring Program

The USGS will maintain the extensometer monitoring station, serve all data through the publicly available National Water Information System (NWIS), and provide annual analysis and reporting of data. The station will be outfitted with continuous real-time monitoring equipment so that short-term compression/expansion motions are know and immediately available via the web. Continuous data will be useful for optimizing performance of the wastewater injection system and will shorten response times for maintenance. The following variables will be monitored and recorded:

- Change in aquifer thickness
- Multiple groundwater levels (regardless of whether or not USGS installs the piezometers)
- Land-surface elevation
- Bedrock-surface elevation
- Performance parameters (e.g., voltage)

All data collected in this project will be stored in USGS data archives and will be publicly available after being quality assured by USGS personnel. A USGS Open-File Report will be published outlining the extensometer installation and initial data-collection results. An annual report summarizing results will also be provided to HRSD. One or more scientific journal articles describing key findings from the study will be published.

The Plan

Under the proposed plan, the USGS will construct an extensometer station at HRSD's in Suffolk. USGS Water Science Center personnel will oversee all aspects of construction and operation of the extensometer system, including drilling and downhole work. The USGS will also deploy its own drill rig, support equipment, and drill crew (see Figure 9.6). A staff geologist with the USGS and a geologist for the Virginia

FIGURE 9.6 USGS drilling rig at Nansemond Wastewater Treatment Plant, Suffolk, VA. (Photograph by F.R. Spellman.)

Department of Environmental Quality (DEQ) will share onsite geologist duties during the approximately 1 month of drilling. Installation of surface reference-frame structures and monitoring equipment will be performed or overseen by USGS staff. The work is anticipated to last from July 2016 through August 2019. Drilling and construction will occur in Year 1, and the USGS will operate the station in Years 1 to 3. Station operation can continue beyond Year 3 by mutual agreement.

The extensometer station will be constructed within the fenced perimeter of the Nansemond Wastewater Treatment Plant in Suffolk, and the final location will be chosen jointly with HRSD. The final location should be 200 to 500 feet from the injection test well and should have paved road access and adequate space for equipment, including a large drilling rig, delivery trucks, a generator, work lights, forklift, dumpsters, and backhoe. A roughly 60-ft by 100-ft area is necessary to house materials, to provide a work area for geologists and drillers, and to dig a pit for the circulation of drilling fluids.

A permanent structure with a 20-ft by 20-ft footprint and a 10-ft height will be built to house the top of the extensometer instrument and the associated monitoring equipment. As currently planned, it will be an insulated metal building with a concrete floor and a window and door, but alternative exteriors can be considered if desired by HRSD. The building will have locked windows and doors.

A cluster of four piezometers is highly recommended to allow monitoring of groundwater levels (pressures) in multiple aquifers. Each piezometer will be housed in a protective casing that protrudes about 3 feet above ground. These piezometers will be permanent structures and will be located near the extensometer building.

THE BOTTOM LINE

HRSD's SWIFT project proposes to add advanced treatment processes to several of its facilities to produce water that exceeds drinking water standards and to pump this clean water into the ground. This will ensure a sustainable source of water to meet current and future groundwater needs through eastern Virginia while improving water quality in local rivers and Chesapeake Bay. The SWIFT project is expected to

- Eliminate HRSD discharge to the James, York, and Elizabeth rivers except during significant storms.
- Restore rapidly dwindling groundwater supplies in eastern Virginia upon which hundreds of thousands of Virginia residents and businesses depend
- Create huge reductions in the discharge of nutrients, suspended solids, and other pollutants to Chesapeake Bay.
- Make available significant allocations of nitrogen and phosphorous to support regional needs.
- Protect groundwater from saltwater contamination and intrusion.
- Reduce the rate of land subsidence, effectively slowing the rate of sea-level rise by up to 25%.
- Extend the life of protective wetlands and valuable developed low-lying lands.

REFERENCES AND RECOMMENDED READING

Eggleston, J. and Pope, J. (2013). *Land Subsidence and Relative Sea-Level Rise in the Southern Chesapeake Bay Region*, Circular 1392. Reston, VA: U.S. Geological Survey.

Holdahl, S.R. and Morrison, N. (1974). Regional investigations of vertical crustal movements in the U.S. using precise relevelings and mareograph data. *Tectonophysics*, 23(4): 373–390.

Hong, B.G., Sturges, W., and Clarke, A.J. (2000). Sea level on the U.S. east coast: decadal variability caused by open ocean wind-curl forcing. *Journal of Physical Oceanography*, 30: 2088–2098.

Jones, B.D. (1980). *Service Delivery in the City: Citizen Demand and Bureaucratic Rules*. New York: Longman.

Mayfield, D. (2016). Aquifer replenishment touted as way to cut cities' cost of bay cleanup. *The Virginian-Pilot*, February 18.

Pope, J.P. and Burbey, T.J. (2004). Multiple-aquifer characteristics from single borehole extensometer records. *Ground Water*, 42(1): 45–58.

Spellman, F.R. (2015). *The Science of Water: Concepts and Applications*, 3rd ed. Boca Raton, FL: CRC Press.

USGS. (2016). *Proposal to Establish an Extensometer Station at the Nansemond Wastewater Treatment Plant in Suffolk, VA*, proposal to Hampton Roads Sanitation District, May 20, 2016. Reston, VA: U.S. Geological Survey (http://www.hrsd.com/pdf/Commission%20Minutes/2016/05-24-16_Draft_Commission_Minutes.pdf).

10 Potomac Aquifer System

> A rock formation or stratum that will yield water in sufficient quantity to be of consequence as a source of supply is called an "aquifer.".... It is water-bearing not in the sense of holding water but in the sense of carrying or conveying water.
>
> **Meinzer (1923)**

INTRODUCTION*

Before the Hampton Roads Sanitation District (HRSD) can commence pumping 130 million gallons per day (mgd) into the Potomac Aquifer System or any other aquifer, it—along with the expert assistance of CH2M, its primary consultant in this matter—must first determine the feasibility of aquifer replenishment by recharging clean water, purified from the advanced treatment of wastewater treatment plant (WWTP) effluent. This chapter provides a description of the essential elements of recharging clean water into the Potomac Aquifer System (PAS) at the following seven Hampton Roads Sanitation District WWTPs: Army Base, Boat Harbor, James River, Nansemond, Virginia Initiative Plant, Williamsburg, and York River (see Figure 10.1). Also, this chapter discusses determining the capacity of individual injection wells at the seven WWTPs, projecting the injection capacity within the existing site area of the seven WWTPs, and characterizing the regional beneficial hydraulic response of the PAS to clean water injection.

The material presented in this chapter utilizes data available in city/county, state, and federal databases, reports, scientific papers, and other literature to characterize the injection capacity of individual wells at each of the WWTPs. Injection well capacities and analytical mathematical modeling were used to estimate the injection capacity of each WWTP based on the plant's flow rate, property size, and the transmissivity of the underlying PAS aquifer.

POTOMAC FORMATION

Given the elevated volume of wastewater requiring disposal, and the importance of minimizing the number of injection wells, the most suitable aquifer units are those that exhibit the highest production capacity. Furthermore, a thick, confining bed composed of impermeable materials such as silt or clay should overlie the aquifer to prevent vertical migration of the injection fluid (injectate) into the surrounding aquifer units. Beneath the HRSD service area, the Cretaceous-age Potomac Formation meets these criteria. The Potomac Formation contains thick sand deposits, forming

---

* Much of this chapter is based on CH2M, *Sustainable Water Recycling Initiative: Groundwater Injection Hydraulic Feasibility Evaluation*, Report No. 1, CH2M, Newport News, VA, 2016.

HRSD Service Area
A Political Subdivision of the Commonwealth of Virginia

Major facilities include the following:

1. Atlantic, Virginia Beach
2. Chesapeake-Elizabeth, Va. Beach
3. Army Base, Norfolk
4. Virginia Initiative, Norfolk
5. Nansemond, Suffolk
6. Boat Harbor, Newport News
7. James River, Newport News

8. Williamsburg, James City County
9. York River, York County
10. West Point, King William County
11. Central Middlesex, Middlesex County
12. Urbanna, Middlesex County
13. King William, King William County

Serving the Cities of
Chesapeake, Hampton,
Newport News, Norfolk,
Poquoson, Portsmouth,
Suffolk, Virginia Beach,
Williamsburg, and the
Counties of Gloucester,
Isle of Wight, James City
King and Queen,
King William, Mathews,
Middlesex and York

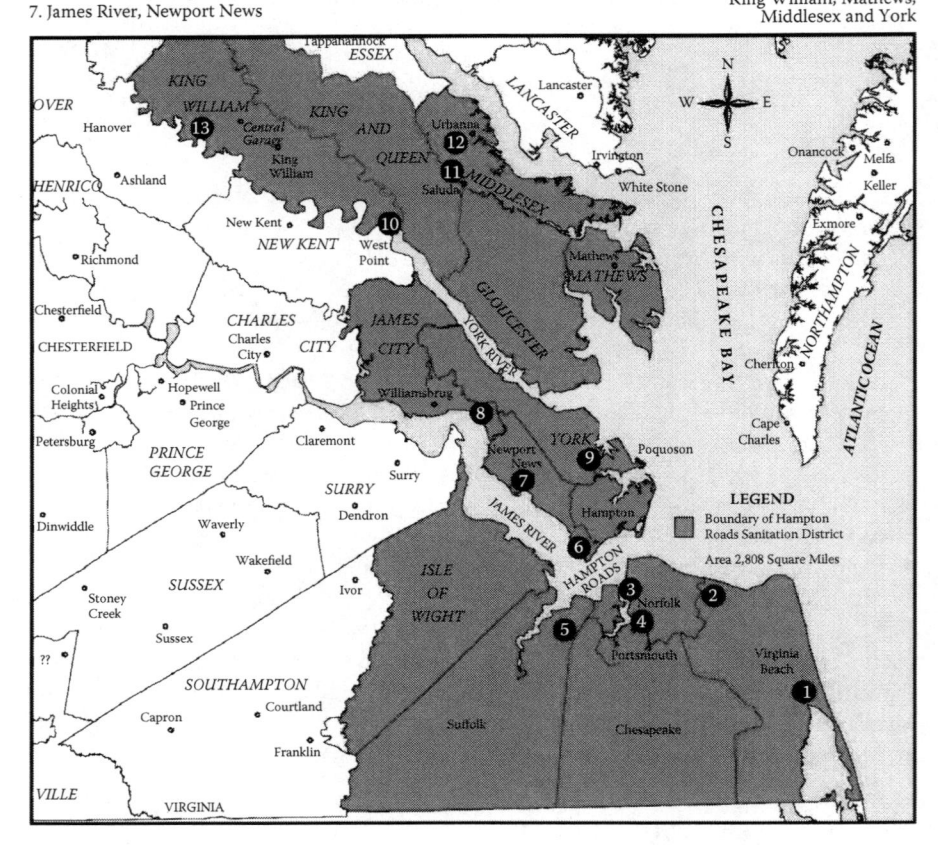

FIGURE 10.1   Map of Hampton Roads Sanitation District service area (2016).

three discrete aquifer units. The modern convention developed by the U.S. Geological Survey (USGS) and the Virginia Department of Environmental Quality (DEQ) is to group the three aquifers as one, named the Potomac Aquifer (McFarland and Bruce, 2006); however, because they behave hydraulically as three distinct units, they must be and are examined separately in this chapter. These three units are referred to here as the Upper Potomac Aquifer (UPA) zone, Middle Potomac Aquifer (MPA) zone, and Lower Potomac Aquifer (LPA) zone (Hamilton and Larson, 1988; Laczniak and Meng, 1988). Each discrete aquifer in the HRSD service area is separated from adjacent aquifers by clay confining beds of measurable thickness, and a cumulative thickness of silt and clay units totaling several hundred feet overlies the top aquifer unit (UPA) of the PAS. Production wells screened in the PAS exhibit significantly greater pumping capacities than wells screened in other aquifers of the Virginia Coastal Plain (Smith, 1999). Some wells can pump at rates approaching 3000 gallons per minute (gpm), or 4.3 mgd. In addition to confinement and production capacity, aquifers in the PAS exhibit deep static water levels, ranging from 80 to 180 feet below grade (fbg), providing available head for injection.

Recent USGS sedimentological studies suggest that the HRSD service area, spanning the York–James Peninsula and Southeastern Virginia (see Figure 10.1), is well situated with regard to the quality of aquifers in the PAS (McFarland, 2013). PAS aquifer sands display greater thickness, coarser grain size, and better sorting in the HRSD service area than units in the northern Virginia Coastal Plain or to the south in northern North Carolina. As a result, aquifers exhibit excellent hydrologic coefficients (e.g., hydraulic conductivity, transmissivity) and, thus, more productive well capacity.

The Potomac Aquifer System outcrops at the ground surface west of King William WWTP (see Figure 10.1). The PAS is recharged by infiltrating precipitation. In the recharge area, the aquifers range in thickness from 70 feet (UPA) to 400 feet (LPA). Further downdip, recharge enters the PAS by leakage through overlying confining beds (Meng and Harsh, 1988).

Individual aquifers thicken and dip to the southeast toward the Atlantic coast line, reaching thicknesses ranging from 170 feet (UPA) to nearly 1000 feet (LPA) at the coast (Teifke, 1973). These thicknesses encompass all sediments contained in the vertical section, including discrete sand beds representing aquifer materials, and interleaving silt and clay lenses from intra-aquifer contained beds. Individual aquifers exhibit a strongly inter-bedded morphology consisting of thin to thick beds of sands, silts, and clays. Beneath Newport News (refer to Figure 10.1), the MPA consists of six discrete sand intervals. Obtaining maximum production (or injection) capacities from wells installed in layered aquifers such as the PAS requires extending the well screen assembly across the maximum thickness of aquifer sand. Screen assemblies can consist of multiple screens and blank sections.

INJECTION WELLS

Typically, when we think about a well, the image of a hole in the ground and some type of device to pump water to the surface for use or for storage comes to mind—a vision perhaps similar to the arrangement shown in Figure 10.2. The well portrayed in the figure is pumping water from the subsurface to the surface for whatever intended purpose. However, wells used for pumping are not what we are concerned with in this book. We are concerned with just the opposite—that is, wells that inject treated injectate and do not pump fluids to the surface. Injection wells are known as Class V wells. There are 22 types of Class V injection wells, as shown in Table 10.1.

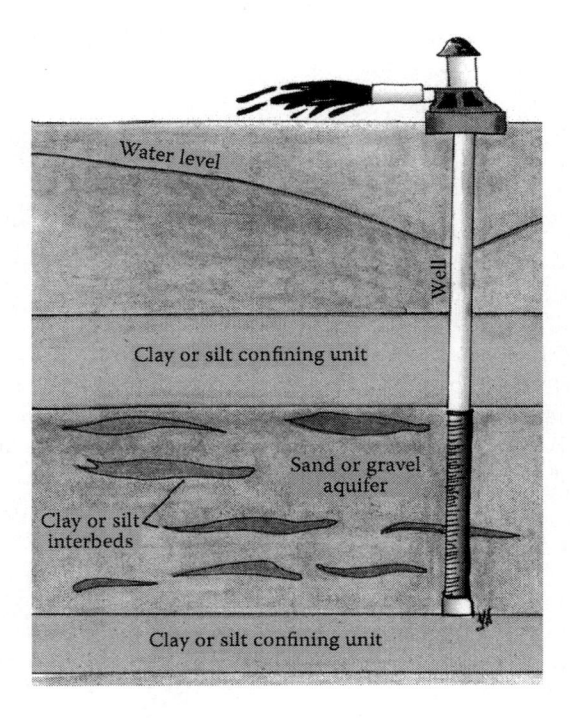

FIGURE 10.2   Standard well used to pump water from subsurface to surface. (Illustration by F.R. Spellman and Kathern Welsh.)

SUBSIDENCE CONTROL WELLS

The focus of this presentation is on the last type of Class V well listed in Table 10.1—subsidence control wells. Subsidence control wells are injection wells whose primary objective is to reduce or eliminate the loss of land surface elevation due to the removal of groundwater providing subsurface support. Subsidence control wells are important to HRSD's SWIFT project. The goal is to inject treated wastewater to drinking water quality into the underground Potomac Aquifer System to maintain fluid pressure and avoid compaction and to ensure that there is no cross-contamination between infected water and underground sources of drinking water. Thus, the injectate must be of the same quality or superior in quality as the existing groundwater supply.

INJECTION WELL HYDRAULICS

With regard to aquifer injection hydraulics, the injection capacity of a well depends on its specific capacity and the pressure (i.e., head) available for injection, a function of the static head, or water level, of the aquifer in which the well is screened. Specific capacity describes the yield (gpm) per unit of head decrease (drawdown) in a pumping well or head increase (drawup) in an injection well. Specific capacity is expressed as gallons per minute per foot (gpm/ft). When injection begins, the water level in the

TABLE 10.1
Class V Underground Injection Wells

Type of Injection Well	Purpose
Agricultural drainage wells	Receive agricultural runoff
Stormwater drainage wells	Dispose of rain water and melted snow
Carwashes without undercarriage washing or engine cleaning	Dispose of wash water from car exteriors
Large-capacity septic systems	Dispose of sanitary waste through a septic system
Food processing disposal wells	Dispose of food preparation wastewater
Sewage treatment effluent wells	Inject treated or untreated wastewater
Laundromats without dry cleaning facilities	Dispose of fluid from laundromats
Spent brine return flow wells	Dispose of spent brine for mineral extraction
Mine backfill wells	Dispose of mining byproducts
Aquaculture wells	Dispose of water used for aquatic sea life cultivation
Solution mining wells	Dispose of leaching solutions (lixiviants)
In situ fossil fuel recovery wells	Inject water, air, oxygen, solvents, combustibles, or explosives into underground or oil shale beds to free fossil fuels
Special drainage wells	Include potable water overflow wells and swimming pool drainage
Experimental wells	Test new technologies
Aquifer remediation wells	Clean up, treat, or prevent contamination of underground sources of drinking water
Geothermal electrical power wells	Dispose of geothermal fluids
Geothermal direct heat return flow wells	Dispose of spent geothermal fluids
Heat pump/air conditioning return flow wells	Reinject groundwater that has passed through a heat exchanger to heat or cool buildings
Saline intrusion barrier wells	Inject fluids to prevent the intrusion of saltwater
Aquifer recharge/recovery wells	Recharge an aquifer
Noncontact cooling water wells	Inject noncontact cooling water
Subsidence control wells	Control land subsidence caused by groundwater withdrawal or over pumping of oil and gas

well rises as a function of the transmitting properties of the receiving aquifer and the efficiency of the well (Warner and Lehr, 1981). Although the transmitting character of the aquifer should remain stable over the service life of an injection well, the available head for injection will decline as injection recharges the aquifer, causing the static water level to rise toward the ground surface.

In this discussion of the planning and study phase of the SWIFT project, it is important to differentiate between production and injection by referring to injection-specific capacity as *injectivity*. Moreover, for our purposes here, the evaluation of specific capacity of local production wells and its conversion to injectivity represent an important variable in determining the capacity of individual injection wells and ultimately the total injection capacity across the affected area, and number of wells required at each WWTP.

DID YOU KNOW?

Specific capacity is one of the most important concepts in well operation and testing. The calculation should be made frequently in the monitoring of well operation. A sudden drop in specific capacity indicates problems such as pump malfunction, screen plugging, or other problems that can be serious. Such problems should be identified and corrected as soon as possible. Specific capacity is the pumping rate per foot of drawdown (gpm/ft):

Specific capacity = Well yield (gpm) ÷ Drawdown (ft)

Problem: If the well yield is 300 gpm and the drawdown is measured to be 20 ft, what is the specific capacity?

Solution: Specific capacity = 300 gpm ÷ 20 ft = 15 gpm per ft of drawdown

To determine injectivity for wells screened in specific aquifer units, specific capacity values were evaluated in the three Potomac Aquifer units with respect to their proximity to HRSD's wastewater treatment plants. The Virginia Department of Environmental Quality (DEQ) supplied a database of production wells that contained a total of 98 wells screened in the UPA, MPA, or LPA, with some wells being screened across sands in two of the three aquifers. In these wells, a technique was employed to determine the specific capacity contributed by each aquifer to the well, based on the length of the screen penetrating the aquifer divided by the total length of the well screen. It proved necessary to separate the aquifers across individual wells for the less represented LPA. Because of its greater depth and water quality, production wells rarely penetrated all of the sand units in the LPA, often screening sand intervals in both the MPA and UPA. Across the study area, the specific capacity of production wells screening the UPA and MPA averaged 35.5 and 32.4 gpm/ft, respectively, essentially equaling each other, whereas wells screened in the LPA exhibited a 40% lower value of 21 gpm/ft. Wells screening the LPA, as identified by the Virginia DEQ, were only located around the Franklin Paper Mill and Franklin City area (see Figure 10.1). Wells screened in the UPA and MPA were better represented across the study area.

As stated previously, the available head for injection represents an important factor in determining injection capacities. Multiplying the injectivity of a well by the available head for injection provides the *injection capacity* of the well. As a driver for this study, local industrial and residential development has resulted in elevated

DID YOU KNOW?

Per 40 CFT 144.121, "No owner or operator shall construct, operate, maintain, convert, plug, abandon, or conduct any other injection activity in a manner that allows the movement of fluid containing any contaminant into underground sources of drinking water, if the presence of that contamination may cause a violation of any primary drinking water regulation under 40 CFR part 142 or may otherwise adversely affect the health of persons."

pumpage, drawing water levels in aquifers of the Potomac Aquifer System downward at rates averaging 1 to 2 feet per year (McFarland and Bruce, 2006). The declining water levels represent greater available head for injection; however, once injection commences, water levels should rebound with a corresponding loss in available head.

INJECTION OPERATIONS

Because of the deep static water levels at most HRSD wastewater treatment plants (Table 10.2), HRSD could inject water under gravity of pressurized conditions. Under gravity conditions, HRSD would fit a foot valve to the base of the injection

TABLE 10.2
Summary of Static Water Levels in Potomac Aquifer System at HRSD Treatment Plants

Wastewater Treatment Plant	Ground Elevation	PAS Aquifer	2005 Static Water Elevation (ft MSL)	2014 Projected Water Level Elevation (ft MSL)	2014 Projected Depth to Water (fbg)	Available Head for Injection (ft)
Army Base	11.0	UPA	−65	−73.28	84.28	107.28
		MPA	−70	−78.28	78.28	101.28
		LPA	−70	−78.28	78.28	101.28
Virginia Initiative Plant	8.5	UPA	−70	−78.28	86.78	109.78
		MPA	−75	−83.28	83.28	106.28
		LPA	−70	−78.28	78.28	101.28
Nansemond	22.5	UPA	−70	−78.28	100.78	123.78
		MPA	−75	−83.28	83.28	106.28
		LPA	−75	−83.28	83.28	106.28
Boat Harbor	4.0	UPA	−65	−73.28	77.28	100.28
		MPA	−60	−68.28	68.28	91.28
		LPA	−65	−73.28	73.28	96.28
James River	18.0	UPA	−58	−66.28	84.28	107.28
		MPA	−50	−58.28	58.28	81.28
		LPA	−55	−63.28	63.28	86.28
York River	5.5	UPA	−50	−58.28	63.78	86.78
		MPA	−45	−53.28	53.28	76.28
		LPA	−45	−53.28	53.28	76.28
Williamsburg	58.5	UPA	−36	−44.28	102.78	125.78
		MPA	−36	−44.28	44.28	67.28
		LPA	−45	−53.28	53.28	76.28

Rate of water level decline   0.92 ft/yr

Source: Adapted from CH2M, *Sustainable Water Recycling Initiative: Groundwater Injection Hydraulic Feasibility Evaluation*, Report No. 1, CH2M, Newport News, VA, 2016.

Note: ft, feet; ft/yr, feet per year; MSL, mean sea level; fbg, feet below grade.

piping to reduce head and maintain a positive pressure in the injection and well-header piping. This design facilitates better control of injection rates compared with cascading water down the injection piping, inducing a vacuum through the system. If HRSD decides to inject under gravity conditions, it would not be necessary to seal the injection head, wellheader piping, and associated fittings. Instead, the design requires monitoring injection levels rising in the well's annular space to prevent it from topping the ground surface.

An option to injecting under gravity conditions could entail sealing the injection head, wellheader piping, and associated fittings, while allowing the injection head to rise above the elevation of the ground surface by maintaining a positive pressure in the annular space of the well. With greater available head for injection, HRSD could achieve higher injection rates while anticipating risking regional water levels inherent with injecting large volumes of water.

The operator is responsible for ensuring that annular pressures stay below an established threshold. Elevated annular pressures can stress the sand filter pack surrounding the well screen by lifting it and initiating the formation of channels that connect the well screen to the surrounding formation materials. Limiting the injection pressure to 10 pounds per square inch (psi) in the well's annular space precludes damage to the filter pack while making available another 23 feet of head for injection.

Regular maintenance is required to operate injection wells to their maximum capacity whenever they are screening fine sandy materials like those found in the Potomac Aquifer System. In aquifer storage and recovery (ASR) wells (see Figures 10.3 and 10.4) screening the Potomac Aquifer System, or similar aquifers at Chesapeake, Virginia, and in North Carolina, Delaware, and New Jersey, fine sandy aquifers have proven susceptible to clogging from total suspended solids (TSS) entrained in the recharge water (McGill and Lucas, 2009). Even high-quality treated water can contain some

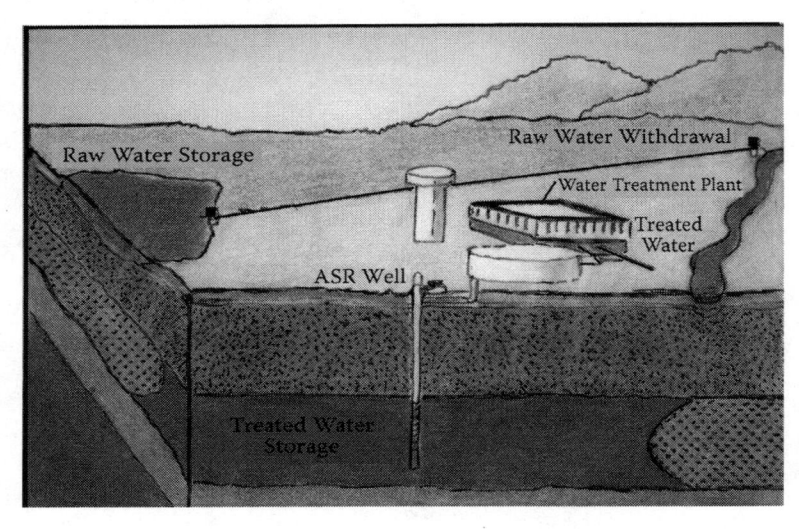

FIGURE 10.3 Treated water storage. (Adapted from USGS, *South Florida Information Access (SOFIA): Projects to Improve the Quantity, Quality, Timing and Distribution of Water*, U.S. Geological Survey, Reston, VA, 2013.)

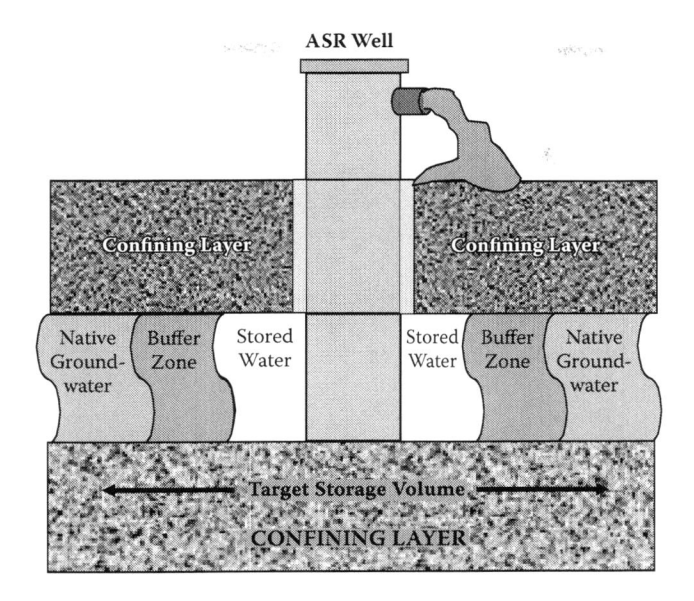

FIGURE 10.4 Aquifer storage and recovery (ASR) well.

amount of TSS that accumulate in the screens and pore spaces of sand filter packs and formations. Clogging reduces the permeability around the screen, filter pack, and aquifer proximal to the well (i.e., wellbore environment), resulting in higher injection levels while reducing injection capacity.

DID YOU KNOW?

The type and quality of injectate and the geology affect the potential for endangering an underground supply of drinking water. The following examples illustrate potential concerns (USEPA, 2016):

- If injectate is not disinfected pathogens may enter an aquifer. Some states allow injection of raw water and treated effluent. In these states, the fate of microbes and viruses in an aquifer is relevant.
- When water is disinfected prior to injection, disinfection byproducts can form *in situ*. Soluble organic carbon should be removed from the injectate before disinfection. If not, a chlorinated disinfectant may react with the carbon to form contaminating compounds. These contaminants include trihalomethanes and haloacetic acids.
- Chemical differences between the injectate and the receiving aquifer may create increased health risks when arsenic and radionuclides in the geologic matrix interact with injectate having a high reduction–oxidation potential.
- Carbonate precipitation in carbonate aquifers can clog wells when the injectate is not sufficiently acidic.

Aquifer storage and recovery wells operating in these states are equipped with conventional well pumps, which allow periodic well backflushing during recharge. Backflushing entails temporarily shutting down recharge and turning on the well pump for a sufficient time to move fine-grained materials from the wellbore environment to the ground surface. Installing a pump in each HRSD injection well for backflushing will slow the accumulation of fine-grained materials in the wellbore environment. To generate sufficient energy to effectively remove solids from the wellbore, the capacities of backflushing pumps should equal or exceed the injection rates planned for the well.

Even with backflushing, progressive clogging can occur and exceed the ability of backflushing to maintain injection wells near their maximum capacity. Each WWTP should contain a sufficient number of wells to compensate for removing wells from service for rehabilitation without compromising the facility's injection capacity.

INJECTION WELL CAPACITY ESTIMATION

In the planning and study stage, the following steps were employed to estimate the capacities of injection wells in the Potomac Aquifer System at each HRSD wastewater treatment plant:

- Estimate specific capacity at individual wells (Specific capacity = Well yield ÷ Drawdown).
- Organize specific capacities by aquifer.
- Separate well screens spanning two aquifers and calculate specific capacity for both.
- In short screen assemblies, normalize screen length to 100 feet.
- Convert specific capacity to injectivity.
- Calculate available head for injection from a USGS synoptic study.
- Combine injectivities of the UPA and MPA or the MPA and LPA in a single injection well.
- Average available head for injection across the UPA and MPA or the MPA and LPA in a single injection well
- Add 23 feet of head to the available injection head to account for maintaining a pressure of 10 psi in the annular space of a well.
- Multiply injectivity by available head for pumping to obtain injection well capacity.
- Limit injection well capacity to 3 mgd (2100 gpm).
- To estimate the number of injection wells per WWTP, divide the plant's effluent rate by the injection well capacity
- To facilitate periodic maintenance, add one injection well for every five per WWTP.

ESTIMATING SPECIFIC CAPACITY AND INJECTIVITY

The Virginia Department of Environmental Quality (DEQ) database of wells and their locations were obtained and evaluated according to their location relative to HRSD's WWTPs at Army Base, Boat Harbor, James River, Nansemond, Virginia

Initiative Plant, Williamsburg, and York River. The database contained static water levels, pumping levels, and stable production rates at most wells, enabling the calculation of specific capacity. Moreover, the Virginia DEQ database identified the Potomac Aquifer System unit spanned by each screen interval in each well. Several wells featured over five well screen intervals. As previously described, maximizing well screen length increases the production capacity of a well. Accordingly, many larger capacity productized wells were equipped with multiple screen intervals and, in many cases, more than one Potomac Aquifer System unit was screened.

Specific capacities were calculated for each well and then grouped according to the aquifer unit(s) in which the well was screened. Specific capacities for wells with screens spanning two aquifers resulted in developing two specific capacities for single wells. Screen assemblies in these wells usually spanned the UPA and MPA or the MPA and LPA. Individual intervals were grouped according to the aquifer that they screened. For wells spanning several aquifers, the specific capacity for a discrete aquifer was then estimated by taking the length of screen spanning the aquifer and dividing it by the total screen length as follows:

$$\text{Total SC of well} \times (\text{SL Aquifer 1/TSL of well}) = \text{SC Aquifer 1}$$

where
    SC = Specific capacity.
    SL = Screen length.
    TSL = Total screen length.

HRSD should design their injection wells with the screens penetrating an entire aquifer's sand thickness to maximize injection capacity. Note, however, that many of the production wells studied in this project used shorter screen assemblies that only partially penetrated the aquifers in which they were installed. Accordingly, to make the estimates compatible with fully penetrating injection wells, the specific capacity for a shortened screen assembly was normalized to a well with 100 feet of screen by applying the following equation:

$$\text{SC Aquifer 1} = (100 \text{ ft of screen/SL}) = \text{SC of Aquifer normalized to 100 ft of screen}$$

A total of 100 feet of screen was recognized as being an average total for assemblies fully penetrating the UPA, MPA, and LPA. Specific capacities for wells exhibiting screens extending over 100 feet across a specific aquifer unit of the PAS were accepted as being representative and were applied without modification.

Based on observations from aquifer storage and recovery wells that function in both injection and pumping modes of operation, the specific capacity and the injectivity of wells installed in the same aquifer usually vary slightly (Pyne, 2005). In a production well, water migrates toward a potentiometric head lowered by pumping in the wellbore. Water moves through an environment constrained by the size and heterogeneity of the porous media into the well. Upon entering the pumping well, the water is no longer impeded by porous media.

By comparison, injected water is driven down the wellbore by elevated head against the resistance of the wellbore environment. As a result of the greater resistance to flow, the injectivity of an ASR well often falls 10 to 50% less than its specific capacity. To recreate this important relationship for the study, converting specific capacity to injectivity involved multiplying specific capacity by a factor of 0.5, an average of the ratio of injectivity to specific capacity in Atlantic Coastal Plain injection-type wells (Pyne, 1995).

Available Head for Injection

The available head for injection also played an important role in determining the capacity of an injection well. The available head in each aquifer at individual WWTPs was determined by obtaining the most recent general view of the whole (synoptic) water level information. The U.S. Geodetic Survey last measured synoptic water levels separated by aquifer units of the Potomac Aquifer System in 2005 (Virginia DEQ, 2006b). More recent work considers all aquifers of the Potomac Aquifer System together.

Note that, with progressively declining water levels attributed to overpumping, potentiometric heads measured in 2005 cannot accurately reflect conditions in 2014. To adjust potentiometric levels to 2015 conditions, hydrographs were examined to quantify the annual conditions in water levels (Virginia DEQ, 2006a). The evaluation revealed that potentiometric levels have declined an average of 0.9 feet per year since 2005 in the Upper Potomac Aquifer, Middle Potomac Aquifer, and Lower Potomac Aquifer. The annual rate of decline in potentiometric levels was multiplied by 10 years and then applied to the potentiometric head from 2005 to project elevations to 2014.

To estimate the depth of water, the projected potentiometric level for 2014 in each aquifer unit was subtracted from elevation of the land surface at each WWTP. A constant head of 23 feet, equal to injecting under a pressure of 10 psi in the well's annular space, was added to the depth of the water level to obtain the available head for injection.

Flexibility for Adjusting Injection Well Capacities

Maximizing the capacity of individual injection wells required screening two aquifer units, either the UPA and MPA or the MPA and LPA, in each well. Because of potential hydraulic inefficiencies inherent with the difference in screen elevations, wells screening the UPA and LPA were not considered for this evaluation. This is important because it provides HRSD with some flexibility for adjusting injection well capacities while installing the injection wellfields. Upon installing the initial injection wells at a WWTP, HRSD could elect to screen all three Potomac Aquifer System units in a single well or revert to screening one aquifer if capacities fail to meet or significantly exceed expectations, respectively. Injection well capacities (Table 10.3) were estimated by adding the injectivities of two aquifer units in a well and averaging the available head for injection between the two aquifers. Then, the resulting injectivity was multiplied by the averaged available head for injection.

TABLE 10.3
Summary of Estimated Injection Well Capacity at HRSD WWTPs

WWTP	2030 Effluent Flow (mgd)	PAS Aquifer	Specific Capacity (gpm/ft)	Water Level (ft bgs)	Injection Capacity	Number of Wells	Total
Atlantic	48	UPA	38.9	104.28	—	—	—
(Virginia		MPA	37.1	102.28	8	8	—
Beach)		LPA	20.9	102.28	6	8	20
Army Base	12	UPA	20.0	107.28	—	—	—
		MPA	5.0	101.28	3	2	—
		LPA	20.9	101.28	3	2	5
Virginia	33	UPA	13.0	109.78	—	—	—
Initiative		MPA	5.0	106.28	2	8	—
Plant		LPA	20.9	101.28	3	6	17
Nansemond	23	UPA	38.9	123.78	—	—	—
		MPA	5.0	106.28	5	4	—
		LPA	20.9	106.28	3	4	10
Boat Harbor	15	UPA	27.5	100.78	—	—	—
		MPA	5.0	91.28	3	3	—
		LPA	20.9	96.28	3	3	7
James River	14	UPA	27.5	107.28	—	—	—
		MPA	34.8	81.28	6	2	—
		LPA	20.9	86.28	6	2	6
Williamsburg	11	UPA	27.5	125.78	—	—	—
		MPA	50.0	67.28	6	2	—
		LPA	20.9	76.28	6	2	5
York River	14	UPA	27.5	86.78	—	—	—
		MPA	34.8	76.28	5	2	—
		LPA	20.9	76.28	5	2	6

Source: Adapted from CH2M, *Sustainable Water Recycling Initiative: Groundwater Injection Hydraulic Feasibility Evaluation*, Report No. 1, CH2M, Newport News, VA, 2016.

NUMBER OF INJECTION WELLS REQUIRED AT EACH WASTEWATER TREATMENT PLANT

Elevated well capacities that exceeded reasonable constructability practices resulted from combining high injectivities from merging screens and large available heads for injection. Constructing relatively deep wells of sufficient diameter to inject at rates approaching 8 mgd (5600 gpm) or equipping the well with a backflush pump capable of the same rate is likely impractical. To reduce well casing, screen, and pumps to dimensions consistent with local well drilling capabilities and operation (such as, equipment available and electrical service), a capacity threshold of 3 mgd was set for the injections wells (Table 10.3). The number of injection wells at each WWTP was determined by dividing the plant's effluent rate by the injection well capacities. At large plants, injection wells were split evenly between the UPA and MPA and the

MPA and LPA combinations. These well totals and aquifer combinations at each WWTP formed the basis for the initial mathematical modeling runs discussed in the modeling section of this work. To accommodate removing wells from service for rehabilitation, one well was added for each five. At plants with smaller effluent flows, one well was added to any total less than five.

HRSD's Virginia Initiative Plant (VIP) and Nansemond WWTP required the largest number of injection wells to replenish the aquifer because of their higher effluent rates and relatively low injectivity in the MPA. Lower injectivities also appear in the UPA adjacent to the Virginia Initiative Plant. Production wells supporting the mapping of specific capacity in the MPA for this project were not present within a 5-mile radius of the Nansemond and Boat Harbor plants. Regionally, specific capacity values in the MPA appear to increase to the southeast with the exception of two production wells located west of Nansemond and Boat Harbor. To maintain a conservative approach to the project, low specific capacity values imparted by these wells were maintained when estimating injection well capacity at Nansemond and Boat Harbor. By comparison, large-specific capacity values in the UPA and MPA around the James River, York River, and Williamsburg plants resulted in elevated injectivities, which yielded elevated hypothetical injection capacities despite the relatively shallow available head for injection.

AQUIFER INJECTION MODELING

Hydrologists, hydrogeologists, and groundwater experts in other professional fields soon learn that groundwater flow models are simplified representations of often highly complex hydrogeological flow systems. Modeling tools are well suited for analyzing aquifer injection experiments. Generally, incorporating as much available hydrogeologic information as possible in the conceptual and numerical models of the flow system is advantageous. This is the approach used by HRSD and its consultant to model HRSD's Sustainable Water Initiative for Tomorrow (SWIFT) project. Hydrogeologic information takes many forms, including maps that show outcropping surfaces of geologic units and faults, cross-sections derived from geophysical surveys and wellbore information that show the likely subsurface location of geologic units and faults, maps of water-table levels, independent point well data, and maps showing the hydraulic properties of the subsurface materials. This information was used to classify the geologic units into hydrogeologic units, which are convenient units with which to define hydrologic properties (Anderman and Hill, 2000). The modeling employed in the SWIFT project helps to project or estimate the capacity of injections of injectate and other important parameters used in the project.

Estimating the capacity of individual injections at each WWTP along with preliminary determinations of the number of wells at each facility are key elements of HRSD's SWIFT project; the significance of these determinations can be seen when the goal is to dispose of nearly 130 mgd into the Potomac Aquifer System. The modeling executed in this section tests the hydraulic interference between injection wells located within the boundaries of each WWTP property. This evaluation will identify whether individual WWTP properties are sufficiently large to contain the projected number of wells required to dispose of the projected effluent volumes.

> **DID YOU KNOW?**
>
> Hydraulic engineers and others are quite familiar with mathematical modeling. With the continuing advancements in computer technology and development of advanced computer engineering programs, engineers rely more and more on mathematical modeling. A mathematical model is an abstract model that uses mathematical language to describe the behavior of a system. A mathematical model has been defined as a representation of the essential aspects of an existing system (or a system to be constructed) which presents knowledge of that system in usable form (Eykhoff, 1974).

MATHEMATICAL MODELING

The estimates of injection well capacity and the appropriate number of wells assigned at each WWTP did not account for hydraulic interference between wells in the same aquifer. For well screens combining the UPA and MPA or the MPA and LPA, hydraulic interference in the MPA exerts the greatest influence on local injection levels. Wastewater treatment plants situated on smaller properties will cause an increase in hydraulic interference, as smaller inter-well spacing is required for fitting the number of injection wells necessary to dispose of effluent. Mathematical modeling techniques were used to quantify the interference between injection wells located at the WWTPs and rebounding water levels in the aquifers receiving effluent. In the following section, analytical groundwater flow modeling is used to evaluate local groundwater mounding at individual WWTPs while injecting effluent (injectate) over 50 years.

GROUNDWATER FLOW MODELING

To evaluate potentiometric levels in the UPA, MPA, and LPA, analytical groundwater flow modeling was applied at each injection well at the seven WWTPs. The modeling study extends the determination of individual injection well capacities by testing the injection rates under the spatial conditions unique to each WWTP property. Although the injection capacities of individual wells may appear feasible at a WWTP based on the head available for injection and the aquifer transmissivities, hydraulic interference between multiple wells can drive injection levels higher in inverse proportion to the available spacing between wells.

At smaller WWTP sites, interfering wells could cause injection levels to exceed 23 feet above the ground surface, the maximum threshold established for injection head at individual wells. Mitigating elevated injection levels can entail screening all three PAS aquifers in a single well or reducing injection rates sufficiently to lower levels below the site injection elevation threshold. Lowering injection rates so that heads fall below site thresholds effectively limits flow injection rates lower than the projected 2040 target flows. Accordingly, groundwater flow models were customized according to property size and the projected number of wells required at each WWTP. The computer program CAPZONE (Bair et al., 1992) was applied to conduct the analytical groundwater flow modeling at each WWTP.

CAPZONE is an analytical flow model that can be used to construct groundwater flow models of two-dimensional flow systems characterized by isotropic and homogeneous confined, leaky-confined, or unconfined flow conditions. CAPZONE computes drawdowns at the intersections of a regularly spaced rectangular grid produced by up to 100 wells using either the Theis equation (see Equation 10.1) for a confined aquifer or the Hantush–Jacob equation for a leaky confined aquifer (see Equation 10.2) (Bair et al., 1992). Unlike the numerical mathematical techniques employed by models such as MODFLOW (comprised of simple algebraic equations that a computer cycles through multiple iterations to solve the flow equation), CAPZONE directly solves the differential flow equation. Subsequently, CAPZONE provides a more exact and conservative solution than models relying on numerical methods. However, analytical groundwater flow models offer less flexibility in simulating the heterogeneous conditions exhibited in natural systems, including multiple layers, variable grid spacing, spatially varying transmissivity, and boundary conditions. At the scale of a single WWTP, where neither the USGS nor Virginia DEQ have characterized heterogeneity beyond single wells, CAPZONE offers a reasonable method for simulating the hydraulic response to injection (or pumping) in the PAS. The Theis equation is simply

$$s = \left(\frac{Q}{4\pi t} \right) W(u) \tag{10.1}$$

$$u = \frac{r^2 S}{4Tt}$$

where
s = Drawdown (change in hydraulic head at a point since the beginning of test).
Q = Discharge (pumping) rate of the well.
t = Time since pumping began (seconds).
$W(u)$ = Well function.
u = Dimensionless time parameter.
r = Distance from the pumping well to the point where drawdown was observed.
S = Storativity of the aquifer around the well.
T = Transmissivity of the aquifer around the well.

The Hantush–Jacob well function for leaky confined aquifers is abbreviated $W(u,r/B)$. The Hantush–Jacob equation can be written in compact notation as follows:

$$s = \left(\frac{Q}{4\pi T} \right) W\left(u, r/B\right) \tag{10.2}$$

where
s = Drawdown in leaky confined aquifer.
Q = Pumping rate.
T = Transmissivity.
u = Dimensionless time parameter.

r = Distance from the pumping well to the point where drawdown was observed.
B = Leakage factor.
$W(u, r/B)$ = Leaky confined well function.

In practice, CAPZONE produces drawdowns/drawups that are then subtracted from water levels that form either a uniform or non-uniform hydraulic gradient. Thus, the analyst can designate a hypothetical gradient of one based on an observed water-level distribution (non-uniform). The non-uniform option has proven particularly useful for injection wellfield analyses at sites where a potentiometric surface exhibits irregularities or deflections that could potentially alter the potentiometric surface geometry.

At HRSD's individual WWTPs, a uniformed hydraulic gradient representing the site's position in the regional potentiometric surface (Figure 10.5) for the PAS, developed by USGS, was input to CAPZONE. The regional gradient was considered locally at each WWTP when estimating ambient groundwater flow direction and hydraulic gradient. In this approach, the regional gradient was considered locally at each WWTP to estimate ambient groundwater flow direction and hydraulic gradient.

To apply CAPZONE, boundaries of the simulated area were defined, and then the area was divided into a grid. The grid and cell dimensions for CAPZONE were unique to each WWTP (Table 10.4), depending on the size of the site. The grid contained up to 75 columns and 75 rows, with a grid node spacing range from 11 to 200 feet at the

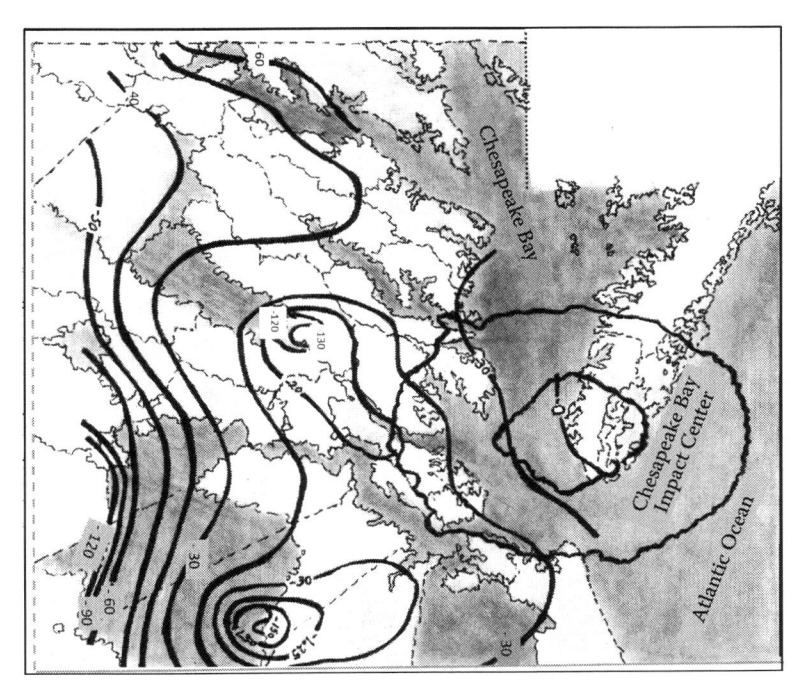

FIGURE 10.5   Potentiometric surface map of the Potomac Aquifer. (Illustration by F.R. Spellman and Kathern Welsh, based on CH2M, *Sustainable Water Recycling Initiative: Groundwater Injection Hydraulic Feasibility Evaluation*, Report No. 1, CH2M, Newport News, VA, 2016.)

TABLE 10.4

CAPZONE Input Parameters

WWTP	No. of Grid Nodes (X)	No. of Grid Nodes (Y)	Spacing (ft-X)	Spacing (ft-Y)	Static Water Level Elevations (ft MSL)	Aquifer Transmissivity (gpd/ft²)
Army Base	60	36	20	20	−75.8/−78.3	203,100/175,000
Boat Harbor	75	75	24	20	−71.6	202,400
James River	75	75	20	20	−62.6	202,400
Nansemond	75	75	54	54	−81.6	242,100
Virginia Initiative Plant	52	34	100	100	−80.8	242,100
Williamsburg	75	75	66.6	66.6	−44.3/−48.8	163,400/153,000
York River	75	75	120	80	−56.0	202,400

Source: Adapted from CH2M, *Sustainable Water Recycling Initiative: Groundwater Injection Hydraulic Feasibility Evaluation*, Report No. 1, CH2M, Newport News, VA, 2016.

Note: Split values represent UPA/MPA vs. MPA/LPA; all other values represent combined UPA/MPA/LPA.

Boat Harbor and York River WWTPs, respectively. Other inputs included coefficients of transmissivity from the results of aquifer testing conducted at production wells in the vicinity of each WWTP and a storage coefficient of 0.0001, typical for a confined aquifer. Hydraulic gradients averaged around 0.0001 ft/ft, with varying directions of groundwater flow based on the site's position within the USGS potentiometric map.

A unique static groundwater elevation was entered for each WWTP site and modified slightly depending on whether injection wells were screened across two or three aquifer units of the PAS. As described by the assessment of local-specific capacity, injection wells were first simulated to screen the two adjoining aquifer units of the three aquifers (UPA/MPA and MPA/LPA) comprising the PAS. This approach entailed adding the transmissivity of the two aquifers together, and obtaining an average static water elevation for the two aquifers. For wells screening the UPA and MPA and the MPA and LPA, discrete simulations were conducted. As the MPA received effluent, whether it was combined with the UPA or LPA to effectively simulate well interference in the aquifer, each simulation involved all the wells. As an example, in simulations involving wells screening the UPA and MPA, wells screening the MPA and LPA received one-half of the total effluent flow.

To obtain the number of injection wells used in each simulation, maximum injection rates were held at 3 mgd per well and divided into the effluent flow that HRSD projected for 2040. The injection wells were spaced (as much as possible) at roughly equal distance around the perimeter of the WWTP. Care was taken to avoid locating wells on existing structures. Locations were not evaluated for the practicality of positioning wells on lawns, parking lots, along fence lines, or other questionable areas that might host an injection well. Consistent with transient model runs conducted by the USGS, CAPZONE simulations were set for a 50-year duration. Simulated injection elevations were compared against the WWTP's threshold elevation (Table 10.5) in feet mean sea level (MSL), as defined by the ground surface elevation, plus 10 psi

TABLE 10.5
Summary of Results of Groundwater Modeling

WWTP	2040 Effluent Flow (mgd)	Achievable Flow (mgd)	UPA/MPA Wells	MPA/LPA Wells	UPA/MPA/ LPA Wells	Threshold Elevation (ft MSL)	Simulated Injection Level Elevation (ft MSL)	Comments
Army Base	12	12	2	2	—	35.0	−3.0/−11.6	Meets 2040 projections
Boat Harbor	16	14	—	—	7	28.0	28.0	Does not meet 2030 or 2040 projections
James River	15	15	—	—	5	42.0	42.0	Meets 2040 projections
Nansemond	28	24	—	—	12	46.5	44.6	Meets 2030 projections but not 2040 projections
Virginia Initiative Plant	33	21	—	—	14	32.5	24.0	Does not meet 2030 or 2040 projections
Williamsburg	13	15	2	3	—	82.5	42.4/52.6	Meets 2040 projections
York River	14	15	—	—	6	29.5	29.5	Meets 2040 projections

Source: Adapted from CH2M, *Sustainable Water Recycling Initiative: Groundwater Injection Hydraulic Feasibility Evaluation*, Report No. 1, CH2M, Newport News, VA, 2016.

Note: Two elevations are shown at WWTPs where injection wells screened the UPA and MPA and the MPA and LPA. One elevation is shown at WWTPs where each injection well screened the UPA, MPA, and LPA.

(23 feet). Simulated injection levels from two aquifers exceeded the threshold elevation, which indicated that the two aquifers could not facilitate effluent flows for the site. Accordingly, the simulation was run again combining all three aquifers of the PAS in each injection well at the WWTP.

Similar to the approach described previously, the transmissivities were added for each aquifer and static water level elevations were averaged. In case the simulated injection levels continued to exceed threshold elevations, the effluent flow rates were reduced in each well until a solution was found where injection rates fell below the WWTP's designated threshold elevation. This approach allowed determination of the sustainable effluent rate for the site.

After the model runs that resolved the sustainable number of wells and effluent rates were completed, sensitivity testing was conducted to quantify the uncertainty in input parameters used to obtain the model solutions. Sensitivity testing was conducted under conditions prevalent at the HRSD's York River Treatment Plant. An additional test was conducted at the York River Treatment Plant to investigate the relationship between well spacing and interference. This testing proved particularly important to stations with elevated projected effluent flows, or smaller stations that could not support the number of wells to inject effluent at the projected 2040 rates. At these stations, HRSD could potentially locate injection wells at offsite locations at distances sufficient to lower the effects of hydraulic interference.

MODELING RESULTS

After the model grid was set up, injection wells were located around the perimeter of each WWTP site, maximizing the number of wells given the site constraints. Through iterations of well layouts and injection rates the 2040 projected demands were tested at each WWTP. Table 10.5 shows the modeling results for each WWTP, including the maximum injection rates for sites that did not meet the 2030 or 2040 demand projections within the existing boundaries.

Army Base Treatment Plant

The Army Base WWTP model was able to meet the 2040 projected demands of 12 mgd using four wells. Two wells were screened in the UPA/MPA and the other two were set in the MPA/LPA, with all four injecting effluent at 3 mgd. The injection head elevation reached a maximum of 9 ft MSL within the UPA/MPA, falling several feet below the threshold elevation of 35 ft MSL. Depending on the aquifer used, the maximum drawup found at the property boundary was approximately 61 to 68 feet above static conditions.

Boat Harbor Wastewater Treatment Plant

Seven injection wells spaced between 200 and 600 feet apart and screening all three PAS aquifers achieved an injection rate totally 14 mgd, falling short of the 16 mgd targeted 2040 flow projections. At 14 mgd, the injection level approached the maximum threshold value of 28 ft MSL. The total injection rate was constrained by the location and layout of the plant. The adjacent highway and the harbor limit the space available for wells, placing some wells at distances as close as 200 feet apart. If HRSD can find locations outside the WWTP boundaries, increasing well spacing to greater than 600 feet, six injection wells should prove sufficient to meet the 2040 flow projections. The maximum drawup found at the property boundary fell approximately 93 feet above static conditions.

James River Treatment Plant

The James River WWTP model was able to meet the 2040 projected flows of 15 mgd using five wells. Unlike the Army Base model, the James River model required that the wells screened all three aquifers in the PAS. Using the three PAS aquifers, the injection head elevation fell below the threshold elevation (42 ft MSL) by almost 3 feet. The maximum drawup found at the property boundary totaled approximately 98 feet above static conditions.

Nansemond Wastewater Treatment Plant

The 2040 projections for the Nansemond WWTP represented the second highest of any of the WWTP sites tested, reaching 28 mgd. The results of the model included using 12 wells screening all three PAS aquifers to inject a maximum rate of 24 mgd, falling short of reaching the 2040 projections. At 24 mgd, the injection head elevation remained 2 feet below the threshold elevation (46.5 ft MLS). The large number of wells required and the limited space available led to the 4-mgd shortfall. The adjacent river and marsh limit the space available within the Nansemond WWTP site for siting injection wells. The maximum drawup found at the property boundary was approximately 122 feet above static conditions.

Virginia Initiative Plant

The Virginia Initiative Plant's 2040 projections were the highest of all of the WWTP sites tested, requiring 33 mgd. The model included 14 UPA/MPA/LPA wells spread out across parcels north and east of the WWTP. The maximum attainable injection rate reached only 21 mgd, falling short of the 2040 projections. Meeting the projected 2040 flow of 33 mgd will require locating wells in offsite locations. The maximum projected drawup found at the property boundary equaled approximately 101 feet above static conditions. In this scenario, drawup was obtained from the boundary of the WWTP with the surrounding open space.

Williamsburg Waste Water Treatment Plant

The Williamsburg WWTP, like the Army Base WWTP, was able to meet 2040 projections using five wells split between the UPA/MPA and MPA/LPA. For Williamsburg simulations, two wells were set in the UPA/MPA and the other three in the MPA/LPA. Given the high threshold elevation at this site (82.5 ft MSL), the WWTP was able to exceed the 2040 demands (13 mgd). At 15 mgd, injected through five wells, the injection elevation reached 53 ft MSL, well below the threshold elevation for the site. The maximum drawup found at the property boundary totaled approximately 77 to 93 feet above static conditions, depending on the aquifer combination (UPA/MPA or MPA/LPA).

York River Treatment Plant

The York River WWTP also successfully met the 2040 projections but required using all three PAS aquifers. The model was able to achieve 15 mgd (exceeding 2040 projections of 14 mgd) using six UPA/MPA/LPA wells, reaching a simulated injection elevation of 29.5 ft MSL, matching the threshold value (29.5 ft MSL). The maximum drawup found at the property boundary fell approximately 79 feet above static conditions.

SENSITIVITY OF AQUIFER PARAMETERS

Thou shalt confess in the presence of sensitivity. Corollary: Thou shalt anticipate criticism.

Leamer (1978)

When reporting a sensitivity analysis, researchers should explain fully their specifications search so that the readers can judge for themselves how that results may have been affected. This is basically an "honesty is the best policy" approach.

Kennedy (2007)

Whenever mathematical models are used, as was the case for HRSD's SWIFT project, there is uncertainty in the inputs applied to get an output. Because of uncertainty in inputs and outputs used in mathematical modeling, a sensitivity analysis is called for and should be part of the entire process. With regard to the SWIFT project, a sensitivity analysis quantifies the doubt in a calibrated or predicted solution caused by uncertainty in the estimates of the aquifer parameters, injection stresses, and groundwater elevations. Basically, what a sensitivity analysis accomplishes is a process of recalculating outcomes under alternative assumptions, and it serves various purposes, including the following (Pannell, 1997):

- In the presence of uncertainty, it tests the strength of the results of the model.
- It amplifies our understanding of the relationships between input and output variables in the system being modeled.
- Further research can reduce uncertainty by identifying the model inputs that cause significant uncertainty in the output.
- Unexpected relationships between inputs and outputs can be revealed by searching for errors in the model.
- It simplifies models.
- It increases and enhances communication and the links between modelers and decision makers.
- It employs Monte Carlo filtering to find regions in the space in input factors to optimum criterion.
- Knowing the sensitivity of parameters saves time because non-sensitive ones can be ignored (Bahremand and De Smedt, 2008).
- It allows the development of better models by identifying important connections among observations, model inputs, and predictions or forecasts (Hill and Tiedeman, 2007; Hill et al., 2007).

TABLE 10.6

Summary of Sensitivity Analyses

Parameter	Head Difference vs. Modeled Solution (ft)		Head Difference vs. Modeled Solution (%)	
	Value +50%	Value –50%	Value +50%	Value –50%
Transmissivity	–26.72	77.00	–90.58	261.02
Simulation duration	2.45	–4.04	8.31	–13.69
Storage coefficient				
0.005	–13.50	—	–45.76	—
0.00005	—	13.60	—	46.10
Injection rate (mgd)	0	2.0	3	4
Maximum injection elevation (ft MSL)	–56	–39.2	–30.9	–22.7
Static water level (ft MSL)	–46	–56.0	–66	—
Maximum injection elevation (ft MSL)	39.6	29.5	19.6	—

Source: Adapted from CH2M, *Sustainable Water Recycling Initiative: Groundwater Injection Hydraulic Feasibility Evaluation*, Report No. 1, CH2M, Newport News, VA, 2016.

For HRSD's SWIFT program, a sensitivity analysis of aquifer parameters was performed on the scenario that simulated injecting 14 mgd at the York River WWTP site. The York River site was chosen for selectivity analysis because it represents a site that accommodated the 2040 flows but required using all three PAS aquifers. Additionally, changes in pumping stresses and groundwater elevations were tested on a single well, eliminating interference from multiple wells. Finally, two wells were simulated to measure interference at varying distances.

For this sensitivity analysis to characterize the uncertainty of the modeled solution, input values for transmissivity, storativity, injection rate, simulation duration, and the groundwater elevations were systematically adjusted to assess how the changes affected groundwater elevations beneath the WWTP site. To quantify the evaluation, the maximum head generated from a sensitivity run was compared against the head from the original modeled solution.

The sensitivity analysis for aquifer transmissivity was performed by changing one parameter value at a time by –50% and +50% of the original parameter (Table 10.6). The storage coefficient, injection rates, and groundwater elevations were tested by increasing and decreasing the values incrementally, not on a percent basis.

Transmissivity

For purpose of clarity, transmissivity is defined here as the capacity of a rock to transmit water under pressure. The coefficient of transmissibility is the rate of flow of water (gallons per day) at the prevailing water temperature through a vertical strip of the aquifer 1 foot wide and extending the full saturated height of the aquifer under a hydraulic gradient of 100%. A hydraulic gradient of 100% indicates a 1-foot drop in head in 1 foot of flow distance.

With regard to HRSD's SWIFT project, the transmissivity values applied to the combined units of the UPA, MPA, and LPA at the York River WWTP were increased and decreased from the values used in the modeled solution. Values used in the sensitivity analysis ranged from 101,200 to 303,600 gpd/ft. The model was more sensitive to decreasing than increasing transmissivity by a factor approaching three times. Reducing transmissivity by 50% increased the maximum groundwater elevation at York River by 77 feet over the modeled solution of 29.5 feet MSL. Conversely, increasing the transmissivity value 50% decreased the mounding by only 27 feet from the modeled solution.

Storage Coefficient

The storage coefficient is the volume of water released from storage in a unit prism of an aquifer when the head is lowered a unit distance. In the HRSD SWIFT model, the aquifer storage coefficient was increased and decreased from the value used in obtaining the modeled solution (0.0005). The storage coefficient used in the sensitivity analysis ranged from 0.00005 to 0.005. Reducing the storage coefficient by an order of magnitude increased the maximum head across the site by 13.6 feet. Conversely, increasing the storage coefficient by an order of magnitude decreased the maximum injection level by 13.5 feet. As the modeled solution fell close to the threshold elevation, adjusting the storage coefficient had significant effects on whether the UPA, MPA, or LPA could exceed the 2040 injection rate.

Injection Rates

The sensitivity to changes in injection rates was tested by comparing the maximum injection levels of a single well at different injection rates. Injection rates within the sensitivity analysis ranged from 0 (static conditions) to 4 mgd. The model was almost identically sensitive to increases and decreases in injection rate. Reducing the rate from 3 mgd to 2 mgd reduced the maximum injection head by 8.3 feet, whereas increasing the rate to 4 mgd increased the injection level by 8.2 feet.

Simulation Duration

The duration of injection activity was adjusted to determine the model's sensitivity to changes in this parameter. An increase of 50% (75 years) over the original simulation (50 years) resulted in a rise in the maximum head value of 2.45 feet. Decreasing the simulation duration by 50% (25 years) lowered the maximum head value by 4.04 feet.

Static Water Levels

The static water level was set at −56 ft MSL in the model solution. The model solution, with 14 mgd of injection into the UPA/MPA/LPA, reached a maximum groundwater elevation of 29.5 ft MSL. The static water level was raised and lowered 10 feet for the sensitivity analysis. The model appeared slightly more sensitive to an increase in the groundwater elevation than a decrease. Increasing the static water level to −46 ft MSL increased the maximum head at the well by 10.1 feet; decreasing the static water level to −66 ft MSL reduced the maximum head by 9.9 feet.

WELL INTERFERENCE

To measure the effect of multiple wells injecting in proximity to each other, two wells were simulated with injection rates of 3 mgd under the subsurface conditions encountered at the York River WWTP for 50 years. The distance between wells was changed incrementally while the maximum groundwater elevation was recorded at each well and at the mid-point between the wells (Table 10.7). The well spacings tested ranged from 500 to 3000 feet. Injection heads at each well ranged from −30.9 ft MSL to −12 ft MSL (Well 1) when spaced 500 feet apart. Between 500 and 3000 feet of spacing, the maximum groundwater elevations in the wells varied about 4 feet. Interferences at Well 1 ranged from almost 19 feet to 15 feet, with Well 2 spaced from 500 to 3000 feet away, respectively. At the midpoint between the wells, the head values ranged from −14.8 to −23.1 ft MSL, for distances of 500 to 3000 feet, respectively. The interference in the PAS changed by 8.3 feet, or slightly less than 2 feet for every 500 feet of separation between the wells. The large amount of interference (greater than 15 feet) caused by a single nearby injection well even at relatively large spacings (3000 feet) appears consistent with the results of the modeling at other WWTPs. WWTPs carrying large injection rates require many wells, each well increasing the injection levels at other wells and in the aquifer.

HAMPTON ROADS REGION GROUNDWATER FLOW

The Virginia Coastal Plain Model (VCPM), a SEAWAT groundwater model, was employed to evaluate the hydraulic response of the PAS to injection operations at HRSD's seven WWTPs (Heywood and Pope, 2009; Langevin et al., 2008). This section presents the results of simulating injection at individual WWTPs and in scenarios with all seven WWTPs injecting simultaneously. SEAWAT is a three-dimensional, variable-density groundwater flow and transport model developed by the USGS based on MODFLOW and MT3DMS (a modular, three-dimensional multispecies transport model). The VCPM groundwater model encompasses all of the coastal plain within Virginia and parts of the coastal plain in northern North Carolina and southern Maryland. The original VCPM was updated for use in the Department of Environmental Quality (DEQ) well permitting process and is now called VAHydro-GW. The VAHydro-GW model is discretized into 134 rows, 96 columns, and 60 layers. The majority of the model cells are square, with the horizontal edges measuring one mile. The upper 48 model layers are each 35 feet. Layer thicknesses for the lower model layers increase to 50 feet after layer 48 (top to bottom) and then 100 feet beneath layer 52.

The model simulates potentiometric water levels in 19 coastal plain hydrogeologic units. The water levels are simulated for each year from 1891 through 2012, based on historic pumping records. The VAHydro-GW also simulates water levels for 50 years beyond 2012. These water levels are based on two scenarios: the

TABLE 10.7
Summary of Interference Modeling

Distance between Wells (ft)	Distance to Midpoint (ft)	Head at Well 1 (ft MSL)	Interference at Well 1 (ft)	Head at Well 2 (ft MSL)	Head at Midpoint between Wells 1 and 2 with Well 1 Active (ft MSL)	Head at Midpoint between Wells 1 and 2 with Both Wells Active (ft MSL)	Interference at Midpoint Between Wells 1 and 2 (ft)
0	0	−30.93	N/A	N/A	N/A	N/A	N/A
500	250	−12.06	18.87	−12.18	−35.24	−14.80	20.44
1000	500	−13.70	17.23	−14.30	−36.83	−18.00	18.83
1500	750	−14.65	16.28	−15.80	−37.77	−19.86	17.91
2000	1000	−15.30	15.63	−15.99	−38.42	−21.19	17.23
2500	1250	−15.85	15.08	−15.88	−38.92	−22.22	16.70
3000	1500	−16.28	14.65	−15.99	−39.33	−23.05	16.28

Source: Adapted from CH2M, *Sustainable Water Recycling Initiative: Groundwater Injection Hydraulic Feasibility Evaluation*, Report No. 1, CH2M, Newport News, VA, 2016.

Note: N/A, not applicable.

total permitted scenario and the reported-use scenario. The total permitted scenario simulates water levels for 50 years beyond 2012 by using the May 2015 total permitted withdrawal rates established for withdrawal permits issued by the DEQ together with the estimates for non-permitted withdrawals (domestic wells, wells in Maryland and North Carolina, wells within unregulated portions of Virginia) based on 2012 estimated use. The total permitted scenario represents the estimated water levels 50 years into the future if all permittees within the coastal plain were to pump at their authorized maximum withdrawal rates for the duration of the 50-year period.

The reported-use scenario simulates water levels for 50 years using pumping rates reported in 2012 for wells permitted by the DEQ, and estimates for non-permitted withdrawals based on 2012 estimated use. For most large permitted systems (greater than 1 mgd), reported pumping rates falling well below their total permitted diversion. The reported-use simulation represents the best available estimate of water levels within the coastal plain aquifers over the next 50 years, if pumping were to continue at the currently reported pumping rates for the permitted wells within the coastal plain.

Virginia regulations have established limits on the amount of drawdown allowed as a result of permitted pumping within the coastal plain. The *critical surface* is defined as the surface that represents 80% of the distance between the land surface and the top of the aquifer. Individual model cells where simulated potentiometric water levels fall below the critical surface are referred to as *critical cells*. Both the reported-use and total permitted simulations show areas of the coastal plain for the Potomac, Virginia Beach, Aquia, Piney Point, and Yorktown–Eastover aquifers, where the predicted water levels at the end of the 50-year simulation end below the critical surface for those aquifers.

For any new or renewing permitted withdrawal, DEQ performs a technical evaluation that involves adding the proposed facility to the total permitted simulation. As a major criterion for permit issuance, the facility cannot create new critical cells in any aquifer due to their proposed withdrawal. The critical cells simulated at the end of the reported-use simulation are not used for permit evaluation or issuance but represent a more plausible estimate of areas where water levels have lowered to crucial levels.

MODEL INJECTION RATES

VAHydro-GW row and column values were assigned to seven of HRSD's proposed injection WWTPs (Table 10.8) by using the well locations (latitude and longitude) to plot the position on a geographic information system (GIS) coverage of the VAHydro-GW finite-difference grid. Each facility was simulated as a single point of injection and consequently assigned to only one row and one column. Each model cell was square with each cell edge measuring 1 mile. As a result, the rates injected through any number of wells were dependent on the individual WWTP. Because of course grid dimensions, the analysis is not intended to evaluate the number of injection wells that are required to dispose of injectate at each facility.

For the initial modeling, the well screen length for each WWTP was assumed to measure between 300 and 350 feet, thus screening across multiple layers of the VAHydro-GW model. Because the VAHydro-GW module utilizes the

TABLE 10.8
VAHydro-GW Injection Well Allocation

WWTP	Row	Column	Layers	Potomac Horizontal Conductivity (ft/day)	Potomac Thickness (ft)	Screen Thickness (ft)	Portion of Potomac Screened (%)
Army Base	103	67	29–37	39.9	1440	350	24
Virginia Initiative Plant	106	67	26–34	37.6	1552	315	20
Nansemond	105	61	22–30	47.9	1392	315	23
Boat Harbor	100	62	27–35	50.3	1250	315	25
James River	92	55	19–27	68.8	1267	315	25
Williamsburg	83	50	18–26	70.3	1171	315	27
York River	83	60	35–43	0.0001	989	315	32

Source: Adapted from CH2M, *Sustainable Water Recycling Initiative: Groundwater Injection Hydraulic Feasibility Evaluation*, Report No. 1, CH2M, Newport News, VA, 2016.

Hydrogeologic-Unit Flow (HUF) package, the model layers are independent of the hydrogeologic units. The HUF package is an alternative internal flow package that allows the vertical geometry of the system hydrogeology to be defined explicitly within the model using hydrogeologic units that can be different than the definition of the model layers. With regard to the model, a model layer may contain multiple hydrogeologic units. In order to ensure that simulated water levels were not artificially influenced by the Potomac confining unit, each injection well was assigned to the uppermost VAHydro-GW model layer filled by the PAS. The remainder of each injection well screen was assigned to lower, adjacent model layers. The VAHydro-GW row, column, and layers assigned to each injection well are listed in Table 10.8. Additionally, the hydraulic conductivity at the top of the PAS is given for each injection cell as well as the total PAS thickness.

MODELING DURATION

In addition to modeling each WWTP operating individually (with the exception of the York River injection facility), all of the proposed facilities were modeled simultaneously at a combined flow of 114.01 mgd. The York River WWTP was not included in the combined simulations because the facility lies within the outer rim of the Chesapeake Bay bolide impact crater. As simulated in the VAHydro-GW, the horizontal hydraulic conductivity for the PAS at the cells within the bolide impact crater equal 0.0001 ft per day. As a result, simulated heads mounded to unrealistically high values when modeling injection at the York River WWTP.

The individual WWTP scenarios and the combined WWTP scenario were simulated by adding the proposed injection rates to the total permitted and reported-use simulations outlined previously, at the beginning of the 50-year predictive portion of those simulations (year 2013). The reported-use and total permitted scenarios were also executed before adding the injection facilities to establish baseline conditions. This presentation refers to these scenarios as the *reported-use baseline* and *total permitted baseline* simulations.

The model runs represent the DEQ's preferred metric for determining the beneficial impacts, if any, of proposed pumping/injection scenarios. The differences between water levels from an injection simulation and water levels from a baseline simulation represent the benefits, or recovery (rebound), resulting from the injection. The results of the injection and baseline simulations were compared at two points, 10 and 50 years into the predictive portion of the simulations.

THE BOTTOM LINE

This section summarizes the findings and conclusions drawn from research, investigations, and modeling procedures conducted by HRSD, USGS, and CH2M for HRSD's SWIFT project. These summarized procedures and findings are based on the evaluation of injection well rates, WWTP injection capacities, and the hydraulic response of the Potomac Aquifer System beneath the HRSD service area to injection operations. The bottom line conclusions that can be drawn from the CH2M (2016b) report are as follows:

- The transmissivities and available head for injection in the Potomac Aquifer System beneath each HRSD WWTP appear to support individual injection well capacities ranging between 3 and 8 mgd.
- By adhering to practical well design standards (e.g., borehole and casing diameter, pumping capacities), injection capacities were capped at 3 mgd for this evaluation.
- To account for the maintenance necessary for injection wells screened in sandy aquifers, one additional injection well was added for every five required to meet the effluent disposal rate at each WWTP.
- Accordingly, the number of injection wells ranged from 5 at Army Base and Williamsburg to 17 at the Virginia Initiative Plant.
- Analytical groundwater flow modeling indicated that of the seven WWTP sties tested, Army Base, James River, Williamsburg, and York River were able to meet the 2040 projected demands, within the site boundaries.
- Only Army Base and Williamsburg met the demands using the original two-aquifer approach.
- Conditions at James River and York River required screening all three of the Potomac Aquifer System aquifers to meet the 2014 demands.
- Sensitivity testing at the York River WWTP revealed that the modeled solution appeared sensitive to all parameters tested (transmissivity, storage coefficient, injection rates, simulation duration, and static water levels).
- The modeled solution exhibited the greatest sensitivity to changes in transmissivity. Changes in static water level resulted in increasing or decreasing the modeled head by the magnitude in the change of the water level.
- An evaluation of hydraulic interference between two wells at the York River WWTP reveal significant (interference greater than 15 feet) between wells spaced even 3000 feet away.
- The results of the analytical modeling show that hydraulic interference exerts a significant influence over the feasibility of replenishing the aquifer at the project 2040 rates.
- Injection was successfully simulated using the VCPM at each wastewater treatment plant except York River. The VCPM simulates very low coefficients of hydraulic conductivity for the Potomac Aquifer System beneath the York River WWTP because of its location inside the outer rim of the Chesapeake Bay bolide impact crater
- Injection at each WWTP resulting in removing most of the critical cells and, regionwide, recovering water levels in the Potomac Aquifer System.
- Water levels in all injection scenarios resulted in simulated water levels that exceeded the land surface across the HRSD service area.

REFERENCES AND RECOMMENDED READING

Anderman, E.R. and Hill, M.C. (2000). *MODFLOW-2000, the U.S. Geological Survey Modular Ground-Water Model—Documentation of the Hydrogeologic-Unit Flow (HUF) Package*, Open-File Report 2000-342. Reston, VA: U.S. Geological Survey (https://pubs.er.usgs.gov/publication/ofr00342).

Bair, E.S., Springer, A.E., and Roadcap, G.S. (1992). *CAPZONE*. Columbus: Ohio State University.

Bahremand, A. and De Smedt, F. (2008). Distributed hydrological modeling and sensitivity analysis in Torysa watershed, Slovakia. *Water Resources Management*, 22(3): 293–408.

CH2M. (2016a). *Sustainable Water Recycling Initiative: Groundwater Injection Hydraulic Feasibility Evaluation*, Report No. 1. Newport News, VA: CH2M.

CH2M. (2016b). *Sustainable Water Recycling Initiative: Groundwater Injection Geochemical Compatibility Feasibility Evaluation*, Report No. 2. Newport News, VA: CH2M.

Eykhoff, P. (1974). *System Identification Parameter and State Estimation*. New York: John Wiley & Sons.

Hamilton, P.A. and Larson, J.D. (1988). *Hydrogeology and Analysis of the Ground-Water-Flow System in the Coastal Plain of Southeastern Virginia*, Water Resources Investigations Report 87-4240. Reston, VA: U.S. Geological Survey.

Hantash, J.E. and Jacob, C.E. (1955). Nonsteady radial flow in an infinite leaky aquifer. *Transactions of the American Geophysical Union*, 36(1): 95–100.

Heywood, C.E. and Pope, J.P. (2009). *Simulation of Groundwater Flow in the Coastal Plain Aquifer System of Virginia*, Scientific Investigations Report 2009-5039. Reston, VA: U.S. Geological Survey(http://pubs.usgs.gov/sir/2009/5039/).

Hill, M. and Tiedeman, C. (2007). *Effective Groundwater Model Calibration, with Analysis of Data, Sensitivities, Prediction, and Uncertainty*. New York: John Wiley & Sons.

Hill, M., Kavetski, D., Clark, M., Ye, M., Arabic, M., Lu, D., Foglia, L., and Mehl, S. (2015). Practical use of computationally frugal model analysis methods. *Groundwater*, 54(2): 159–170.

Kennedy, P. (2007). *A Guide to Econometrics*, 5th ed. Hoboken, NJ: Blackwell.

Laczniak, R.J. and Meng III, A.A. (1988). *Ground-Water Resources of the York–James Peninsula of Virginia*, Water Resources Investigations Report 88-4059, Reston, VA: U.S. Geological Survey.

Langevin, C.D., Thorne, Jr., D.T., Dausman, A.M. et al. (2008). *SEWAT Version 4: A Computer Program for Simulation of Multi-Species Solute and Heat Transport*. Reston, VA: U.S. Geological Survey (https://pubs.usgs.gov/tm/tm6a22/).

Leamer, E. (1978). *Specification Searches: Ad Hoc Inferences with Nonexperimental Data*. New York: John Wiley & Sons.

McFarland, E.R. (2013). *Sediment Distribution and Hydrologic Conditions of the Potomac Aquifer in Virginia and Parts of Maryland and North Carolina*, Scientific Investigations Report 2013-5116. Reston, VA: U.S. Geological Survey.

McFarland, E.R. and Bruce, T.S. (2006). *The Virginian Coastal Plain Hydrogeologic Framework*, U.S. Geological Survey Professional Paper 1731. Reston, VA: U.S. Geologic Survey.

McGill, K. and Lucas, M.C. (2009). Mitigating Specific Capacity Losses in Aquifer Storage and Recovery Wells in the New Jersey Coastal Plain, paper presented at New Jersey American Water Works Association Annual Conference, Atlantic City, NJ, March 31.

Meinzer, O.E. (1923). The Occurrence of Ground Water in the United States with a Discussion of Principles, PhD dissertation, Department of Geology and Paleontology, Ogden Graduate School of Science, University of Chicago.

Meng, A.A. and Harsh, J.F. (1988). *Hydrogeologic Framework of the Virginia Coastal Plain*, U.S. Geological Survey Professional Paper 1404-C. Reston, VA: U.S. Geological Survey.

Pannell, D.J. (1997). Sensitivity analysis of normative economic models: theoretical framework and practical strategies. *Agricultural Economics*, 16: 139–152.

Pyne, D.G. (1995). *Groundwater Recharge and Wells*. Ann Arbor, MI: Lewis Publishers.

Pyne, D.G. (2005). *Aquifer Storage and Recovery: A Guide to Groundwater Recharge Through Wells*. Gainesville, FL: ASR Press.

Smith, B.S. (1999). *The Potential for Saltwater Intrusion in the Potomac Aquifers of the York–James Peninsula*, Water-Resources Investigations Report 98-4187. Reston, VA: U.S. Geological Survey.

Theis, C.V. (1935). The relation between the lowering of the piezometric surface and the rate and duration of discharge of a well using ground-water storage. *Eos,* 16(2): 519–524.

Teifke, R.H. (1973). Stratigraphic units of the Lower Cretaceous through Miocene series. In: *Geologic Studies, Coastal Plain of Virginia*, Bulletin 83, pp. 1–78. Charlottesville: Virginia Division of Mineral Resources.

USEPA. (2016). *Aquifer Recharge and Aquifer Storage and Recovery*. Washington, DC: U.S. Environmental Protection Agency (https://www.epa.gov/uic/aquifer-recharge-and -aquifer-storage-and-recovery).

USGS. (2013). *South Florida Information Access (SOFIA): Projects to Improve the Quantity, Quality, Timing and Distribution of Water*. Reston, VA: U.S. Geological Survey (https://sofia.usgs.gov/publications/reports/doi-science-plan/waterkissokee.html).

Virginia DEQ. (2006a). *Status of Virginia's Water Resources*. Richmond: Virginia Department of Environmental Quality, Office of Water Supply.

Virginia DEQ. (2006b). *Virginia Coastal Plain Model 2005 Withdrawals Simulation*. Richmond: Virginia Department of Environmental Quality, Office of Water Supply.

Warner, D.L. and Lehr, J. (1981). *Subsurface Wastewater Injection: The Technology of Injecting Wastewater into Deep Wells for Disposal*. Berkeley, CA: Premier Press.

11 Native Groundwater and Injectate Compatibility

> Ted Henifin's jaw-dropping, eyebrow-raising idea was first proposed in 2015, and last month the sanitation district [HRSD] general manager kicked off its pilot phase to stop what some scientists have called a nightmare in super slow motion.
>
> **Fears (2016)**

Not without some controversy, Hampton Roads Sanitation District's general manager, Ted Henifin, is putting into place a "jaw-dropping, eyebrow-raising project" in an attempt "to stop ... a nightmare in super slow motion"—in other words, land subsidence. Critics similarly doubted Leonardo da Vinci's audacious far-thinking, Newton's calculus, the Wright brothers' attempts to fly like a bird, Einstein's theories, General Patton's march against the bad guys, and Jonas Salk's vaccine for polio. Two things seem obvious to the author: Critics number in the mega-millions, while innovators, risk-takers, far-thinkers, people with grit, people with backbone and vision like Henifin are almost as rare as the dodo bird. Moreover, it is only the innovators who think outside the stovepipe; the rest go up the chute, undampered.

INTRODUCTION*

Bear with the author's presentation of a very simplistic view of what this chapter is all about. Envision two 1-liter glass beakers. One of these beakers is filled halfway with clean, safe drinking water. The other beaker is filled halfway with salty seawater. A normal person wanting to quench their thirst would obviously prefer to drink from the beaker of clean, safe water, leaving the beaker of salty seawater alone. Now, suppose we pour the contents of one of the beakers into the other to mix the two different contents. Is the new 1-liter mixture of clean, safe drinking water and salty seawater something that any of us would want to drink? Not likely, although some people would gargle with this mixture. Anyway, the point being made here is that by mixing clean, safe drinking water with salty seawater we have changed or adulterated the mixture.

The Hampton Roads Sanitation District (HRSD) does not intend to mix clean, safe drinking water with salty seawater or any other contaminant. The intent is not to adulterate native groundwater in any way; rather, the goal is to inject, replenish, and recharge the Potomac Aquifer's native groundwater supply with purified, safe water from the advanced treatment of wastewater effluent. To ensure that the treated

---

* Much of this chapter is based on CH2M, *Sustainable Water Recycling Initiative: Groundwater Injection Geochemical Compatibility Feasibility Evaluation*, Report No. 2, CH2M, Newport News, VA, 2016.

wastewater that is injected is of the same quality as the native groundwater contained in the Potomac Aquifer, HRSD and its consultant, CH2M, conducted a feasibility study. This feasibility study evaluated the geochemical compatibility of recharging clean water (injectate), native groundwater, and injectate interactions with minerals in the Potomac Aquifer System (PAS) aquifers. Three discrete injectate chemistries originate from the advanced water treatment processes of reverse osmosis (RO), nanofiltration (NF), and biological activated carbon (BAC).

The focus of the study was on a single wastewater treatment plant where conditions (such as geography, flow, geology, injectate quality, and groundwater quality) best represent the HRSD system. This determination was based on a large number of permutations involved with comparing three injectates with native groundwater chemistries from the three PAS aquifers beneath HRSD's seven wastewater treatment plants (WWTPs) (see Figure 11.1) and then applying the injectate chemistries to aquifer minerals in the three aquifers. HRSD's York River Treatment Plant was selected for this evaluation and for the subsequent pilot study. In addition to displaying fairly representative conditions, the property surrounding the York River Treatment Plant is sufficiently spacious to accommodate a WWTP upgraded with advanced water treatment processes and an injection wellfield.

HRSD'S WATER MANAGEMENT VISION

As mentioned earlier, the Hampton Roads region is faced with a variety of challenges related to management of the region's water supply and receiving water resources. These challenges involve technical, financial, and institutional complexities that invite the exploration of using non-traditional approaches that provide benefits on a larger scale beyond what the current wastewater treatment and disposal model can achieve. Aquifer replenishment can protect and enhance the region's groundwater supplies, in addition to reducing the potential damage caused by the discharge of nutrients to the lower James River and the Chesapeake Bay and perhaps slowing or arresting relative sea-level rise in the region.

GEOCHEMICAL CHALLENGES FACING SWIFT PROJECT

In Chapter 10, evaluations and analytical groundwater flow modeling revealed that HRSD will recharge between 77 and 131 million gallons per day (mgd) to the Potomac Aquifer System using over 60 injection wells with maximum capacities approaching 3.0 mgd per well, at seven WWTPs. Groundwater flow modeling revealed that the injection wells will screen combinations of two or all three PAS aquifers zones, according to the hydrologic conditions found at individual WWTPs. Both physical and geochemical challenges can emerge when recharging clean water into aquifers composed of reactive metal-bearing minerals and potentially unstable clay minerals, while also containing brackish native groundwater, typical conditions found in the PAS. Physical and chemical reactions are important with respect to both well facility operation and aquifer water quality. Potential damaging effects can result from

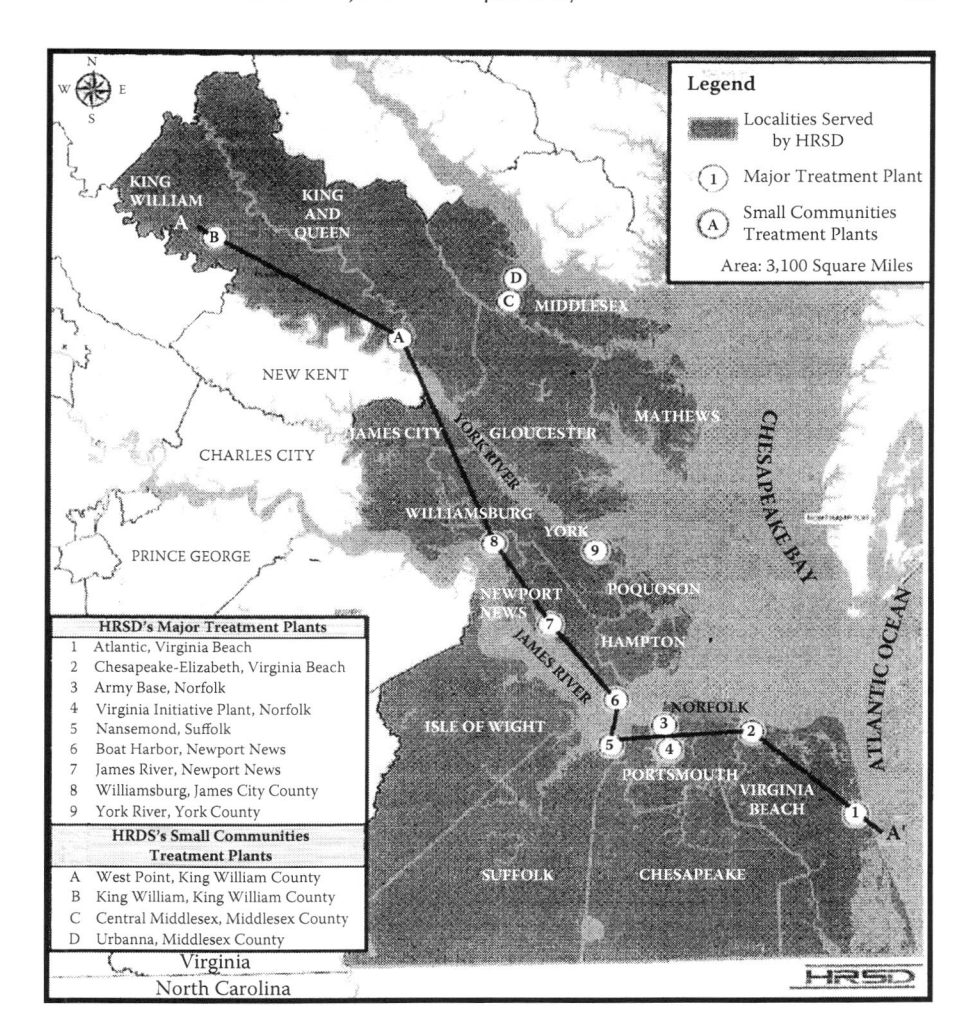

FIGURE 11.1 Location map of HRSD and section line A–A'. (From CH2M, *Sustainable Water Recycling Initiative: Groundwater Injection Geochemical Compatibility Feasibility Evaluation*, Report No. 2, CH2M, Newport News, VA, 2016).

- *Water to water mixing of injectate water with native groundwater*—Impacts are in the form of pore spaces becoming plugged with solids or precipitated metals, reduced permeability local to the wellbore, and eventually lower injectivity.
- *Interaction of the injectate/native groundwater mix with the aquifer matrix*—Impacts are in the form of damage to water-sensitive clays, precipitation or dissolution of metal-bearing minerals, and potential release of metals troublesome to injection activities (iron and manganese) or water quality issues (arsenic).

REDUCTION IN INJECTIVITY

The following two factors affect the injectivity of an injection well:

1. Physical plugging
2. Mineral precipitation

Physical Plugging

The injection rate per unit of head buildup (drawup) in an injection well is known as its injectivity, expressed in gallons per minute per foot (gpm/ft) of drawup (Warner and Lehr, 1981). When injection begins, the water level in the well rises as a function of the transmitting properties of the receiving aquifer and the efficiency of the well. Although the transmitting character of the aquifer should remain stable over the service life of an injection well, the available head for injection will decline as the injectate recharges the aquifer. This causes the static water level to rise toward the ground surface. The efficiency of a well will decrease with time, depending on the quality of the injectate, particularly its total suspended solids (TSS) content, which, in water, is commonly expressed as a concentration in terms of milligrams per liter and typically is described as the amount of filterable solids in a wastewater sample. More specifically, the term *solids* mean any material suspended or dissolved in water and wastewater. Although normal domestic wastewater contains a very small amount of solids (usually less than 0.1%), most treatment processes are designed specifically to remove or convert solids to a form that can be removed or discharged without causing environmental harm. However, 100% removal is unlikely. Even the most purified injectate can contain small amounts of TSS. If left to accumulate in the borehole environment (wellbore), solids can clog the screen, filter pack, and aquifer proximal to the well, which reduces the injectivity of the well, increasing drawup and eventually lowering the injection capacity of the well (Pyne, 2005). TSS can originate from scale or dirt in piping, treatment residuals, and reactions in the injectate that result in solids precipitation. One of the more common reactions occurs when oxygen dissolved in the injectate reacts with dissolved iron or manganese, precipitating ferric or manganese oxides and turning a dissolved component of the injectate into a source of solids.

Mineral Precipitation

In addition to physical plugging, chemical reactions between the injectate and the native groundwater or between the injectate and aquifer mineralogy can precipitate metal-bearing oxides and hydroxides. These reactions often arise from injectate that contains dissolved oxygen (DO). Considering the relatively small surface areas around the wellbore, precipitating metal-bearing minerals can clog pore spaces and reduce permeability and well injectivity. An important part of the research and planning involved with HRSD's SWIFT project is determining precisely the type and composition of injectate from the reverse osmosis (RO), nanofiltration (NF), or the biological activated carbon (BAC) processes to estimate its potential for plugging the wellbore.

GEOCHEMICAL CONCERNS

Beyond problems associated with physically plugging pore spaces around the borehole, several geochemical reactions can negatively affect injection well operations due to

- Clay minerals damage
- Metals precipitation
- Mineral dissolution

Damaging Clay Minerals

The term *clay* is applied to materials having a particle size of less than 2 micrometers (μm) (25,400 μm = 1 inch) and to a complex group of poorly defined hydrous silicate minerals that contain primarily aluminum, along with other cations (potassium and magnesium) according to the exact mineral species. Displaying a platy or tabular structure, clay minerals exhibit an extremely small grain size and typically adsorb water to their particle surfaces. In aquifer sand, trace amounts (less than 10%) of clay are found as components of the aquifer's interstitial spaces, coating framework particles such as quartz grains or lining or filling pore spaces, or occurring as a weathering product of feldspars.

Damage to clays occurs with the disruption of their mineral structure. The damage can arise when injecting water of significantly different ionic strength than the native groundwater, a concern when injecting dilute freshwater into an aquifer containing brackish or saline native groundwater (Drever, 1988). The dilute water contains significantly fewer cations and has a weaker charge than brackish native groundwater. When displacing the brackish water in the diffuse double layer between clay particles, the weaker charge can induce repulsive forces dispersing the particles, fragmenting the clay structure while mobilizing the fragments into flowing pore water. The particles can eventually accumulate in smaller pores and physically plug the pore space, thus reducing the permeability of the aquifer.

Damage can also arise when injectate displays differing cation chemistry than the native groundwater and the clay minerals (Langmuir, 1997). Exchanging cations can disrupt clay mineral structure particularly when their atomic radius exceeds the radius of the exchanged cation. The larger cation fragments the tabular structure, shearing off the edges of the mineral. Plate-like fragments break off the main mineral particle and migrate with flowing groundwater. Like the damage incurred by water of differing ionic strength, migrating clay fragments will pile up in pore spaces, physically plugging passageways and reducing aquifer permeability. Unlike the accumulation of TSS in the wellbore, formation damage by migrating clays develops in the aquifer away from the wellbore, making its removal difficult by backflushing or even invasive rehabilitation techniques.

Mineral Precipitation

Metal-bearing minerals can precipitate in the aquifer away from the well. These reactions typically occur when the injectate contains dissolved oxygen (DO) at concentrations exceeding anoxic levels (<1.0 mg/L) but can also occur if the pH of the injectate exceeds 9.0. As surface areas in the aquifer increase geometrically away from the well, mineral precipitation does not create as great a concern as the same reactions at the borehole wall.

Mineral Dissolution

Injectate reactions with minerals in the aquifer matrices can dissolve minerals leaching their elemental components (Stuyfzand, 1993). Injectate containing DO above anoxic concentrations will react with common, reduced metal-bearing minerals such as pyrite (FeS_2) and siderite ($FeCO_3$) to release iron and other metals such as manganese that occupy sites in the mineral structure. Iron and manganese can precipitate as oxide and hydroxide minerals if they contact injectate-containing DO. Oxidation of arseniferous pyrite can release arsenic, creating a water quality concern in the migrating injectate.

WATER QUALITY AND AQUIFER MINERALOGY

Note: Because it would require an unwieldy number of permutations (63) to assess the injection of three injectate chemicals into three discreet aquifers at seven WWTPs, the chemical compositions of three injectate types and the native groundwater in the three PAS aquifer zones beneath the York River Treatment Plant were chosen for discussion here. Targeting the York River Treatment Plant for the discussion that follows is not an issue and does not skew data, because the York River Treatment Plant exhibits effluent and local aquifer characteristics typical of conditions across the Hampton Roads Sanitation District.

Mass-balance relationships between raw water entering the plant and modeling of the advanced water treatment process were used to determine injectate chemistry. As no wells installed in the PAS aquifer zones currently exist at the York River Treatment Plant, water quality data from the area around the site were obtained from the National Water Information System (NWIS) database, which is maintained by the U.S. Geological Survey (USGS). The NWIS database provides samples collected by USGS personnel from local municipal, irrigation, and industrial supply wells, along with designated monitoring wells.

Table 11.1 shows that the York River WWTP water chemistry constituents included major and trace metals, anions, nutrients, nonmetallic trace constituents, and field chemistry measurements. The constituents listed in the table were used for geochemical evaluation of PHREEQC modeling, a type of speciation modeling used in situations where the possibility of mineral dissolution or precipitation needs to be known, as in water treatment, aquifer storage and recovery, artificial recharge, and well injection. Water quality data are discussed in relation to the wells closest to York River WWTP screening the three PAS aquifer zones. The search radius extended a maximum of 5 miles from the York River WWTP. Data are also listed based on treatment via advanced water treatment (AWT): biological activated carbon (BAC), nanofiltration (NF), and reverse osmosis (RO). BAC is the combination of ozonation and granular activated carbon (GAC) and is usually referred to as a biologically enhanced activated carbon process. Nanofiltration is a membrane filtration method that uses nanometer-sized cylindrical throughpores that pass through the membrane at 90°. Nanofiltration membranes have pore sizes from 1 to 10 nm, smaller than those

TABLE 11.1
Water Chemistry Constituents for Geochemical Evaluation and PHREEQC Modeling

Constituent	Injectate/ GW Units	BAC	NF	RO	UPA NWIS 7A	MPA NWIS 7B	UPA NWIS 7C
pH	SU	7.8	7.8	7.8	8.2	8	7.7
DO	mg/L	1.3	5	5	0.5	0.5	0.5
Ammonia	mg/L	1.3	0.87	0.23	0.01	0.01	0.01
Nitrate	mg/L	3.2	3.1	0.48	0.01	0.01	0.01
Phosphate	mg/L	0.5	0.03	0.01	0.01	0.01	0.5
Chloride	mg/L	212	125	9	494	1200	2950
Calcium	mg/L	62	19	17	4.5	13	51
Iron^{2+}, Fe(II)	mg/L	0.7	0.01	0.01	0.04	0.4	0.4
Iron^{3+}, Fe(III)	mg/L	0.001	N/A	0.01	0.01	0.02	0.06
Maganese^{2+}	mg/L	0.0001	0.0001	0.001	0.05	0.02	0.12
Magnesium	mg/L	10	2.5	0.05	2.5	5.8	17
Potassium	mg/L	13	7.7	0.78	13.5	15	25
Sodium	mg/L	103	58	7.2	516	870	1700
Sulfate	mg/L	44	1.8	0.4	59	73	146
Silica	mg/L	31	26	18	21	46	36
Alkalinity	mg/L	79	57	40	46	370	248
TDS	mg/L	615	262	46	1280	2780	4580
Ionic strength	Unitless	8.1E-03	5.01E-03	1.97E-03	2.03E-02	4.16E-02	8.25E-02

Source: CH2M, *Sustainable Water Recycling Initiative: Groundwater Injection Geochemical Compatibility Feasibility Evaluation*, Report No. 2, CH2M, Newport News, VA, 2016.

Notes: GW, groundwater; BAC, biological activated carbon; NF, nanofiltration; RO, reverse osmosis; UPA, Upper Potomac Aquifer; MPA, Middle Potomac Aquifer; NWIS, National Water Information System; SU, standard unit; DO, dissolved oxygen; TDS, total dissolved solids.

used in microfiltration and ultrafiltration but just larger than for reverse osmosis. In reverse osmosis, solutions of differing ion concentration are separated by a semipermeable membrane. Typically, water flows from the chamber with lesser ion concentration into the chamber with the greater ion concentration, resulting in hydrostatic or osmotic pressure. In RO, enough external pressure is applied to overcome this hydrostatic pressure, thus reversing the flow of water. This results in the water on the other side of the membrane becoming depleted in ions and demineralized.

Two methods were used to identify potential minerals in the PAS aquifers that could react with the injectate. First, thermodynamic equilibrium models were applied to identify the potential mineral suite in each aquifer. The models were run using water chemistry analyses obtained from the NWIS database for the PAS zones around the York River WWTP area. The models project potential minerals that occur in equilibrium with water chemistry. Second, mineralogical analysis of cores collected at the City of Chesapeake's aquifer storage and recovery (ASR) facility were examined to gain information on the mineralogy of the PAS. Cores were collected from the PAS zones in the City of Chesapeake. The composition of the PAS should remain fairly consistent across the HRSD service area; however, the grain size and sorting (texture) decline proceeding down the stratigraphic dip in the Virginia Coastal Plain (Teifke, 1973). Consistent with the changes in texture, the percentage of fines (texture) increases downdip. As the City of Chesapeake lies over 20 miles downdip from HRSD's York River WWTP, data from cores should portray more conservative aquifer properties than actually occur below the York River.

INJECTATE WATER CHEMISTRY

Injectate chemistry was estimated by modeling water quality entering the York River WWTP and through the advanced water treatment (AWT) processes. Effluent chemistry was estimated for the reverse osmosis, nanofiltration, and biological activated carbon advanced water treatment processes (see Table 11.1).

Reverse Osmosis

Injectate modeled for treatment by the reverse osmosis process featured a pH of approximately 7.8 after adjustment with lime ($CaOH_2$), dilute total dissolved solids (TDS) (46 mg/L), and, correspondingly, a low ionic strength (0.0015). Cations and anions in the influent were reduced following the treatment process, resulting in concentrations of cations such as potassium (<1 mg/L), magnesium (<1 mg/L), and sodium (<10 mg/L) falling below their method detection limits (MDLs), with similarly lower concentrations of anions such as phosphate (0.01 mg/L), chloride (<10 mg/L), and sulfate (<1 mg/L). At the York River WWTP, RO-treated waste displayed a calcium bicarbonate water type. Metals such as iron, manganese, and arsenic exhibited low concentrations, all near MDLs. RO had a limited effect on DO concentrations in the injectate, which ranged around 5 mg/L.

Nanofiltration

Injectate derived from the nanofiltration process exhibited a pH of approximately 7.8, moderate TDS (262 mg/L), and corresponding ionic strength (0.005). Cations and anions are reduced following the treatment process, but unlike reverse osmosis they displayed measureable concentrations of major cations such as potassium (7.7 mg/L), magnesium (2.5 mg/L), and sodium (58 mg/L) and anions including phosphate (0.03 mg/L), chloride (125 mg/L), and sulfate (1.8 mg/L). Nanofiltration injectate displayed sodium chloride chemistry. Concentrations of metals including iron, manganese, and arsenic fell below MDLs. NF injectate also displayed near-saturated concentrations of DO around 5 mg/L.

Biological Activated Carbon

Injectate originating from biological activated carbon treatment displayed a pH of approximately 7.8, slightly brackish TDS (615 mg/L), and corresponding ionic strength (0.009). Cations and anions exhibited less reduction following the treatment process compared with nanofiltration and reverse osmosis. Cation concentrations of potassium (13 mg/L), magnesium (10 mg/L), and sodium (103 mg/L) exceeded the concentrations yielded by the membrane (reverse and nanofiltration) treatments. Concentrations of anions such as phosphate (0.5 mg/L), chloride (212 mg/L), and sulfate (44 mg/L) also appeared correspondingly higher. Unlike reverse osmosis and nanofiltration, DO concentrations fell to 1.3 mg/L after treatment using biological activated carbon. Biological activated carbon injectate featured sodium chloride water chemistry. Iron and arsenic displayed concentrations higher than the membrane treatment options at 0.002 and 0.73 mg/L, respectively. Iron concentrations above 0.1 mg/L created significantly large amounts of TSS that could quickly clog an injection well.

NATIVE GROUNDWATER

Native groundwater quality from the Upper Potomac Aquifer (UPA), Middle Potomac Aquifer (MPA), and Lower Potomac Aquifer (LPA) zones was obtained from nested observation wells maintained by the USGS and NWIS, located 5 miles west of the York River WWTP. At this location, observation wells installed were screened in the UPA from 527 to 537 feet below grade (fbg), in the MPA from 820 to 830 fbg, and in the LPA from 1205 to 1215 fbg.

Upper Potomac Aquifer Zone

The Upper Potomac Aquifer (UPA) featured a slightly alkaline pH (8.2), brackish TDS (1280 mg/L), anoxic water (DO less than 1.0 mg/L), and an ionic strength of 0.02. The groundwater exhibited low amounts of nutrients, with concentrations of ammonia, nitrates, and phosphate falling below 0.01 mg/L. Chloride concentrations approached 500 mg/L, and sodium concentrations appeared similarly elevated (516 mg/L). Concentrations of other cations including calcium (4.5 mg/L), and potassium (13.5 mg/L) were comparatively low. Iron concentrations fell around its MDLs (0.01 to 0.04 mg/L), whereas manganese concentrations approached the drinking water maximum contaminant level (MCL) of 0.05 mg/L. Water from the UPA displayed a sodium chloride water chemistry.

Middle Potomac Aquifer Zone

The Middle Potomac Aquifer (MPA) also featured a slightly alkaline pH (8.0), brackish TDS (2780 mg/L), anoxic groundwater (DO less than 1.0 mg/L), and an ionic strength of 0.04. Similar to the UPA, concentrations of nutrients fell below 0.01 mg/L. Concentrations of anions—chloride (1200 mg/L), alkalinity (370 mg/L), and sulfate (73 mg/L)—exceeded concentrations encountered in the UPA. Sodium concentrations appeared similarity elevated at 870 mg/L, and concentrations of other cations, such as calcium, magnesium, and potassium, fell below 15 mg/L. Iron concentrations were near MDLs (0.01 to 0.04 mg/L), and manganese concentrations were 0.02 mg/L. Similar to the UPA, groundwater in the MPA exhibited sodium chloride water chemistry.

Lower Potomac Aquifer Zone

The Lower Potomac Aquifer (LPA) displayed a circumneutral pH (7.7) brackish TDS (4580 mg/L), and anoxic groundwater (DO less than 1.0 mg/L). Similar to the other PAS aquifers, concentrations of nutrients fell below 0.01 mg/L, although phosphate concentrations, at 0.5 mg/L, appeared notably higher than in the other PAS aquifers. Concentrations of chloride (2950 mg/L) and sulfate (146 mg/L) anions exceeded concentrations encountered in the UPA and MPA. Sodium concentrations were similarly elevated (1700 mg/L). Concentrations of other cations such as potassium (25 mg/L) and calcium (51 mg/L) increased over the concentrations encountered in the other PAS zones. Yet, iron and manganese in the LPA mimicked groundwater from the other aquifers with concentrations at MDLs and 0.02 mg/L, respectively. Similar to the other PAS aquifers, groundwater from the LPA displayed sodium chloride water chemistry.

GEOCHEMICAL ASSESSMENT OF INJECTATE AND GROUNDWATER CHEMISTRY

This section discusses chemical assessment of the injectate (that is, the effluent from the advanced water treatment processes of reverse osmosis, nanofiltration, and biological activated carbon) that potentially is to be injected as injectate into native groundwater. Obviously, as pointed out with the example of combining beakers of pure water and seawater to come up with something that is not potable, HRSD does not want a similar outcome with injection of its treated wastewater into the Potomac Aquifer's native groundwater. The goal is to determine the appropriate injectate. Keeping the desired outcome and goal in mind, the evaluation of the chemistry of the injectate water and native groundwater from the PAS revealed the following:

- Reverse osmosis and nanofiltration displayed ionic strengths differing by over one order of magnitude from the native groundwater in the PAS.
- Biological activated carbon displayed ionic strengths within the same order of magnitude as the PAS.
- Reverse osmosis exhibited a different cationic chemistry than groundwater from the PAS.

Influence of Ionic Strength

The ionic strength of RO diluted, treated water appeared lower than groundwater in the three PAS zones by at least one order of magnitude. By comparison, the ionic strength displayed by nanofiltration differed from the LPA by over one order of magnitude. The ionic strength of biological treated carbon, although lower than the PAS aquifers, fell within the same order of magnitude. The low ionic strength of reverse osmosis compared to the PAS groundwater represents a concern for injection operations, particularly for reverse osmosis' potential to disperse clay minerals. Clay dispersion is an electrokinetic process (Gray and Rex, 1966; Meade, 1964; Reed, 1972), where an electrostatic attraction between negatively charged clay particles is opposed by the tendency of ions to diffuse uniformly throughout an aqueous solution. One of the most important factors leading to the dispersion of clay minerals involves a change in the double-layer thickness of a clay particle. A double layer of ions lies adjacent to the clay mineral surface or between the mineral's structural layers because a negative charge attracts cations toward the surface. Because the fluid must maintain electrical neutrality, a more diffuse layer of anions surrounds the cations.

As in brackish water, when the concentration of ions is large, the double layer around the particle or between the clay's structure layers gets compressed to a smaller thickness. Compressing the double layer causes particles to coalesce, forming larger aggregates. This process is called *clay flocculation*. When the ionic concentration of a fluid invading the aquifer is significantly lower than the native groundwater, the diffuse double layer expands, forcing clay particles and the structural layers within clay minerals apart. The expansion prevents the clay particles from moving closer together and forming an aggregate. The tendency toward dispersion is measured in clay minerals by their zeta potential (i.e., where colloids with high zeta potential are electrically stabilized while colloids with low zeta potentials tend to coagulate or flocculate) according to the following relationship:

$$Z = (4\pi\delta q)/D$$

where
Z = Zeta potential.
δ = Thickness of the zone of influence surrounding the charged particle.
q = Charge on the clay particle before attaching cations.
D = Dielectric constant of the liquid.

For any solution and clay mineral, reducing the zeta potential involves lowering the thickness of the zone of influence. Substituting small, double- or triple-charged cations such as Ca^{2+} or Al^{3+}, respectively, in place of large singly charged and hydrated ions such as Na^{2+} lowers the zeta potential, permitting clay particles to coalesce. This behavior explains the tendency for sodium to cause clay dispersion, whereas calcium and aluminum induce its flocculation.

Aquifer sands containing more complex clay minerals, such as mixed-layer clays and smectite group clays that display small particle sizes but large surface charges, usually exhibit the greatest sensitivity to freshwater (Brown and Silvey, 1977). As little as 0.4% smectite in a sand body has been shown to reduce the aquifer's hydraulic conductivity by 55% after exposure to fresher water (Hewitt, 1963). Mixed-layer clays and smectite encountered in cores from the UPA and MPA at the City of Chesapeake exhibited only trace abundance (less than 1% to 4% of the whole rock composition of the sand).

In the 1970s, the USGS tested an aquifer storage and recovery facility in Norfolk, VA, that exhibited greater than 50% reduction in injectivity only 150 minutes after beginning injection operations (Brown and Silvey, 1977). The aquifer storage and recovery well was installed in the UPA and screened nearly 85 feet of sand in the unit. The USGS employed nuclear, electrical, and mechanical geophysical logging techniques to evaluate the origin of the injectivity losses and discriminate among the causes of physical plugging documented at other sites, such as TSS loading. Injectivity losses caused by physical plugging from TSS loading typically occur at discrete zones through the well screen. In contrast, geophysical logging of the aquatic storage and recovery test well at Norfolk showed hydraulic conductivity losses distributed evenly across the entire screen. Also, compared to physical plugging by TSS, which responds positively to mechanical and chemical rehabilitation techniques, the USGS was unable to restore even a fraction of the well's original injectivity.

The USGS used a solution of calcium chloride ($CaCl_2$ greater than 10,000 mg/L) to treat the wellbore and proximal aquifer to arrest the declining injectivity. The double-charged calcium cation forms a stronger particle and interlayer bond than the monovalent sodium cation. Using a concentrated solution ensures calcium exchanges for sodium at the maximum number of sites. After applying the treatment at Norfolk, the injectivity of the aquifer storage and recharge test well remained stable over two more test cycles before the project ended. Concentrated solutions containing the trivalent aluminum proved to be effective in stabilizing clay minerals prone to dispersion in the presence of dilute injectate (Civan, 2000). Applying a calcium or aluminum chloride treatment to the Potomac Aquifer System before initiating injection operations offers a viable alternative for stabilizing clay minerals *in situ*, precluding formation damage and injectivity loss, should regulators select reverse osmosis as the most viable method for protecting local water users. These treatments could also benefit injection operations using nanofiltration or biological activated carbon as the preferred injectate.

Cation Exchange

In addition to differing ionic strengths, reverse osmosis, as calcium carbonate water, differs from the sodium chloride chemistry encountered in the UPA, MPA, and LPA. As previously described, double-charged calcium ions should benefit the long-term stability of clay minerals where calcium exchanges for sodium. However, calcium exhibits a large ionic radius that can damage clay minerals when entering the position left by the sodium, fragmenting the edges of the minerals and mobilizing the fragments in the aquifer environment.

Iron and Manganese

None of the injectates or native groundwater from the PAS aquifers appears to exhibit problematic concentrations of iron or manganese. Iron concentrations in reverse osmosis and nanofiltration effluent typically occurred below method detection levels (MDLs). During injection operation, iron and manganese contained in the injectate or native groundwater can precipitate oxide and hydroxide minerals when exposed to DO. Formation of these minerals presents a problem if they precipitate close to the wellbore, which is a zone featuring small surface areas sensitive to physical plugging. Accordingly, the absence of iron and manganese in injectate or native groundwater benefits injection operations.

LITHOLOGY OF THE POTOMAC AQUIFER SYSTEM

The lithology and minerals comprising the PAS aquifers are described in this section, starting with the general composition across the study area and then focusing on cores collected from the UPA and MPA near the City of Chesapeake.

Lithology

As previously mentioned, the Potomac Aquifer System consists of three discrete aquifer zones (Upper Potomac Aquifer, Middle Potomac Aquifer, and Lower Potomac Aquifer) named for their position in the section. Deposited in river (fluvial) and shallow marine environments, the aquifers consist of coarse to fine sands with occasional gravel, interbedded with thin gray to pale green clays (Teifke, 1973). The aquifers are separated by clay beds of thicknesses exceeding 20 feet; however, thinner clay beds transect the sand units in the Middle Potomac Aquifer and the Lower Potomac Aquifer. Because of the abundance of clay beds, the Middle Potomac and Lower Potomac often consist of multiple, stacked units requiring repeated screen and blank combinations for supply wells installed in these aquifers.

Sands are comprised primarily of quartz (Meng and Harsh, 1988), often reaching amounts exceeding 90% by weight, forming the predominant framework mineral. Accessory minerals include orthoclase, muscovite, glauconite, and locally lignite. Trace minerals mostly occupy the interstitial spaces in the sands and include biotite, pyrite, siderite, magnetite, and clays.

CITY OF CHESAPEAKE AQUIFER STORAGE AND RECOVERY FACILITY CORE SAMPLES

In 1989, at the City of Chesapeake's aquifer storage and recovery facility, 10 core samples were collected from the UPA and MPA at depths ranging from 560 to 835 fbg. The cores were submitted to Mineralogy, Inc., a laboratory specializing in mineralogical assays, for the following analyses:

- Specific gravity
- Porosity
- Permeability
- X-ray diffraction
- Cation exchange capacity (CEC)
- Grain size distribution
- Energy dispersive chemical analysis
- Scanning electron microscopy

Potomac Aquifer System sediments found at the City of Chesapeake locations, even though they are located approximately 35 miles apart, should display characteristics similar to those underlying the York River WWTP. Because aquifer characteristics such as grain size, sorting, textural maturity, porosity, and permeability decline moving downdip, PAS sediments at the York River WWTP should display characteristics better suited to injection operations than at the City of Chesapeake.

Core samples from the UPA and MPA were composed of coarse to very coarse-grained sands, in a medium-grained matrix. Aquifer sands appeared conglomeratic and unsorted; however, as unconsolidated sands they displayed open pore spaces yielding good porosity (21 to 34%) and air permeability. Grain size diminished with depth. Samples from the deeper portions of the MPA exhibited a medium-grain size with a larger percentage of fine sands than shallower samples. Most clay minerals were found in interstitial spaces of the aquifers and showed a high degree of crystallinity, suggesting that they formed after deposition and burial (i.e., were authigenic).

Sands from the UPA and MPA consisted of 84 to 89% quartz (Table 11.2) with 8 to 12% potassium and plagioclase feldspar, thus classifying the sands as subarkosic or lithic arkosic. Trace amounts of calcite and dolomite (less than 10%) were detected in

TABLE 11.2

Results of X-Ray Diffraction Analysis (City of Chesapeake Cores)

						Minerals					
Sample	Depth (fbg)	Description	Quartz (%)	Plagioclase (%)[b]	Orthoclase (%)[b]	Calcite (%)[b]	Dolomite (%)[b]	Siderite[a] (%)[b]	Kalolinite[a] (%)[b]	Illite/Mica[a] (%)[b]	Smectite[a] (%)[b]
1	560.4	UPA	89	2	9	trc	1	N/A	1	1	trc
2	580.4	UPA	89	2	8	trc	trc	N/A	trc	1	trc
3	595.4	Intra-aquifer clay bed	69	1	9	trc	trc	19	1	1	trc
4	623.4	UPA	88	trc	7	1	1	N/A	1	1	1
5	665.4	UPA	89	1	7	trc	2	N/A	trc	trc	1
6	685.4	Confining bed	79	2	2	—	—	trc	8	5	4
7	715.4	MPA	88	7	4	trc	1	N/A	trc	1	trc
8	765.4	MPA	85	4	7	trc	trc	N/A	2	1	trc
9	820.2	MPA	86	6	6	trc	trc	—	1	1	trc
10	835.3	MPA	84	3	9	trc	trc	—	2	1	1

Source: CH2M, *Sustainable Water Recycling Initiative: Groundwater Injection Geochemical Compatibility Feasibility Evaluation*, Report No. 2, CH2M, Newport News, VA, 2016.

Notes: UPA, Upper Potomac Aquifer; MPA, Middle Potomac Aquifer; fbg, feet below grade; trc, tag–redox coulometry.

[a] Clay mineral.

[b] Percentage of whole rock composition as determined through analysis by x-ray diffraction.

every sample of the aquifer sands. Clay minerals, including kaolinite, illite/mica, and smectite, accounted for up to 4% of the same samples. The iron carbonate mineral siderite ($FeCO_3$) was encountered in a confining bed sample (595 fbg) at amounts up to 19%. Siderite was also encountered in an aquifer core (685.2 fbg) in trace amounts.

Permeabilities in air (intrinsic permeability) ranged from 1280 to 5900 millidarcies (Table 11.3). Generally, intrinsic permeability and porosity values declined with depth, so the greatest permeabilities were encountered in cores from the UPA. Intrinsic permeability displayed minimal anisotropy, with horizontal and vertical values from the same core yielding nearly equal permeabilities.

Cation exchange capacity (CEC) refers to the number of exchangeable cations per dry weight that a soil can hold, at a given pH, and are available for exchange with the soil-water solution which is influenced by the amount and type of clay and the amount of organic matter (Drever, 1982). CEC serves as a measure of soil fertility, nutrient retention capacity, and the capacity to protect groundwater from cation contamination. The CEC of minerals contained in confining beds often controls the cation chemistry in the adjacent aquifers by exchanging cations across the contact between the units. For injection purposes, knowing the CEC of the aquifer and confining bed materials can help assess how these materials will react with recharge water displaying a specific cation ionic chemistry. The CEC is expressed as milliequivalents of hydrogen per 100 grams of dry soil (meq/100 g). Table 11.4 presents the CEC values of some clay minerals.

TABLE 11.3
Porosity and Permeability to Air, City of Chesapeake Cores

Sample No.	Depth[a] (fbg)	Description	Permeability (millidarcies) Horizontal	Vertical	Porosity (%)
1	560.4	UPA	4280	3800	31.0
2	580.4	UPA	5200	4800	33.6
3	595.4	Intra-aquifer clay bed	5100	5900	34.0
4	623.4	UPA	2570	1350	28.0
5	665.4	UPA	4200	4230	31.8
6	685.4	Confining bed	1370	1335	30.7
7	715.4	MPA	3640	3600	33.0
8	765.4	MPA	2200	2175	30.3
9	820.2	MPA	2410	2410	29.1
10	835.3	MPA	1320	1280	21.5

Source: CH2M, *Sustainable Water Recycling Initiative: Groundwater Injection Geochemical Compatibility Feasibility Evaluation*, Report No. 2, CH2M, Newport News, VA, 2016.

Notes: UPA, Upper Potomac Aquifer; MPA, Middle Potomac Aquifer; fbg, feet below grade.

[a] Core lengths = 0.4 feet. Reported depths represent base of cores.

TABLE 11.4

Cation Exchange Capacities of Some Clay Minerals

Mineral	Cation Exchange Capacity (meq/100 g)	Mineral	Cation Exchange Capacity (meq/100 g)
Smectite	80–150	Kaolinite	1–10
Vermiculite	120–200	Chlorite	<10
Illite	10–40		

Source: Drever, J.M., *The Geochemistry of Natural Waters*, Prentice Hall, Upper Saddle River, NJ, 1982.

All ten samples at the Chesapeake site were analyzed for CEC. Sodium represented the most dominant exchangeable cation followed by magnesium, calcium, and potassium. The confining bed sample at 685.2 fbg displayed the most elevated CEC at 12.5 meq/100 g of core. Aquifer sand samples from the UPA and MPA exhibited CEC values for sodium ranging from 0.7 to 3.9 meq/100 g of core. Sodium, a monovalent ion in the exchange position of clays, will not benefit from injection operations.

Despite the dominance of sodium, CEC values from cores from the City of Chesapeake were low, suggesting that the clays should display minimal tendency to exchange cations. In environments showing more elevated CEC values, divalent ions such as calcium or magnesium in the injectate can exchange with sodium, temporarily disrupting the clay's atomic structure. Over the long term, replacing sodium with a divalent ion will strengthen the atomic structure of the clay mineral, eventually transitioning it to a stable smectite.

Mineralogy–Geochemical Modeling

The thermodynamic equilibrium model PHREEQC (Parkhurst, 1995) was used to gain a greater understanding of the stability of the clay minerals in the Potomac Aquifer System aquifers beneath the York River WWTP, based on the native groundwater and injectate chemistries. As previously described, the stability of clay minerals can control the success of injection operations in sandy aquifers such as the Potomac Aquifer System. Thermodynamic equilibrium models consist of computer programs using a relatively sophisticated set of equations (Davies, 1962; Debye and Huckel, 1923; Truesdell and Jones, 1974) to stimulate the chemical equilibrium of a solution under natural or laboratory conditions and to simulate the effects of chemical reactions. These models perform the following types of calculations:

- Correct all equilibrium constants to the temperature of the specific sample.
- Calculate speciation, the distribution of chemical species by element, by solving a matrix of equations.
- Calculate activity coefficients of each chemical species.

- Calculate the state of saturation for potential mineral species that occur in equilibrium with the samples water chemistry. These calculations identify potential mineral species and determine whether they will dissolve or precipitate under the changing conditions consistent with aquifer storage and recovery operations.
- Perform calculations related to oxidation–reduction processes.

Thermodynamic equilibrium computer models are a powerful tool for predicting chemical behavior in a natural system. Manual manipulation of the same equations performed by these programs are time consuming and prone to calculation errors.

STABILITY OF CLAY MINERALS

PHREEQC was employed for evaluating the stability of clays contained in the $CaO-Al_2O_3^-SiO_2^-H_2O$; $NaO-Al_2O_3SiO_2^-H^2O$; and $K_2O-Al_2O_3^-SiO_2^-H_2O$ mineral systems. Minerals contained in these systems represent clays and their weathering products (gibbsite, kaolinite) commonly found in sediments of the PAS. The objectives of these simulations included determining how native groundwater chemistries fall into the stability fields of clay minerals and identifying potential instabilities. Along with ambient clay stabilities, PHREEQC simulates how clay can evolve during the exchange of cations; however, the program does not address instability arising from introducing injectate of a differing ionic strength. The chemistries of groundwater from the UPA, MPA, and LPA, as well as potential injectate waters from the York River WWTP, were plotted on stability diagrams for three systems describing common clay minerals: $CaO-Al_2O_3^-SiO_2^-H_2O$; $NaO-Al_2O_3SiO_2^-H^2O$; and $K_2O-Al_2O_3^-SiO_2^-H_2O$. Common clay minerals including smectite, beidellite, montmorillonite, illite, and the gibbsite (weathering product) were oversaturated in recharge water samples, which suggests a tendency to precipitate over time. When waters of incompatible ionic strength or differing cations are injected, damage to clay minerals can arise; however, this was not a concern during injection operations in the Potomac Aquifer System because precipitation of clay minerals represented a relatively minimal matter regarding permeability loss. Moreover, the precipitation of clay minerals requires significant amounts of geologic time rather than the relatively short service life of an injection facility.

SIMULATED INJECTATE AND WATER INTERACTIONS

Along with characterizing clay mineral stability during injection operations, PHREEQC was also employed to assess reactions originating from mixing the three injectate types—reverse osmosis, nanofiltration, and biological activated carbon—with native groundwater from each of the PAS aquifers, as well as reactions between the injectate and reactive minerals in the PAS aquifers. Mixing reactions occur when injectate interfaces with native groundwater. As injection operations proceed, injectate drives the mixing interface further into the PAS aquifers. Surface areas in an aquifer undergoing injection are small around the injection wellbore, but increase geometrically with distance away from well. The larger surface areas away from the injection wells help buffer reactions that cause permeability losses. Thus, these reactions diminish in importance as injection operations progress. Reactions between injectate and reactive minerals raise the following concerns:

- Permeability losses with the precipitation of iron or manganese oxide minerals
- Leaching of environmentally problematic constituents such as iron, manganese, and arsenic, along with other metals, depending on the ambient mineralogy

MIXING

Mixing could arise during injection operations:

- Mixing between the injectate and native groundwater
- Mixing between groundwater from the UPA, MPA, and LPA in the injection wellbore

Mixing Injectate and Native Groundwater

As mentioned, mixing reactions prove most troublesome around the injection wellbore. One common reaction involves the precipitation of oxide minerals when injectate containing dissolved oxygen contacts dissolved iron or manganese entrained in the injector or native groundwater. Although each injectate from the advanced water treatment processes contained measurable concentrations of dissolved oxygen, dissolved iron and manganese concentrations (with the exception of biological activated carbon) were absent in the injectate and native groundwater. Despite the absences of iron and manages other minerals can also precipitate during mixing.

PHREEQC modeling was employed to simulate mixing between the injectate and native groundwater chemistries at a 1:1 ratio in order to evaluate the mixing between the differing water types. The modeling was also used to evaluate reactions between the native groundwater chemistries in the UPA, MPA, and LPA when mixed in an injection wellbore, simulating an injection well screening the three Potomac Aquifer System aquifers.

During the mixing simulations, important reactions were tracked including the potential precipitation of metal oxide, hydroxide, sulfate, and carbonate minerals along with dissolution of silicates, including clays. Because of the similar bulk chemistry of the three injectates and native groundwater from the UPA, MPA, and LPA, potential mineral suites identified by PHREEQC in the mixed water and their saturation indices were repeated across the nine mixtures. The *saturation index* (SI) of a mineral determines whether the mineral occurs in equilibrium (Langmuir, 1997) with mixed water chemistry (SI = 0.0); is undersaturated (SI < 0.0) and, if present, should dissolve; or is supersaturated (SI > 0.0) and should precipitate. Estimation of saturation indices are usually not exact, often varying over ±0.3 units, depending on the composition of the solution. The SI provides a guideline on how minerals will behave in a water sample.

Of the common minerals and their weathering products identified in nine combinations of mixed water, quartz was the most common in sands of the Potomac Aquifer System, returning a slightly oversaturated SI = 0.66 to 0.87 for all simulations. Oversaturation of quartz and near-equilibrium saturation indices for less crystalline forms of silica such as chalcedony (SI = −0.06 to 0.29) and cristobalite (SI = 0.02 to 0.37) indicate that feldspars are dissolving, releasing silica. Gibbsite $(Al(OH)_3)$ appeared oversaturated in the mixed water consistent with the dissolution of feldspar and precipitation of residual byproducts in a weathered environment.

Other minerals potentially reacting in the mixed water included the carbonates, including calcite, aragonite, and siderite. Calcite $(CaCo_3)$ appeared uniformly undersaturated (SI = −1.42 to −0.24) in all the mixed waters suggesting it will not precipitate. Calcite, in an undersaturated state, benefits injection operations by not precipitating, blocking pore spaces, and reducing the permeability of the aquifer. Aragonite, an isomorph of calcite, shows similar indices, ranging from −1.56 to −0.39. The iron carbonate mineral siderite $(FeCO_3)$ displayed SI values similar to those for calcite and aragonite, with strongly undersaturated indices ranging from −8.09 to −11.73.

Other important minerals included gypsum and jarosite, a weathering product of iron-bearing mineral sand sulfides. Similar to the carbonates, with one exception, gypsum and jarosite exhibited undersaturated indices ranging from −2.21 to −3.37 and −5.62 to 0.33, respectively. As a single exception, biological activated carbon mixing with groundwater from the LPA resulted in near-equilibrium SI for jarosite.

Mixing in the Injection Wellbore

With an open conduit extending between the three PAS aquifers at the York River WWTP, groundwater will mix in the injection wellbore before the start of injection operations. Once injection operations begin, injectate will displace the groundwater away from the wellbore so mixing groundwater will no longer present an issue. An examination of static water level elevations at the nested National Water Information System wells near the York River WWTP suggests a vertically downward hydraulic gradient of 0.085 feet/foot (ft/ft) occurring between the UPA and MPA, whereas an upward gradient of 0.031 ft/ft appears between the LPA and MPA. The differing gradient directions impose converging flow in the wellbore. Accordingly, groundwater will flow in through intervals screening the UPA and LPA and out through the screen against the MPA, promoting the mixing of the three water types in the wellbore rather than stratification.

In the geochemical simulation, groundwater from the UPA, MPA, and LPA was mixed at even proportions between the three units. To maintain a conservative approach to the simulations, ferrous iron (Fe(II)) concentrations at the UPA, MPA, and LPA were assumed to occur at 0.1 mg/L, the MDL for iron. Concentrations in the mixed water oxidized from Fe(II) to ferric iron (Fe(III)) remained at a concentration of 0.1 mg/L.

No deleterious reactions associated with mixing in the aquifer storage and recovery wellbore were detected through modeling. The pH of the mixed water declined slightly to 7.83. The SI for calcite and siderite appeared near equilibrium at −0.08 and −0.04, respectively. This suggests that these minerals should neither dissolve nor precipitate in the mixed water. The mixed water displayed sodium chloride chemistry similar to groundwater from the three PAS aquifers. Aragonite, gypsum, and jarosite displayed unsaturated SIs, indicating that the mineral should not precipitate in the mixed water. The weathering product of clay minerals, gibbsite remained saturated roughly the same as the individual groundwater chemistries.

INJECTATE AND AQUIFER MINERAL REACTIONS

In addition to reactions between dilute injectate and clay minerals, dissolved oxygen in the injectate can react with reduced metal-bearing minerals, releasing metals and other constituents that can compromise the quality of water disposed in the PAS aquifers. These reactions should not affect injection operations but can result in environmental concerns prompting the attention of regulators. Analysis of cores from the aquifer storage and recovery facility at the City of Chesapeake encountered microcrystalline siderite in the interstices of aquifer sands and as larger crystalline forms in adjoining confining beds. Pyrite is another reduced, metal-bearing mineral common to the Virginia Coastal Plain aquifers, including the PAS (McFarland and Bruce, 2006; Meng and Harsh, 1988). Although not detected in core samples from the City of Chesapeake, pyrite and siderite typically occur together in sediments subject to flooding by marine and freshwater systems (Postma, 1982), typical of the near coastal environment in which the formation bearing the PAS aquifers was deposited. Accordingly, pyrite was considered in the geochemical modeling evaluation. The primary metal in siderite and pyrite is Fe(II), but both minerals can also contain cadmium and manganese. Additionally, pyrite occasionally contains varying amounts of arsenic (Evangelou, 1995). Dissolved oxygen in the three injectates should range between 1.3 and 5 mg/L, providing a source for oxidizing reactions.

Siderite Dissolution

Dissolved oxygen reacts with siderite to release Fe(II) and CO_3^{2-} (carbon trioxide). Upon encountering dissolved oxygen, Fe(II) oxidizes to Fe(III), which acts as a strong oxidant, continuing the dissolution of siderite. At equilibrium, a small amount of siderite can release large amounts of Fe(II) into the surrounding pore water. PHREEQC was employed to simulate potential reactions between the injectates and siderite. In this reaction, 1 mole of siderite was reacted with reverse osmosis, nanofiltration, and biological activated carbon, each containing dissolved oxygen concentrations ranging from 1 to 7 mg/L in 1.0-mg/L increments. Resulting Fe(II) at dissolved oxygen concentrations of 7 mg/L ranged from 90 to 130 mg/L for reverse osmosis and biological activated carbon, respectively, with nanofiltration exhibiting concentrations between reverse osmosis and biological activated carbon. Even acting with only 1 mg/L dissolved oxygen, siderite produced Fe(II) concentrations ranging from 50 to 90 mg/L for reverse osmosis and biological activated carbon, respectively. Bicarbonate concentrations increased from 95 mg/L for reverse osmosis at 1 mg/L dissolved oxygen to over 200 mg/L at 7 mg/L dissolved oxygen for biological activated carbon. A portion of the total bicarbonate was comprised of carbonic acid, lowering the simulated pH of the pore water from 7.8 to 6.91 for reverse osmosis at 7 mg/L dissolved oxygen. The pH of reverse osmosis, nanofiltration, and biological activated carbon injectate dropped below 7.15 when reacting siderite with a dissolved oxygen of only 1 mg/L; reverse osmosis, the most dilute injectate, exhibited the lowest capacity to buffer the pH during the reaction between siderite and dissolved oxygen.

Pyrite Oxidation

Although pyrite was not encountered in the cores collected at the City of Chesapeake, its appearance elsewhere in the PAS and its deleterious reactions when encountering dissolved oxygen make evaluating the mineral an important part of any injection feasibility study. Reacting dissolved oxygen with pyrite releases Fe(II) and bisulfide ions (Evangelou, 1995). Upon encountering dissolved oxygen, Fe(II) oxidizes to Fe(III), which also acts as a strong oxidant, continuing the oxidation of pyrite. The bisulfide ion (S_2^{2-}) further reacts with dissolved oxygen to form sulfuric acid (H_2SO_4), lowering the pH of the surrounding pore water.

PHREEQC was used to simulate an operational injection scenario to predict the chemistry of effluent containing varying amounts of dissolved oxygen, exposed to pyrite in the PAS aquifer matrices. Pyrite was equilibrated with reverse osmosis, nanofiltration, and biological activated carbon containing concentrations ranging from 1.0 to 7.0 mg/L in 1.0-mg/L increments. Similar to the siderite simulations, the injectate chemistries were equilibrated with 1 mole of pyrite. Thus, simulations provide conservative results that overestimated the concentrations of iron, sulfate, and arsenic in the recovered water. Where present in the Atlantic Coastal Plain aquifers, pyrite comprises less than 1% of the whole rock composition, or 0.05 to 0.1 moles. Modeling results showed that Fe(II) concentrations increased from 3 to over 10 mg/L, while sulfate increased at twice this rate. Sulfate concentrations simulated with biological activated carbon were elevated above the other effluent types by its initial concentration of 44 mg/L. Similar to the simulations with siderite, reverse osmosis exhibited the greatest decline in pH after reacting with pyrite, with the pH declining from 7.8 to less than 6.8.

The iron concentrations simulated from the modeling are considered conservative, as a large portion of Fe(II) released by the oxidation of pyrite will precipitate as hydrous ferric oxide (HFO). The HFO typically precipitates on the pyrite mineral surface, progressively reducing its reactivity (i.e., passivation). Moreover, these surfaces can adsorb Fe(II) migrating in the aquifer environment. In the absence of pyrite, these simulations illustrate potential groundwater quality problems that can emerge from effluent containing dissolved oxygen in the presence of this reactive mineral.

Arsenic

The release of arsenic from pyrite was simulated with PHREEQC. As a substitution for sulfur, arsenic concentrations were estimated at 1% by weight of the mass of pyrite. Similar to previous simulations, 1 mole of pyrite was equilibrated with reverse osmosis, nanofiltration, and biological activated carbon containing dissolved oxygen concentrations ranging from 1 to 7 mg/L. Applying this approach, arsenic concentrations increased from 31 to nearly 58 μg/L in reactions with reverse osmosis, and from 58 to 85 μg/L during reactions with biological activated carbon. Similar to relationships between pyrite and iron, arsenic concentrations simulated with PHREEQC represent conservative conditions. HFO surfaces in aquifer settings display a strong affinity for adsorbing the oxyanions of arsenic and lowering its concentrations in groundwater. Moreover, continuing oxidation of pyrite precipitates HFO on the mineral surface, passivating the mineral and diminishing the concentration of constituent released during oxidation reactions.

Mitigating Pyrite Oxidation

At aquifer storage and recovery facilities recharging beneath the Atlantic Coastal Plain, siderite and pyrite dissolution is addressed by increasing the pH of the injectate by adding potassium and (caustic) sodium hydroxide. Increasing the injectate pH raises it above the solubility limit of iron, buffering the dissolution of iron-bearing minerals. Hydroxyl ions in sodium and potassium hydroxide will react with Fe(II) released from siderite or pyrite, oxidizing Fe(II) and Fe(III). It precipitates HFO on the surface of these minerals, which then passivates the minerals to future reactions in the aquifer environment. In addition to isolating the minerals, HFO surfaces display excellent adsorption properties, adsorbing metals migrating in the aquifer environment including arsenic and Fe(II). Iron adsorbs as a surface precipitate on HFO, while these surfaces exhibit an affinity for adsorbing arsenic at pH values encountered in groundwater environments (Dzomback and Morel, 1990).

PHREEQC was employed to simulate adjusting the pH of reverse osmosis, nanofiltration, and biological activated carbon injectate containing dissolved oxygen from 7.8 to 8.5 in the presence of pyrite. Consistent with other simulations, dissolved oxygen was varied between 1 and 7 mg/L. Fe(II) concentrations approaching 10 mg/L during reactions with only dissolved oxygen fell to less than 1.0×10^{-7} mg/L in simulations with injectate pH adjusted to 8.5. Fe(II) was nearly completely oxidized to Fe(III), which precipitated as $Fe(OH)_3$. The modeling results illustrate how well adjusting the pH of the injectate can control Fe(II) concentrations. As equilibrium simulations, the modeling did not account for the reactions passivating siderite and pyrite over time which also reduce Fe(II) concentrations in groundwater.

THE BOTTOM LINE

Modeling and geochemical evaluations of injectate, native groundwater, mixing, and the reactions between injectate and aquifer minerals in the Potomac Aquifer System beneath HRSD's York River Treatment Plant have generated several conclusions. The bottom line conclusions that can be drawn from the CH2M (2016b) report are as follows:

- The chemistry of reverse osmosis, a potential injectate, differed significantly from the chemistry of native groundwater exhibited by the UPA, MPA, and LPA.
  - Reverse osmosis displayed a dilute ionic strength that differed by over one order of magnitude from the chemistry encountered in the UPA and MPA, but approached two orders of magnitude when compared against the LPA.
  - Reverse osmosis displayed a calcium bicarbonate water chemistry, but groundwater from the three PAS aquifers uniformly exhibited sodium chloride chemistry.

- The low ionic strength of reverse osmosis compared to groundwater from the PAS is a concern for injection operations, particularly for its potential to disperse clay minerals. Once dispersed, clay particles migrate through connected pores in the aquifer until accumulating and blocking narrowed pores, thus reducing aquifer permeability and ultimately injection well capacity.
- A USGS-sponsored aquifer storage and recovery facility tested at Norfolk in the 1970s used an injectate similar in ionic strength and cation chemistry to reverse osmosis. The injection capacity of the aquifer storage and recovery well declined by 50% after only 4 hours of operation and dropped 75% over several days. The USGS was not able to restore the capacity of the aquifer storage and recovery well, despite applying several, for the time, state-of-the-art rehabilitation techniques.
- Cores collected at the City of Chesapeake's aquifer storage and recovery facility exhibited trace concentrations of smectitic clays dispersed throughout the interstices of every sample collected in aquifer sands. Smectites possess a complex lattice expanding structure vulnerable to dispersion or swelling when exposed to dilute water.
- When considering the varied cation chemistry between reverse osmosis and groundwater in the PAS aquifers, the doubly charged calcium ion should benefit the long-term stability of clay minerals where calcium exchanges for sodium. However, calcium, when hydrated, exhibits a large ionic radius that can damage clay minerals upon entering the position vacated left by sodium, fragmenting the edges of the mineral and mobilizing the fragments in the aquifer environment.
- Conversely, cores from the City of Chesapeake project, analyzed for cation exchange capacity, displayed little tendency to exchange, which is a benefit of injection operations.
- Geochemical modeling of potential clay minerals in the PAS aquifers produced a similar result, with the stability of clay minerals improving over time during injection operations.
- Given the concerns with reverse osmosis as a source of injectate, and the problems experienced at the City of Norfolk's aquifer storage and recovery facility, HRSD should consider eliminating reverse osmosis from further evaluation on the SWIFT project.
- The ionic strength of nanofiltration and biological activated carbon injectate fell within one order of magnitude of the groundwater chemistries originating from the PAS aquifers. Nanofiltration and biological activated carbon as a source of injectate are of significantly less concern for dispersing water-sensitive clays in the PAS aquifers during injection operations.
- Applying a calcium or aluminum chloride treatment to the PAS aquifers before initiating injection operations offers a viable alternative for stabilizing clay minerals *in situ*, precluding formation damage and injectivity losses, should regulators select reverse osmosis as the most viable method for protecting local water users. These treatments could also benefit injection operations using nanofiltration or biological activated carbon as the preferred injectate.

- Biological activated carbon exhibits 0.7 mg/L iron, which presents a considerable source of TSS in the injectate and a strong physical plugging agent in injection wells. HRSD will need to remove iron from biological activated carbon effluent before employing it as an injectate.
- Geochemical modeling runs that simulated mixing between reverse osmosis, nanofiltration, and biological activated carbon and groundwater from the PAS aquifers showed no evidence of deleterious reactions that might clog the injection wells or surrounding aquifer such as precipitating oxide, hydroxide, carbonate, or sulfate minerals, or the dissolution of silicate minerals.
- Geochemical modeling that simulated mixing between the three groundwaters in an injection well screening the UPA, MPA, and LPA displayed no evidence of deleterious reactions that might clog the injection well or surrounding aquifers.
- Cores collected at the City of Chesapeake contained the iron carbonate mineral siderite at amounts ranging from 0.5 to 19% of the whole rock composition. In reactions with injectate containing dissolved oxygen, siderite released up to 130 mg/L Fe(II).
- Although not a concern for injection operations, dissolving siderite can compromise the quality of the disposed water, prompting attention from state and federal regulators.
- Adjusting the pH of the injectate water with a source of hydroxyl such as sodium or potassium hydroxide can help lower Fe(II) concentrations. During model runs simulating reactions between injectate containing varying amounts of dissolved oxygen and pH of 8.5 with pyrite, Fe(II) concentrations fell below 10E-7 mg/L. Fe(II) oxidized to Fe(III), which precipitated as Fe(III)-oxide and Fe(III)-hydroxide minerals.

REFERENCES AND RECOMMENDED READING

Anderman, E.R. and Hill, M.C. (2000). *MODFLOW-2000, the U.S. Geological Survey Modular Ground-Water Model—Documentation of the Hydrogeologic-Unit Flow (HUF) Package*, Open-File Report 2000-342. Reston, VA: U.S. Geological Survey (https://pubs.er.usgs.gov/publication/ofr00342).

Bahremand, A. and De Smedt, F. (2008). Distributed hydrological modeling and sensitivity analysis in Torysa watershed, Slovakia. *Water Resources Management*, 22(3): 293–408.

Bair, E.S., Springer, A.E., and Roadcap, G.S. (1992). *CAPZONE*. Columbus: Ohio State University.

Brown, D.L. and Silvey, W.D. (1977). *Artificial Recharge to a Freshwater-Sensitive Brackish-Water Sand Aquifer*, U.S. Geological Survey Professional Paper 939. Reston, VA: U.S. Geological Survey.

CH2M. (2016a). *Sustainable Water Recycling Initiative: Groundwater Injection Hydraulic Feasibility Evaluation*, Report No. 1. Newport News, VA: CH2M.

CH2M. (2016b). *Sustainable Water Recycling Initiative: Groundwater Injection Geochemical Compatibility Feasibility Evaluation*, Report No. 2. Newport News, VA: CH2M.

Civan, F. (2000). *Reservoir Formation Damage: Fundamentals, Modeling, Assessment, and Mitigation*. Houston, TX: Gulf Publishing.

Davies, C.W. (1962). *Ion Association*. Washington, DC: Butterworths.

Debye, P. and Huckel, E. (1923). On the theory of electrolytes. I. Freezing point depression and related phenomena. *Physikalische Zeitschrift*, 24(9): 185–206.

Drever, J.M. (1982). *The Geochemistry of Natural Waters*. Upper Saddle River, NJ: Prentice Hall.

Drever, J.M. (1988) *The Geochemistry of Natural Waters*, 2nd ed. Upper Saddle River, NJ: Prentice Hall.

Dzombak, D.D. and Morel, F.M.M. (1990). *Surface Complex Modeling*. New York: John Wiley & Sons.

Evangelou, V.P. (1995). *Pyrite Oxidation and Its Control*. New York: CRC Press.

Eykhoff, P. (1974). *System Identification Parameter and State Estimation*. New York: John Wiley & Sons.

Fears, D. (2016). Hampton Roads' solution to stop the land from sinking? Wastewater. *The Washington Post*, October 20.

Gray, D.H. and Rex, R.W. (1966). *Formation Damage in Sandstones Caused by Clay Dispersion and Migration*. La Habra, CA: Chevron Research Company.

Hamilton, P.A. and Larson, J.D. (1988). *Hydrogeology and Analysis of the Ground-Water-Flow System in the Coastal Plain of Southeastern Virginia*, Water Resources Investigations Report 87-4240. Reston, VA: U.S. Geological Survey.

Hantash, J.E. and Jacob, C.E. (1955). Nonsteady radial flow in an infinite leaky aquifer. *Transactions of the American Geophysical Union*, 36(1): 95–100.

Hewitt, E.J. (1963). Mineral nutrition of plants in culture media. In: *Plant Physiology*, Vol. III (Steward, F.C., Ed.), pp. 97–133. New York: Academic Press.

Heywood, C.E. and Pope, J.P. (2009). *Simulation of Groundwater Flow in the Coastal Plain Aquifer System of Virginia*, Scientific Investigations Report 2009-5039. Reston, VA: U.S. Geological Survey(http://pubs.usgs.gov/sir/2009/5039/).

Hill, M. and Tiedeman, C. (2007). *Effective Groundwater Model Calibration, with Analysis of Data, Sensitivities, Prediction, and Uncertainty*. New York: John Wiley & Sons.

Hill, M., Kavetski, D., Clark, M., Ye, M., Arabic, M., Lu, D., Foglia, L., and Mehl, S. (2015). Practical use of computationally frugal model analysis methods. *Groundwater*, 54(2): 159–170.

Laczniak, R.J. and Meng III, A.A. (1988). *Ground-Water Resources of the York–James Peninsula of Virginia*. Water Resources Investigations Report 88-4059. Reston, VA: U.S. Geological Survey.

Langevin, C.D., Thorne, Jr., D.T., Dausman, A.M. et al. (2008). *SEWAT Version 4: A Computer Program for Simulation of Multi-Species Solute and Heat Transport*. Reston, VA: U.S. Geological Survey (https://pubs.usgs.gov/tm/tm6a22/).

Langmuir, D. (1997). *Aqueous Environmental Geochemistry*. Upper Saddle River, NJ: Prentice Hall.

Leamer, E. (1978). *Specification Searches: Ad Hoc Inferences with Nonexperimental Data*. New York: John Wiley & Sons.

McFarland, E.R. (2013). *Sediment Distribution and Hydrologic Conditions of the Potomac Aquifer in Virginia and Parts of Maryland and North Carolina*. Scientific Investigations Report 2013-5116. Reston, VA: U.S. Geological Survey.

McFarland, E.R. and Bruce, T.S. (2006). *The Virginian Coastal Plain Hydrogeologic Framework*, U.S. Geological Survey Professional Paper 1731. Reston, VA: U.S. Geologic Survey.

McGill, K. and Lucas, M.C. (2009). Mitigating Specific Capacity Losses in Aquifer Storage and Recovery Wells in the New Jersey Coastal Plain, paper presented at New Jersey American Water Works Association Annual Conference, Atlantic City, NJ, March 31.

Meade, R.H. (1964). *Removal of Water and Rearrangement of Particles During the Compaction of Clayey Sediments*, U.S. Geological Survey Professional Paper 497-B. Reston, VA: U.S. Geological Survey.

Meng, A.A. and Harsh, J.F. (1988). *Hydrogeologic Framework of the Virginia Coastal Plain*. U.S. Geological Survey Professional Paper 1404-C. Reston, VA: U.S. Geological Survey.

Pannell, D.J. (1997). Sensitivity analysis of normative economic models: theoretical framework and practical strategies. *Agricultural Economics*, 16: 139–152.

Parkhurst, D.D. (1995). *User's Guide to PHREEQC—A Computer Program for Speciation, Reaction-Path, Advective-Transport, and Inverse Geochemical Calculations*, Water-Resources Investigations Report 95-422. Reston, VA: U.S. Geological Survey.

Postma, D. (1982). Pyrite and siderite formation in brackish and freshwater swamp sediments. *American Journal of Science*, 282: 1154–1183.

Pyne, D.G. (1995). *Groundwater Recharge and Wells*. Ann Arbor, MI: Lewis Publishers.

Pyne, D.G. (2005). *Aquifer Storage and Recovery: A Guide to Groundwater Recharge through Wells*. Gainesville, FL: ASR Press.

Reed, M.G. (1972). Stabilization of formation clays with hydroxy-aluminum solutions. *Journal of Petroleum Technology*, 24(7): 860–864 (https://www.onepetro.org/journal-paper/SPE-3694-PA).

Smith, B.S. (1999). *The Potential for Saltwater Intrusion in the Potomac Aquifers of the York–James Peninsula*, Water-Resources Investigations Report 98-4187. Reston, VA: U.S. Geological Survey.

Stuyfzand, P.J. (1993). Hydrochemistry and Hydrology of the Coastal Dune Area of the Western Netherlands, PhD thesis, Vrije Universiteit, Amsterdam, Netherlands.

Teifke, R.H. (1973). Stratigraphic units of the Lower Cretaceous through Miocene series. In: *Geologic Studies, Coastal Plain of Virginia*, Bulletin 83, pp. 1–78. Charlottesville: Virginia Division of Mineral Resources.

Theis, C.V. (1935). The relation between the lowering of the piezometric surface and the rate and duration of discharge of a well using ground-water storage. *Eos*, 16(2): 519–524.

Truesdell, A.H. and Jones, B.F. (1974). WATEQ, a computer program for calculating chemical equilibria of natural waters. *U.S. Geological Survey Journal of Research*, 2: 233–274.

USEPA. (2016). *Aquifer Recharge and Aquifer Storage and Recovery*. Washington, DC: U.S. Environmental Protection Agency (https://www.epa.gov/uic/aquifer-recharge-and-aquifer-storage-and-recovery).

Virginia DEQ. (2006a). *Status of Virginia's Water Resources*. Richmond: Virginia Department of Environmental Quality, Office of Water Supply.

Virginia DEQ. (2006b). *Virginia Coastal Plain Model 2005 Withdrawals Simulations*. Richmond: Virginia Department of Environmental Quality, Office of Water Supply.

Warner, D.L. and Lehr, J. (1981). *Subsurface Wastewater Injection: The Technology of Injecting Wastewater into Deep Wells for Disposal*. Berkeley, CA: Premier Press.

12 Feasibility of Advanced Water Purification Processes

It is common practice when treating wastewater to churn out water that is cleaner than water found in the local waterways that it is ultimately outfalled into. It is ultimately sent on to the ocean with no downstream use—in other words, "one and done" usage. Why? Why waste such a valuable resource? Why not recycle and reuse it?

INTRODUCTION

Hampton Roads Sanitation District (HRSD) is faced with a variety of future challenges related to the treatment and disposition of wastewater in its region of responsibility; this region includes much of the southern Chesapeake Bay, with its many major tributaries and surrounding communities. HRSD envisions that it can protect and enhance the region's groundwater supplies by reusing highly purified wastewater through advanced treatment and subsequent injection into the region's groundwater aquifers. Those of us who understand the natural water cycle, the urban water cycle, the proper use of various advanced wastewater treatment processes to purify the water, and HRSD's commitment to providing absolute excellence for its ratepayers and all who live in the Hampton Roads region, as well as to restoring and sustaining the Chesapeake Bay—we understand that HRSD's vision not only has merit but is also necessary. It is necessary because the groundwater supply within the Potomac Aquifer is dwindling and is in danger of contamination from saltwater intrusion. Treated wastewater from HRSD's wastewater treatment plants outfalls nutrients into the Bay which contribute to dead zones and other environmental issues. Finally, HRSD understands that withdrawing native groundwater from the underlying aquifers contributes to land subsidence in the region. Land subsidence plus global sea-level rise contribute to relative sea-level rise and if not abated will soon (within less than 150 years) result in the inundation of many Hampton Roads major cities and other low-lying areas in the region.

The previous chapter pointed out that HRSD and its contractor along with the U.S. Geological Survey (USGS) are working in unison to implement steps and modeling to ensure the compatibility of treated injectate with native groundwater. It is also important to make sure that the chemical match between injectate and native groundwater results in water that is safe for consumption. This is where advanced water treatment (AWT) comes into play. Wastewater treated only to conventional standards is probably safe enough for pipe-to-pipe connection but suffers from the public perception of the "yuck factor." It should be pointed out, though, that when HRSD's plans were published in the surrounding Hampton Roads area the overwhelming majority of the populace voiced no objection to HRSD's plans. However, people with wells drawing water from the Potomac Aquifer were worried that HRSD's Sustainable Water Initiative for Tomorrow (SWIFT) project might contaminate their water source with toilet water. This is exactly what HRSD is working hard to prevent by implementing advanced water treatment, which is in addition to normal wastewater treatment and filtration. This chapter describes the three treatment processes of reverse osmosis, nanofiltration, and biological activated carbon, as well as the pilot studies used to determine which process or processes would be best suited to facilitate HRSD's goal of producing and injecting the safest water possible.

BY THE BOOK ONLY, PLEASE!

The goal of most public service entities is to perform their functions by the book. The book in most cases is the written volume that contains the applicable regulations—the so-called laws of the land—that apply to their activities. For operations that can directly or indirectly affect the environment, the applicable regulations are generally federally based and enforced (by the U.S. Environmental Protection Agency, for example). However, it is interesting to note that injection of reclaimed water into an aquifer that is used as a potable water supply is referred to as indirect potable reuse (IPR), and regulations have not been developed by the U.S. Environmental Protection Agency (EPA) for potable reuse projects; therefore, states in which IPR is being practiced or is being actively considered have developed state-specific potable reuse regulations. With regard to indirect potable water reuse compliance, it is the state regulator knocking at the door, not the Feds.

THOSE PLAYING BY THE BOOK FOR INDIRECT POTABLE REUSE*

California and Florida have developed regulations governing the practice of indirect potable reuse (IPR). Other states allow IPR, but have established project-specific requirements on a case-by-case basis (e.g., Virginia, Texas). Because Virginia has not developed IPR regulations but does allow IPR, the state will likely look to successful full-scale IPR projects within the state (e.g., Upper Occoquan Service Authority, Loudoun Water) and other states' IPR regulations for guidance in regulating HJRSD's proposed direct injection IPR project. Table 12.1 presents the treated water quality requirements for other IPR projects and associated regulations that Virginia could reference with respect to an HRSD IPR project:

- *Total organic carbon (TOC)*—Total organic carbon, the amount of carbon found in an organic compound, is often used as a nonspecific indicator of water quality. Applying California's strict TOC limit of 0.5 mg/L would most likely require the use of reverse osmosis, which could increase

---

* The following material is adapted from CH2M, *Sustainable Water Recycling Initiative: Advanced Water Purification Process Feasibility Evaluation*, Report No. 3, CH2M, Newport News, VA, 2016.

TABLE 12.1
Example Treated Water Quality Requirements for Indirect Potable Reuse

Parameter	Virginia's Occoquan and Dulles Area Watershed Policies (Surface Water Augmentation)	TCEQ Policy for El Paso, TX (Direct Injection)	Florida IPR Regulations for Direct Injection	California's IPR Regulations for Direct Injection	USEPA's IPR Guidelines for Direct Injection
Relevant IPR projects	Upper Occoquan Service Authority; Centreville, VA Broad Run Water Reclamation Facility (WRF); Loudoun County, VA	Hueco Bolson Recharge Project; El Paso, TX	N/A	West Basin Water Recycling Plant; Los Angeles, CA Los Alamitos Seawater Intrusion Barrier; Long Beach, CA	N/A
TOC	COD < 10 mg/L (~3 mg/L TOC)	None	<3 mg/L; TOC < 0.2 mg/L	0.5 mg/L	≤2 mg/L (of wastewater origin)
Pathogens	Multiple barriers required; $E.\ coli$ < 2 cfu/100 mL	None, but multiple barriers required	Multiple barriers required; total coliform < 4 cfu/100 mL	LRV from raw wastewater to finished water: 12 log for enteric viruses; 10 log for $Giardia$ and $Cryptosporidium$	Multiple barriers required; total coliform below detection limit
Nitrogen	TKN < 1 mg/L; TN < 4 mg/L (Broad Run WRF only)	NO_x-N < 10 mg/L	TN < 10 mg/L	TN < 10 mg/L	None
TDS	None	<1000 mg/L	None	Reverse osmosis treatment required	None
Misc.	TSS < 1 mg/L Turbidity < 0.5 NTU TP < 0.1 mg/L	Turbidity < 1 NTU	Turbidity < 2–2.5 NTU	Reverse osmosis and advanced oxidation process treatment required for CECs	Turbidity < 2 NTU

Source: CH2M, *Sustainable Water Recycling Initiative: Advanced Water Purification Process Feasibility Evaluation*, Report No. 3, CH2M, Newport News, VA, 2016.

Note: Not all parameters are listed; for example, other requirements such as compliance with all drinking water maximum contaminant levels (MCLs), travel time, disinfection residual, and such are required in some states and locations. N/A, not applicable; TOC, total organic carbon; COD, chemical oxygen demand; TOC, total organic carbon; TKN, total Kjehldahl nitrogen; TN, total nitrogen; TDS, total dissolved solids; TSS, total suspended solids; CECs, contaminants of emerging concern.

HRSD's project costs significantly. Conversely, application of the TOC and chemical oxygen demands (COD) limits (used to measure the amount of organic compounds in water) used at other IPR facilities (approximately 2 to 3 mg/L TOC) could allow implementation of more sustainable alternative treatment technologies. For example, the use of ozone and activated carbon operating in biological and adsorption modes has been studied for potable reuse projects and such an approach can often produce water with TOC less than 3 mg/L.

- *Pathogens*—California's strict log reduction values (LRVs) are more challenging to meet than requirements at other locations; thus, additional disinfection-based treatment technologies might be necessary and project costs could increase significantly if these LRVs are adopted by Virginia.
- *Nitrogen*—The USEPA's maximum contaminant level (MCL), the legal threshold limit on the amount of substance that is allowed in public water systems, for nitrate in drinking water is 10 mg/L as nitrogen (N). Therefore, total nitrogen (TN) or nitrite/nitrate (if the predominant nitrogen species) is typically limited in IPR applications to 10 mg/L to achieve MCL compliance with the nitrate limit or to prevent conversion of ammonia (NH_3) to nitrite or nitrate in the aquifer.
- *Total dissolved solids (TDS)*—The USEPA's secondary MCL for total dissolved solids, a measure of the combined content of all inorganic and organic substances contained in a liquid in molecular or colloidal suspended form, is 500 mg/L. Secondary MCLs are not enforceable; therefore, compliance is not required. Drinking water customers, though, may complain of objectionable taste when TDS levels exceed 500 mg/L, depending on the ionic makeup of the TDS. The TDS concentration in the Potomac Aquifer is suspected to be high (>750 mg/L; see Table 12.2), but TDS removal may not be necessary, although further investigation is warranted.

TABLE 12.2

Estimated TDS Concentrations (mg/L) in the Potomac Aquifer at HRSD's WWTP Locations

WWTP	Upper Potomac	Middle Potomac	Lower Potomac
Boat Harbor	750	1000	3500
Army Base	1500	2500	15,000
Virginia Initiative Plant	1000	2600	15,000
Nansemond	750	800	5000
James River	1000	1500	10,000
York River	5000	5000	10,000
Williamsburg	1000	1500	3000

Source: CH2M, *Sustainable Water Recycling Initiative: Advanced Water Purification Process Feasibility Evaluation*, Report No. 3, CH2M, Newport News, VA, 2016. Data from Focazio et al. (1993).

ADDITIONAL DRINKING WATER CONSIDERATIONS

All IPR projects are required to comply with drinking water MCLs, although this requirement is typically not difficult to meet because most modern wastewater treatment plants comply with drinking water MCLs, except for Tennessee, where nitrification and denitrification are not practiced. Contaminants of emerging concern (CECs), such as personal care products and pharmaceuticals (PCPPs), are any synthetic or naturally occurring chemical or any microorganism that is not commonly monitored in the environment but has the potential to enter the environment and cause known or suspected adverse ecological and/or human health effects. Personal care products and pharmaceuticals represent a very broad, diverse collection of thousands of chemical substances, including prescription and over-the-counter therapeutic drugs, fragrances, cosmetics, sunscreen agents, diagnostic agents, nutrapharmaceuticals, and biopharmaceuticals, among many others. These emerging contaminants have attracted significant media attention in recent years because of improvements in analytical techniques allowing measurement of these chemicals at a parts-per-trillion level (ppt). Although some impact on the ecology has been noted at a few wastewater treatment plant (WWTP) discharge locations due to endocrine-disrupting compounds, no impact on human health has been observed. However, at this point in time the best we can say about PCPPs and their effect on the environment and human health is that we do not know what we do not know about them.

What we do know about the impact of endocrine disruptors is that there is a growing body of evidence suggesting that humans and wildlife species have suffered adverse health effects after exposure to endocrine-disrupting chemicals, or environmental endocrine disruptors. In this book, environmental endocrine disruptors are defined as exogenous agents that interfere with the production, release, transport, metabolism binding, action, or elimination of natural hormones in the body responsible for the maintenance of homeostasis and the regulation of developmental processes. The definition reflects a growing awareness that the issue of endocrine disruptors in the environment extends considerably beyond that of exogenous estrogens and includes antiandrogens and agents that act on other components of the endocrine system such as the thyroid and pituitary glands (Kavlock et al., 1996). Disrupting the endocrine system can occur in various ways. Some chemicals can mimic a natural hormone, fooling the body into over-responding to the stimulus (e.g., a growth hormone that results in increased muscle mass) or responding at inappropriate times (e.g., producing insulin when it is not needed). Other endocrine-disrupting chemicals can block the effects of a hormone from certain receptors. Still others can directly stimulate or inhibit the endocrine system, causing overproduction or underproduction of hormones. Certain drugs are used to intentionally cause some of these effects, such as birth control pills. In many situations involving environmental chemicals, an endocrine effect may not be desirable.

In recent years, some scientists have proposed that chemicals might inadvertently be disrupting the endocrine systems of humans and wildlife. Reported adverse effects include declines in populations, increases in cancers, and reduced reproductive function. To date, these health problems have been identified primarily in domestic or wildlife species with relatively high exposures to organochlorine compounds, including DDT and its metabolites, polychlorinated biphenyls (PCBs) and dioxides, or naturally occurring plant estrogens (phytoestrogens). However, the relationship between human diseases of the endocrine system and exposure to environmental contaminants is poorly understood and scientifically controversial.

Although domestic and wildlife species have demonstrated adverse health consequences from exposure to environmental contaminants that interact with the endocrine system, it is not known if similar effects are occurring in the general human population, but again there is some evidence of adverse effects in populations with relatively high exposures. Several reports of declines in the quality and quantity of sperm production in humans over the last five decades and the reported increase in incidences of certain cancers (breast, prostate, testicular) that may have an endocrine-related basis have led to speculation about environmental etiologies (Kavlock et al., 1996). There is increasing concern about the impact of the environment on public health, including reproductive ability, and some controversy has arisen from reviews claiming that the quality of human semen has declined (Carlson et al., 1992). However, little notice has been paid to these warnings, possibly because the suggestions have been based on data for selected groups of men recruited from infertility clinics, from among semen donors, or from candidates for vasectomy. Furthermore, the sampling of publications used for such reviews has not been systematic, thus implying a risk of bias. Because a decline in semen quality may have serious implications for human reproductive health, it is of great importance to determine whether the reported decrease in sperm count reflects a biological phenomenon or is due to methodological errors.

Data on semen quality collected systematically from reports published worldwide indicate clearly that sperm density declined appreciably from 1938 to 1990, although we cannot conclude whether or not this decline is continuing today. Concomitantly, the incidence of some genitourinary abnormalities including testicular cancer and possibly also maldescent (faulty descent of the testicle into the scrotum) and hypospadias (abnormally placed urinary meatus) has increased. Such remarkable changes in semen quality and the occurrence of genitourinary abnormalities over a relatively short period are probably due more to environmental rather than genetic factors. Some common prenatal influences could be responsible both for the decline in sperm density and for the increase in cancer of the testis, hypospadias, and cryptorchidism (one or both testicles fail to move to scrotum). Whether estrogens or compounds with estrogen-like activity or other environmental or endogenous factors damage testicular function remains to be determined (Carlson et al., 1992). Even though we do not know what we do not know about endocrine disruptors, it is known that the normal functions of all organ systems are regulated by endocrine factors, and small disturbances in endocrine function, especially during certain stages of the life cycle such as development, pregnancy, and lactation, can lead to profound and lasting effects. The critical issue is whether sufficiently high levels of endocrine-disrupting chemicals exist in the ambient environment to exert adverse health effects on the general population.

Current methodologies for assessing, measuring, and demonstrating human and wildlife health effects (e.g., the generation of data in accordance with testing guideline) are in their infancy. The USEPA has developed testing guidelines and the Endocrine Disruption Screening Program (discussed later), which is mandated to use validated methods for screening the testing chemicals to identify potential endocrine disruptors, to determine adverse effects and dose–response, to assess risk, and ultimately to manage risk under current laws. The best way in which to end this brief discussion is to provide a statement made by someone who really knows and understands endocrine disruptors and the potential impact on humans, wildlife, and the environment in general:

> Large numbers and large quantities of endocrine-disrupting chemicals have been released in the environment since World War II. Many of these chemicals can disturb development of the endocrine system and of the organs that respond to endocrine signals in organisms indirectly exposed during prenatal and/or early postnatal life; effects of exposure during development are permanent and irreversible.
>
> **Colborn et al. (1993)**

Regardless of our current lack of definitive evidence on the impact of PCPPs on human health, multiple barriers and relatively low TOC limits (0.5 to 3 mg/L) have been established for most IPR projects to limit the presence and concentration of CECs. In addition, some states require specific treatment (e.g., advanced oxidation) to ensure oxidation of a large portion of CECs.

Other nonregulated water quality parameters are often considered when implementing an IPR project, including the following:

- *Nitrosamines*—Nitrosamines are chemical compounds of the chemical structure $R^1N(–R^2)–N=O$; they are suspected carcinogens and have recently been shown to form during wastewater treatment, primarily through the disinfection process when chloramines react with N-nitrosodimethylamine (NDMA) precursors such as secondary, tertiary, and quaternary amines present in the water. California has established a notification level for NDMA of 10 nanograms per liter (ng/L), and the USEPA is considering regulating some of the nitrosamines under the Safe Drinking Water Act as part of the Contaminant Candidate List 3 process, which may result in limitations being applied to IPR projects.
- *Total hardness*—Total hardness in WWTP secondary effluent can often exceed levels generally deemed aesthetically acceptable by the public, unless treatment or blending is specifically provided.

INJECTATE WATER QUALITY CONCERNS

Table 12.3 presents some preliminary recommended water quality ranges for the injection of reclaimed water into the Potomac Aquifer. These water quality ranges are based on the aforementioned human health regulatory information. Note that further investigation of some of these parameters is necessary to more clearly define the anticipated limits (e.g., TDS, pH). In addition, discussion with Virginia regulators will be required to clearly define other regulatory requirements (e.g., TOC,

TABLE 12.3
Preliminary Recommended Water Quality for Injection of Reclaimed Water into the Potomac Aquifer

Parameter	Recommended Range	Notes
pH	To be determined	Further investigation on aquifer geochemical conditions is required.
Pathogens	Multiple pathogen barriers; nondetect for pathogen indicators (e.g., *E. coli*, total coliform)	Discussion with Virginia regulators is required to further establish how pathogen removal will be regulated (e.g., indicator organism approach, log removal value)
TOC	0.5–3 mg/L	Discussion with Virginia regulators is required to further establish the likely limit. Use of COD limit instead of TOC should also be considered.
Drinking water MCLs	Compliance with all MCLs	—
TN	<10 mg/L	—
NMDA	<10 mg/L	Treatment can be provided to meet this goal, but withdrawal at strategic locations in the WWTP process (e.g., prior to disinfection) may also limit the NDMA concentration, thus requiring no treatment for NDMA.
TDS	<750 mg/L	USEPA's secondary MCL is 500 mg/L, but the Potomac Aquifer is suspected to exceed 750 mg/L in some locations. More investigation is necessary to accurately determine the TDS at various locations within the Potomac Aquifer, especially given the cost impact of having to including TDS reduction as part of the advanced treatment train.
Total hardness	50–150 mg/L as $CaCO_3$	Water with a total hardness above 150 mg/L as $CaCO_3$ is generally considered a "hard" water; however, this is not a firm limit. Some customers find water with hardness above this value acceptable. In addition, reclaimed water with significantly different cationic characteristics or ionic strengths than the native groundwater can cause destabilization of clays in the aquifer. Therefore, careful comparison of ion concentrations in reclaimed water to groundwater must be conducted.
CECs	Multiple organics barriers; implement an advanced oxidation process	Discussion with Virginia regulators is required to further establish how trace organics and CECs will be addressed.

Source: CH2M, *Sustainable Water Recycling Initiative: Advanced Water Purification Process Feasibility Evaluation*, Report No. 3, CH2M, Newport News, VA, 2016.

Note: The recommended water quality goals shown are preliminary. More investigation and discussion with Virginia regulators and the public are required to establish firm water quality requirements. TOC, total organic carbon; MCLs, maximum contaminant levels; TN, total nitrogen; NDMA, *N*-nitrosodimethylamine; TDS, total dissolved solids; CECs, contaminants of emerging concern.

pathogens). Ensuring compatibility between the aquifer and the injected water is also required to avoid scale formation that may plug injection wells or release undesirable compounds form the soil matrix (e.g., arsenic).

ADVANCED WATER TREATMENT PROCESSES

Advanced treatment provided at indirect potable reuse plants varies but is typically focused on providing multiple barriers for the removal of pathogens and organics. Nitrogen and TDS removal is provided at some locations where necessary. Table 12.4 shows most of the indirect potable reuse projects that have been implemented in the United States. The table has been sorted according to the type of potable reuse application (i.e., direct aquifer injection, aquifer recharge with surface spreading, and surface water augmentation). The first five projects shown in this table are direct injection projects that match the proposed HRSD concept. Water extracted from direct injection and surface spreading projects that recharge groundwater is not typically treated again prior to distribution into the potable water system; however, water from surface augmentation projects is typically treated again at water treatment plants because of water treatment requirements stipulated by the USEPA's Surface Water Treatment Rule (SWTR). For example, Fairfax County's Griffith Water Treatment Plant provides coagulation, sedimentation, ozone oxidation, biological activated carbon filtration, and chlorine disinfection for water extracted from the Occoquan Reservoir that is augmented by the Upper Occoquan Service Authority's indirect potable reuse plant.

As shown in Table 12.4, the treatment provided for indirect potable reuse projects is typically a combination of multiple barriers for the removal of pathogens and organics. Multiple barriers for pathogens are typically provided through a combination of coagulation, flocculation, sedimentation, lime clarification, filtration (granular or membrane), and disinfection (chlorine, ultraviolet, or ozone). Multiple barriers for organics removal are typically provided through a combination of advanced treatment processes (e.g., reverse osmosis, granular activated carbon, ozone in combination biological activated carbon), although conventional treatment processes (e.g., coagulation, softening) also provide removal at some locations. All potable reuse plants listed in Table 12.4 include robust organics removal processes involving the use of granular activated carbon (GAC), reverse osmosis, or soil aquifer treatment (SAT), which are effective barriers to bulk and trace organics and represent the backbone of the potable treatment process. SAT is the controlled application of wastewater to earthen basins in permeable soils at a rate typically measured in terms of meters of liquid per week. The purpose of a soil aquifer treatment system is to provide a receiver aquifer capable of accepting liquid intended to recharge shallow groundwater, and system design and operating criteria are developed to achieve that goal. However, there are several alternatives with respect to the utilization or final fate of the treated water (USEPA, 2006):

- Groundwater recharge
- Recovery of treated water for subsequent reuse or discharge
- Recharge of adjacent surface streams
- Seasonal storage of treated water beneath the site with seasonal recovery for agriculture

TABLE 12.4
Treatment Technologies Employed at Operational Potable Reuse Plants

Project	Type of Potable Reuse Application	Year	Capacity (mgd)	Advanced Treatment Processes
Hueco Bolton Recharge Project; El Paso, TX	Groundwater recharge via direct injection and spreading basins	1985	10	Lime + GMF + ozone + BAC + Cl$_2$
West Basin Water Recycling Plant; Carson, CA	Groundwater recharge via direct injection	1993	12.5	MF + RO + UVAOP
Scottsdale Water Campus; Scottsdale, AZ	Groundwater recharge via direct injection	1999	20	MF + RO + Cl$_2$
Los Alamitos Seawater Intrusion Barrier; Long Beach, CA	Groundwater recharge via direct injection	2006	3	MF + RO + UV disinfection
Groundwater Replenishment System; Orange County, CA	Groundwater recharge via direct injection and spreading basins	2008	70	MF + RO + UVAOP + SAT (spreading basins for a portion of the flow)
Montebello Forebay, Groundwater Recharge District; Los Angeles, CA	Groundwater recharge via spreading basins	1962	44	GMF + Cl$_2$ + SAT (spreading basins)
Chino Basin Groundwater Recharge Project; Chico, CA	Groundwater recharge via spreading basins	2007	18	GMF + O$_2$ + SAT (spreading basins)

(continued)

Upper Occoquan Service Authority; Centreville, VA	Surface water augmentation	1978	54	High lime clarification + recarbonation
Clayton County Water Authority; Clayton County, GA	Surface water augmentation	1985	18	Cl_2 + UV disinfection + SAT (wetlands)
Gwinnett County, GA	Surface water augmentation	2000	60	Coagulation/flocculation/sedimentation + UF + ozone + BAC
Broad Run Water Reclamation Facility; Loudoun County, VA[a]	Surface water augmentation	2008	11	MBR + GAC + UV

Source: CH2M, *Sustainable Water Recycling Initiative: Advanced Water Purification Process Feasibility Evaluation*, Report No. 3, CH2M, Newport News, VA, 2016.

Note: GMF, granular media filtration; BAC, biological activated carbon; MF, microfiltration; RO, reverse osmosis; UV, ultraviolet; UVAOP, ultraviolet advanced oxidation process; SAT, soil aquifer treatment; MBR, membrane bioreactor; GAC, granular activated carbon.

[a] Indirect potable reuse projects practicing surface water augmentation provide additional treatment at a downstream surface water treatment plant. For example, in Virginia, Fairfax County's Griffith WTP uses water from the Occoquan Reservoir (downstream of the discharge). The treatment process at the Griffith WTP includes substantial treatment through coagulation, sedimentation, ozone, BAC, and Cl2. A similar level of treatment is provided at Fairfax County's Corbalis WTP, which is located downstream of the Broad Run WRF discharge.

The SAT process typically includes application of the reclaimed water using spreading basins and subsequent percolation through the vadose zone. SAT provides significant removal of both pathogens and organics through biological activity and natural filtration. However, because the Potomac Aquifer is confined, it is not possible to utilize SAT for treatment through the vadose zone to recharge the aquifer for the HRSD SWIFT project. On the other hand, movement of reclaimed water through the aquifer after direct injection will provide significant treatment benefits, including excellent removal of pathogens. Advanced water treatment plants based on reverse osmosis and granular activated carbon are often utilized at locations where SAT treatment through the vadose zone is not feasible because these processes can be implemented at most locations.

Advanced treatment trains based on reverse osmosis and granular activated carbon were developed for the HRSD groundwater recharge project using the historical WWTP effluent water quality data and the preliminary aquifer recharge water quality goals discussed previously. Three treatment trains were developed from this analysis: (1) reverse osmosis (RO)-based train, (2) nanofiltration (NF)-based train, and (3) granular activated carbon (GAC)-based train (see Figure 12.1). Consideration for each of these treatment trains includes the following:

1. *RO-based train*—Reverse osmosis has become common for potable water reuse projects in California and many international locations (e.g., Singapore, Australia) because of its effective removal of total dissolved solids (TDS), total organic carbon (TOC), and trace organics. California regulations require the use of RO for direct injection reuse projects or a comparable alternative with regulatory approval. RO creates a waste (concentrate) stream that can be difficult and costly to dispose of, especially at inland locations. Most locations where RO has been implemented are located near the ocean where disposal of RO concentrate is convenient and much less costly than at inland locations.

2. *NF-based train*—Although similar to the RO-based train, the NF-based train operates at significantly lower pressure and generates a less saline concentrate, which results in significant cost savings. This process does not meet California's IPR regulatory requirements but it does provide excellent treatment with significant removal of pathogens and organics. This train offers partial TDS removal by providing a high level of removal of divalent ions (e.g., calcium, magnesium) and moderate removal of monovalent ions (e.g., sodium chloride). NH_3 and NO_x-N removal is much lower with NF compared to RO, which results in lower TN concentration in the concentrate.

3. *GAC-based train*—The GAC-based train is a modernized version of full-scale operational IPR plants that have successfully been in operation for decades in Virginia, Texas, and Georgia. GAC adsorption is used as the backbone process for organics removal, and other treatment has been added for multiple barriers to pathogens and organics. Flocculation and sedimentation provide removal of solids, pathogens, organics, and

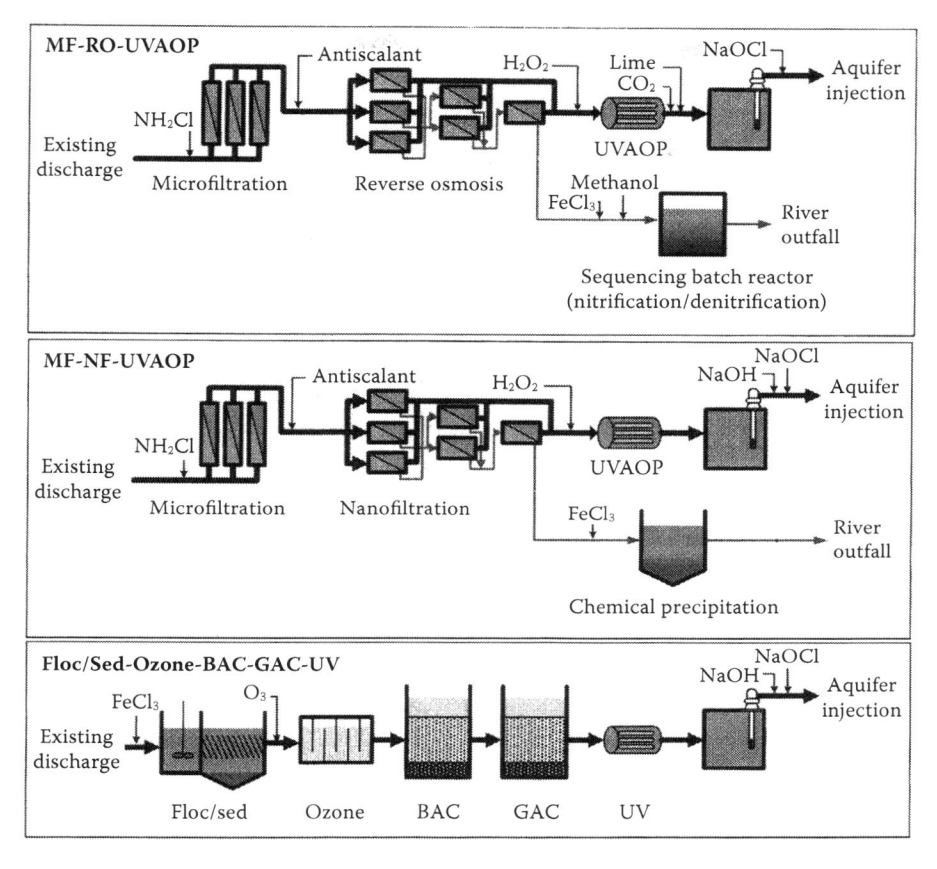

FIGURE 12.1 Process flow diagrams for three advanced wastewater treatment processes. (From CH2M, *Sustainable Water Recycling Initiative: Advanced Water Purification Process Feasibility Evaluation*, Report No. 3, CH2M, Newport News, VA, 2016.)

phosphorus. Ozone provides disinfection of pathogens and oxidation of organics, including oxidation of CECs and high-molecular-weight organic matter to smaller organic fractions that can be assimilated by biological activity present on GAC media, which is referred to as biological activated carbon (BAC) filtration. This treatment train does not provide any TDS removal and, therefore, does not generate a TDS-enriched waste that might require further treatment prior to discharge.

TREATMENT PLANT EFFLUENT WATER QUALITY

HRSD provided historic effluent water quality data for seven WWTPs to identify specific water quality challenges requiring treatment. The WWTPs analyzed include Army Base (AB), Boat Harbor (BH), James River (JR), Nansemond (NP), Virginia Initiative Plant (VIP), and York River (YR).

Data Sources for Evaluation

The following three primary data sources were used in the evaluation:

- *2013/2014 water quality data*—Detailed water quality data were provided for each WWTP effluent for October 2013 through September 2014. The data were provided as raw data in tables. With few exceptions, these data included the following parameters at the stated frequency: chloride (1x/week), calcium (2x/week), magnesium (2x/week), potassium (2x/week), sodium (2x/week), total alkalinity (3x/week), 5-day biochemical oxygen demand (BOD$_5$; 5x/week), pH (1x/day), turbidity (1x/day), ammonia-nitrogen (NH$_3$-N; 2x/week), NO$_x$-N (4x/week), orthophosphate (3x/week), total Kjehldahl nitrogen (TKN; 4x/week), total phosphorus (TP; 5x/week), and total suspended solids (TSS; 5x/week).
- *2011, 2012, and 2013 water quality data*—Detailed influent and effluent water quality data were provided for each WWTP from the 2011, 2012, and 2013 HRSD wastewater characteristics studies (no James River data were provided for 2013). The data provided were presented as minimum, average, and maximum values. The data included flow (continuous), temperature (3x/day), pH (12x/day), total alkalinity (1x/week), BOD$_5$ (5x/week), TSS (5x/week), turbidity (5x/week), fecal coliform (5x/week), TKN (frequency not reported), NO$_x$-N (frequency not reported), and TP (frequency not reported). Data were also provided for influent chloride (1x/week) and influent sulfate (4x/week), selected heavy metals (1x/year), and a variety of organics (volatile, base/acid, pesticides, total trihalomethanes [TTHMs]; 1x/year).
- *2014 total dissolved solids data*—Effluent TDS data were provided for each WWTP for January through September 2014. The data were provided as raw data in tables, and data points were provided once per week.

Effluent COD and TOC data collected by HRSD on a weekly basis from February 2015 through April 2015 were also used in the evaluation.

Data Evaluation

Evaluation of effluent quality from each of the seven treatment plants involved in the HRSD's SWIFT project included identification of the strength of each data source and was qualitatively documented as excellent, good, or limited. Excellent data included detailed 2013/2014 raw data and minimum, average, and maximum annual data from 2011 through 2013. Good data included a full dataset from only one of the sources. Limited data included data that were only collected once per year. Table 12.5 shows the average WWTP effluent water quality for select parameters and the strength of each data source. In addition, the data in Table 12.5 were analyzed to identify any problematic parameters that would require a specific treatment process or additional treatment to meet probable regulatory requirements or aesthetic issues. Specific challenges revealed in Table 12.5 include the following:

TABLE 12.5
Average Effluent Water Quality Data by WWTP

Criteria	Unit	AB	BH	JR	NP	VIP	WB	YR	Data Quality
Flow	mgd	10.9	14.03	13.3	16.5	30.0	8.9	12.1	Excellent
Temperature	°C	22.0	21.7	22.2	22.2	22.4	23.8	19.9	Excellent
pH	SU	7.0	7.2	6.8	7.0	6.7	7.1	7.19	Excellent
TDS (measured)	mg/L	**1393**	**599**	392	**712**	**853**	**596**	**575**	Good
Total hardness	mg/L	143	**161**	99	66	**181**	97	**194**	Calculated
NH_3	mg/L	**25.2**	**16.0**	0.52	1.3	0.5	1.3	1.3	Good
NO_x-N	mg/L	0.1	0.4	3.1	3.8	6.8	5.4	3.2	Excellent
TKN	mg/L	**28.4**	**22.9**	2.6	2.6	1.7	2.7	2.4	Excellent
Organic N	mg/L	3.3	6.9	2.1	1.3	1.3	1.4	1.1	Calculated
TN	mg/L	**28.5**	**23.3**	5.7	6.4	**8.5**	**8.1**	5.6	Calculated
TP	mg/L	1.0	0.70	0.82	1.3	0.58	1.1	0.56	Excellent
OP	mg/L	0.55	0.15	0.56	0.59	0.19	0.55	0.56	Good
BOD_5	mg/L	9.0	5.3	4.0	5.4	3.7	3.5	3.7	Excellent
sCOD	mg/L	59.0	34.5	30.0	39.0	43.5	35.0	25.0	Good
DOC	mg/L	9.9	9.8	9.0	11.2	8.6	10.1	9.5	Good
TSS	mg/L	8.9	6.2	4.6	6.4	5.2	4.3	1.6	Excellent
Fecal coliform	#/100 mL	9.2	5.5	3.0	3.2	6.6	8.4	1.4	Excellent
Total TTHM	µg/L	3.2	8.8	2.8	**82.4**	35.6	**64.7**	49.3	**Limited**
NDMA	ng/L	N/A	N/A	N/A	N/A	N/A	33	N/A	**Limited**
Turbidity	NTU	4.8	4.5	2.5	3.9	4.0	2.6	1.4	Excellent
Alkalinity	mg/L	126	145	84	160	59	164	107	Excellent
Chloride[a]	mg/L	393	206	76	220	426	101	179	Excellent
Calcium	mg/L	29	44	34	18	33	32	62	Good
Magnesium	mg/L	17	12	3.6	5	24	4.0	10	Good
Potassium	mg/L	17	20	13	23	20	16	13	Good
Sodium	mg/L	152	128	68	193	246	142	103	Good
Sulfate[a]	mg/L	45	54	32	39	104	39	44	Limited
Iron	mg/L	1.2	1.7	2.0	3.0	2.1	1.6	0.7	Limited

Source: CH2M, *Sustainable Water Recycling Initiative: Advanced Water Purification Process Feasibility Evaluation*, Report No. 3, CH2M, Newport News, VA, 2016.

Note: Bold cells indicate areas of concern. SU, standard unit; AB, Army Base; BH, Boat Harbor; JR, James River; NP, Nansemond; VIP, Virginia Initiative Plant; YR, York River; TDS, total dissolved solids; TKN, total Kjehldahl nitrogen; TN, total nitrogen; TP, total phosphorus; OP, orthophosphate; BOD_5, biochemical oxygen demand; sCOD, soluble chemical oxygen demand; DOC, dissolved organic carbon; TSS, total suspended solids; TTHM, total trihalomethanes; NMDA, *N*-methyl-d-aspartate; N/A, not applicable.

[a] Based on influent data.

- *Total dissolved solids*—The drinking water secondary MCL for TDS is 500 mg/L. The average effluent TDS from each WWTP except James River exceeds 500 mg/L; the Army Base plant (1292 mg/L) and Virginia Initiative Plant (853 mg/L) have notably high TDS concentrations.

- *Ammonia*—Army Base and Boat Harbor plants have an average effluent NH_3 of 25.2 and 16.0 mg-N/L, respectively, although the Army Base plant was recently upgraded to biological nutrient removal (BNR) and now produces effluent with low ammonia and total nitrogen concentrations that are comparable to HRSD's other biological nutrient removal (BNR) plants. Total nitrogen (TN) concentrations in excess of 10 mg/L are typically not allowed for groundwater recharge into potable aquifers; therefore, additional nitrogen treatment would likely be required at the Boat Harbor plant.
- *Total trihalomethanes*—Elevated TTHM levels were recorded at Nansemond (82.4 µg/L) and Williamsburg (64.7 µg/L). TTHM levels at the other plants ranged from 3 µg/L to 50 µg/L. More TTHM data should be collected as the data sources used were limited. The drinking water primary MCL for TTHMs is 80 µg/L.
- *Total hardness*—Hardness in water is caused by the presence of certain positively charged metallic ions in solution in the water. The most common of these hardness-causing ions are calcium and magnesium; others include iron, strontium, and barium. The two primary constituents of water that determine the hardness of water are calcium and magnesium. If the concentration of these elements in the water is known, the total hardness of the water can be calculated. To make this calculation, the equivalent weights of calcium, magnesium, and calcium carbonate must be known; the equivalent weights are given below:

Equivalent Weights

Calcium (Ca)   20.04
Magnesium (Mg)   12.15
Calcium carbonate ($CaCO_3$)   50.045

Calcium and magnesium ions are the two constituents that are the primary cause of hardness in water. To find total hardness (mg/L as $CaCO_3$), we simply add the concentrations of calcium and magnesium ions (mg/L as $CaCO_3$), using Equation 12.1:

$$\text{Total hardness} = \text{Calcium hardness} + \text{Magnesium hardness} \qquad (12.1)$$

When total hardness has been calculated, it is sometimes used to determine another expression of hardness—carbonate and noncarbonate. When hardness is numerically greater than the sum of bicarbonate and carbonate alkalinity, that amount of hardness equivalent to the total alkalinity (both in units of mg/L as $CaCO_3$) is called the *carbonate hardness*; the amount of hardness in excess of this is the *noncarbonate hardness*. When the hardness is numerically equal to or less than the sum of carbonate and noncarbonate alkalinity, all hardness is carbonate hardness and noncarbonate hardness is absent. Again, the total hardness is comprised of carbonate hardness and noncarbonate hardness:

$$\text{Total hardness} = \text{Carbonate hardness} + \text{Noncarbonate hardness} \quad (12.2)$$

During the evaluation, total hardness data were not specifically provided but were calculated using the detailed calcium and magnesium data. Total hardness in drinking water systems is often limited to 150 mg/L as $CaCO_3$ or less to avoid customer complaints. The Boat Harbor plant (161 mg/L), Virginia Initiative Plant (181 mg/L), and York River Treatment Plant (194 mg/L) all showed average effluent data above this value. Total hardness concentrations in potable water in the surrounding area were less than the secondary MCL of 500 mg/L.

- *Dissolved organic carbon (DOC) and soluble chemical oxygen demand (sCOD)*—These are important parameters to measure for potable reuse plants because advanced treatment goals and regulatory requirements are often developed for these constituents. The average DOC and sCOD concentrations in the effluent from the seven WWTPs ranged from 8.6 to 11.2 mg/L and 25 to 49 mg/L, respectively, values that are within the typical range for WWTPs practicing biological nutrient removal. TOC concentrations above 10 mg/L become increasingly difficult to treat to recommended levels for certain advanced treatment trains so additional DOC sampling and bench and pilot testing are recommended to confirm adequate treatment performance.
- *Heavy metals, VOCs, synthetic organic chemicals, and other organics*—The 2011, 2012, and 2013 dataset included minimum, average, and maximum data for heavy metals, volatile organic chemicals, synthetic organic chemicals, and other organics. Although the datasets were limited (typical frequency of once per year), the data were compared to the applicable drinking water MCLs in order to identify any potential contaminants of concern. The recorded data that were compared to drinking water MCLs included antimony, arsenic, beryllium, cadmium, chromium, copper, lead, mercury, selenium, thallium, benzene, TTHMs, carbon tetrachloride, chlorobenzene, 1,2-dichloroethane, 1,1-dichloroethylene, 1,2-dichloropropene, ethylbenzene, toluene, 1,1,1-trichloroethane, 1,1,2-trichloroethane, trichloroethane, and vinyl chloride. There were some additional parameters with regulated MCLs that did not have any data provided for these WWTPs. Consequently, additional regular sampling for all MCLs, USEPA Contaminant Candidate List 3 parameters, and other chemicals of concern at each WWTP is recommended. In addition, a comprehensive study identifying industrial and commercial facilities that discharge to the wastewater collection system is advisable to identify potential locations where chemical contamination could occur that could negatively affect finished water quality or treatment processes if not already known. An outreach program aimed at limiting discharge of contaminants of concern or additional water quality monitoring at specific locations may have to be implemented if aquifer replenishment is ultimately pursued and implemented. In addition to the average WWTP

TABLE 12.6

Selected 99th-Percentile Water Quality Data by WWTP

Criteria	Unit	AB	BH	JR	NP	VIP	WB	YR
NO$_x$-N	mg/L	1.4	**16.6**	5.0	5.6	10.5	9.3	5.0
BOD	mg/L	**18.0**	**17.0**	**30.5**	13.5	**19.0**	8.5	**17.8**
TSS	mg/L	15.3	14.5	**40.2**	18.4	23.1	8.9	4.0
TDS	mg/L	**2303**	**927**	506	929	**1510**	764	741

Source: CH2M, *Sustainable Water Recycling Initiative: Advanced Water Purification Process Feasibility Evaluation*, Report No. 3, CH2M, Newport News, VA, 2016.

Note: Bold cells indicate areas of concern. AB, Army Base; BH, Boat Harbor; JR, James River; NP, Nansemond; VIP, Virginia Initiative Plant; YR, York River.

effluent data shown in Table 12.5, selected 99th-percentile effluent data from the 2013/2014 dataset were also analyzed to determine peak loadings from the WWTP that could be problematic for various treatment processes. Peak loadings can be accounted for either by selecting a treatment process that is designed for the maximum values or by providing a large enough equalization volume of primary or secondary effluent to attenuate the loading. Table 12.6 shows the selected parameters of concern based on the 99th-percentile data:

- *Nitrate/nitrite-nitrogen*—Average effluent NO$_x$-N levels were well under the nitrate MCL of 10 mg-N/L; however, 99th-percentile data at the Virginia Initiative Plant (10.5 mg/L) and at Williamsburg (0.3 mg/L) show that NO$_x$-N levels could periodically exceed the nitrate MCL, which could require NO$_x$-N-specific treatment or additional storage. The 99th-percentile NO$_x$-N concentration at Boat Harbor was also high, but biological nutrient removal (BNR) is not currently practiced at this plant. When BNR is implemented at Boat Harbor, the variability of the NO$_x$-N data should be reevaluated.
- *Biochemical oxygen demand and total suspended solids*—High 99th-percentile 5-day biochemical oxygen demand (BOD$_5$) and TSS levels suggest occasional plant upsets that could be problematic for filtration (granular or membrane). This could require increased storage or treatment or automated monitoring to divert flow away from the AWT plant during high BOD and TSS loadings.
- *Total dissolved solids*—WWTP effluent TDS values are not expected to fluctuate significantly; yet, Army Base, Virginia Initiative Plant, and Boat Harbor each show 99th-percentile values significantly higher than the average. Periodically high TDS values could violate treatment goals if reverse osmosis is not selected and would require additional storage or provision for divisions.

ADVANCED TREATMENT PRODUCT WATER QUALITY

INORGANIC WATER QUALITY

Using the historical water quality data presented previously and the expected performance of each unit process based on professional judgment, mass balance calculations for key inorganic parameters were performed for each treatment train at seven of HRSD's WWTPs. Summary tables for each treatment train are provided in Tables 12.7, 12.8, and 12.9. The detailed mass balance calculations revealed the following:

- The RO-based treatment process provided the lowest concentration of all water quality parameters (see Table 12.7); however, treatment to this level may not be necessary in all cases. For example, the finished water TDS concentration was about 50 mg/L, well below the secondary MCL (500 mg/L) and the minimum background TDS in the Potomac Aquifer (~750 mg/L). The very low TDS reverse osmosis permeate may increase mobilization of trace metals in the aquifer, which is undesirable.
- The NF-based treatment process provided excellent water quality, as shown in Table 12.8. The NF process removes very little nitrogen; therefore, the TN concentration in the finished water exceeded the recommended upper range (10 mg/L) at Boat Harbor and approached the 10-mg/L limit at two other WWTPs (VIP and Williamsburg). Nitrification and denitrification improvements at these WWTPs may be necessary to ensure regular compliance with the recommended TN limit. Alternatively, NF membranes that have higher nitrogen removal can be considered; however, their use will result in a higher TDS concentrate stream.

TABLE 12.7

Projected Average Finished Water from RO-Based Treatment Train: Inorganics

Parameter	Unit	Recommended Range	AB	BH	JR	NP	VIP	WB	YR
Average flow	mgd	N/A	9.26	11.90	11.30	14.02	25.49	7.56	10.28
Alkalinity	mg/L	40–150	40	40	40	40	40	40	40
TDS	mg/L	0–750	87	46	37	52	61	47	46
Hardness	mg/L as $CaCO_3$	50–150	43	43	42	43	43	42	44
TN	mg/L	0–10	4.84	3.95	0.91	1.01	1.31	1.27	0.89
NO_x-N	mg/L	0–8	0.02	0.06	0.47	0.57	1.02	0.81	0.48
TP	mg/L	0–1	0.005	0.0035	0.004	0.01	0.00	0.01	0.00

Source: CH2M, *Sustainable Water Recycling Initiative: Advanced Water Purification Process Feasibility Evaluation,* Report No. 3, CH2M, Newport News, VA, 2016.

Note: AB, Army Base; BH, Boat Harbor; JR, James River; NP, Nansemond; VIP, Virginia Initiative Plant; WB, Williamsburg; YR, York River; TDS, total dissolved solids; TN, total nitrogen; TP, total phosphorus.

TABLE 12.8

Projected Average Finished Water from NF-Based Treatment Train: Inorganics

Parameter	Unit	Recommended Range	AB	BH	JR	NP	VIP	WB	YR
Average flow	mgd	N/A	9.26	11.90	11.30	14.02	25.49	7.56	10.28
Alkalinity	mg/L	40–150	67	77	45	85	31	87	57
TDS	mg/L	0–750	642	280	191	333	415	283	262
Hardness	mg/L as CaCO₃	50–150	65	76	68	57	68	67	47
TN	mg/L	0–10	<10	**15.7**	4.8	5.5	**7.9**	**7.2**	48
NO$_x$-N	mg/L	0–8	0.10	0.40	3.1	3.8	6.7	5.3	3.2
TP	mg/L	0–1	0.05	0.04	0.04	0.06	0.03	0.06	0.03

Source: CH2M, *Sustainable Water Recycling Initiative: Advanced Water Purification Process Feasibility Evaluation*, Report No. 3, CH2M, Newport News, VA, 2016.

Note: Bold cells indicate areas of concern; modifications to WWTP operations may be required and/or additional treatment at the advanced wastewater treatment plant (AWTP). AB, Army Base; BH, Boat Harbor; JR, James River; NP, Nansemond; VIP, Virginia Initiative Plant; WB, Williamsburg; YR, York River; TDS, total dissolved solids; TN, total nitrogen; TP, total phosphorus.

TABLE 12.9

Projected Average Finished Water from GAC-Based Treatment Train: Inorganics

Parameter	Unit	Recommended Range	AB	BH	JR	NP	VIP	WB	YR
Average flow	mgd	N/A	10.90	14.00	13.30	16.50	30.00	8.89	12.09
Alkalinity	mg/L	40–150	80	99	38	132	13	118	79
TDS	mg/L	0–750	**1422**	616	420	734	**918**	623	615
Hardness	mg/L as CaCO₃	50–150	143	**161**	99	66	**181**	97	**194**
TN	mg/L	0–10	<10	**23**	5.7	6.4	**8.5**	**8.1**	5.6
NO$_x$-N	mg/L	0–8	0.10	0.40	3.1	3.8	<5	**5.4**	3.2
TP	mg/L	0–1	0.50	0.50	0.50	0.50	0.50	0.50	0.50

Source: CH2M, *Sustainable Water Recycling Initiative: Advanced Water Purification Process Feasibility Evaluation*, Report No. 3, CH2M, Newport News, VA, 2016.

Note: Bold cells indicate areas of concern; modifications to WWTP operations may be required and/or additional treatment at the advanced wastewater treatment plant (AWTP). AB, Army Base; BH, Boat Harbor; JR, James River; NP, Nansemond; VIP, Virginia Initiative Plant; WB, Williamsburg; YR, York River; TDS, total dissolved solids; TN, total nitrogen; TP, total phosphorus.

- The GAC-based treatment process provided excellent water quality, as shown in Table 12.9. Specific considerations related to this process include the following:

- Although some incidental nitrification may occur in the biological activated carbon filters, the process is not intended nor typically designed to remove nitrogen. Therefore, nitrogen removal should be considered at the upstream WWTPs, where nitrification and denitrification improvements would be necessary to ensure regular compliance with the recommended TN limit.
- No TDS are removed through this process. The Army Base and VIP plants showed elevated TDS levels that regularly exceeded 750 mg/L. Upstream mitigation, such as reducing infiltration and inflow in areas with high TDS or eliminated industrial discharge high in TDS, may be required at these locations if a TDS limit of 750 mg/L is established.
- Hardness removal with chemical precipitation may be required at three plants (Boat Harbor, VIP, and York River), although more investigation is necessary to determine whether or not the total hardness (161 to 194 mg/L as $CaCO_3$) at these plants is acceptable from aesthetic and aquifer geochemistry perspectives. If not, the proposed flocculation–sedimentation process shown for the GAC-based treatment train could be modified to a chemical softening process for those plants with elevated hardness.

ORGANIC WATER QUALITY

Bulk Organics

The application of robust treatment barriers for the removal of organics has historically been a central tenet of implementing full-scale potable reuse projects to address the presence of unknown organic compounds of chronic health concern that may be present in the secondary effluent—a significant and pressing example of the old "we do not know what we do not know" syndrome. As presented in Table 12.1, regulations and permits for potable reuse projects have been developed by establishing limits on bulk organic parameters, such as COD and TOC, which act as surrogates for organic compounds of wastewater origin. Virginia established a COD limit of 10 mg/L for the Occoquan and Dulles area watershed policies, which apply to the Upper Occoquan Sewage Authority (UOSA) IPR project (constructed in 1978) and the Broad Run Water Reclamation Facility (BRWRF) (constructed in 2008), respectively. California and Florida have established TOC limits of 0.5 mg/L and 3 mg/L, respectively, in their IPR regulations.

An advanced water treatment plant's finished water COD and TOC concentration that would need to comply with the established permit limit is dependent on the initial concentration in the WWTP effluent and the specific treatment processes employed at the advanced water treatment plant. Table 12.10 shows the estimated finished water TOC concentrations from the three proposed advanced water treatment plant treatment trains (i.e., RO, NA, and BAC) when treating effluent from each of HRSD's WWTPs. The calculations use full-scale advanced water treatment plant effluent TOC and DOC sampling and treatment process pilot testing. The following can be concluded from the information in the table:

TABLE 12.10

Estimated Total Organic Carbon (TOC) Concentration in Wastewater Treatment Plant (WWTP) Effluent and Advanced Wastewater Treatment Plant (AWTP) Finished Water

Location	AB	BH	JR	NP	VIP	WB	YR
WWTP effluent dissolved organic carbon (DOC)	9.9	9.8	9.0	11.2	8.6	10.1	9.5
AWTP reverse osmosis (RO)	0.3	0.3	0.3	0.3	0.3	0.3	0.3
AWTP nanofiltration (NF)	1.0	1.0	0.9	1.1	0.9	1.0	1.0
AWTP granular activated carbon (GAC)	2.5	2.5	2.3	2.8	2.2	2.6	2.4

Source: CH2M, *Sustainable Water Recycling Initiative: Advanced Water Purification Process Feasibility Evaluation*, Report No. 3, CH2M, Newport News, VA, 2016.

Note: Preliminary TOC goal: 0.5 to 3 mg/L. The following DOC removal percentages were used in the table based on full-scale treatment performance data: 97% for RO; 90% for NF; 35% for flocculation + sedimentation; 40% for ozone + BAC; and 40% for GAC adsorption. AB, Army Base; BH, Boat Harbor; JR, James River; NP, Nansemond; VIP, Virginia Initiative Plant; WB, Williamsburg; YR, York River.

- Compliance with a California-based TOC limit of 0.5 mg/L could only be achieved by implementing an RO-based treatment train.
- Compliance with a Florida-based UOSA-type permit (3 mg/L TOC and 10 mg/L COD) could likely be achieved at most WWTPs by any of the three proposed treatment trains or by a hybrid treatment train that combined partial RO treatment with GAC-based treatment.
- GAC-based advanced water treatment plants will require regular replacement or regeneration of the GAC media for consistent TOC removal. Pilot testing is necessary to determine the GAC regeneration frequency requirements.
- Measurement on a regular basis of TOC and DOC in the final effluent from each WWTP is recommended to accurately determine TOC removal requirements.

Trace Organics

Earlier, personal care products and pharmaceutical (PCPPs) were mentioned, along with other contaminants of emerging concern (CECs). Additional questions about CECs continue to be raised with regard to the potential for chronic human health effects related to the thousands of organic chemicals that may end up in wastewater effluent at trace levels (mg/L). Furthermore, the efficacy of conventional water treatment processes that may end up treating source waters that have some effluent contribution is typically low. Each advanced treatment process considered in this discussion differs in its effectiveness at removing CECs. Research has shown that RO-based, NF-based, and GAC-based potable reuse treatment trains provide multiple unit processes that are effective barriers to a wide range of CECs. The RO- and NF-based treatment trains provide substantial removal through membranes

(RO/NF) and ultraviolet advanced oxidation (UVAOP), while the GAC-based treatment train provides significant removal through ozone–biological activated carbon (BAC) and granular activated carbon (GAC). Table 12.11 shows representative removals by advanced treatment processes for a variety of CECs as determined through recent research and monitoring of full-scale treatment facilities. These processes are redundant in the removal of some CECs (i.e., provide multiple barriers to their passage) and are complementary in the removal of others. For example, both ozone and GAC are effective barriers to the anticonvulsant drug carbamazepine, but only GAC acts as an effective barrier to the flame-retardant TCEP. No one process provides complete removal of all compounds, but RO generally provides the best removal of a wide range of compounds. However, these compounds are not destroyed or transformed by RO but instead are transferred to the RO concentrate (at a higher concentration); thus, their presence in the concentrate must be considered, particularly when the concentrate is discharged to a receiving water body.

At the present time, treatment for all CECs does not appear to be a differentiator among potable reuse treatment trains. Although health effects of many CECs—either alone or as mixtures—have not been demonstrated at the ng/L concentrations typically detected in wastewater effluent, the proposed treatment trains do reduce the concentrations of many of these chemicals to a significant degree. Meanwhile, the USEPA is prioritizing and studying a number of chemicals through their candidate contaminant list program.

MASS BALANCE FOR REVERSE OSMOSIS SYSTEMS*

To this point in the book we have discussed the benefits of reverse osmosis operating systems. It is important to point out, however, that along with the good there is the not so good; that is, RO systems have their advantages but they also have a few disadvantages. The one disadvantage pointed out and discussed here is the major one—that is, concentrate disposal. Where is the concentrated wastestream to be disposed of?

Mass Balance Principle

To gain a better understanding of membrane disposal issues and techniques we begin with a discussion of mass balance. The simplest way to express the fundamental engineering principle of mass balance is to say, "Everything has to go somewhere." More precisely, the *law of conservation of mass* says that when chemical reactions take place matter is neither created nor destroyed. What this important concept allows us to do is track materials (concentrates)—that is, pollutants, microorganisms, chemicals, and other materials—from one place to another. The concept of mass balance plays an important role in reverse osmosis system operations (especially in desalination)

---

* The following material is adapted from Spellman, F.R., *Reverse Osmosis: A Guide for the Non-Engineering Professional*, CRC Press, Boca Raton, FL, 2015.

TABLE 12.11

Removal of CECs by Advanced Water Treatment Programs

CEC	CEC Class	Percent Reduction				
		Ozone[a] (O$_3$; TOC = 1)	UVAOP[b] (671 mJ/cm^2; 5 mg/L H$_2$O$_2$)	NF[c] (Dow NF270 after 1500 hr)	RO[c,d] (ESPA2 RO)	GAC[b] (Super Darco)
Atendolol	High blood pressure medicine	98%	ND	85%	88%, >98.9%	ND
Atrazine	Pesticide	70%	90%	85%	98%, 56.5%	>95% at 41,000 BVs
Carbamazepine	Psychoactive medication	99%	97%	90%	98%. >99.5%	>95% at 54,000 BVs
DEET	Pesticide	93%	89%	83%	98%, ND	>95% at 36,000 BVs
Diclofenac	Analgesics	98%	>99%	ND	ND, >96.9%	>95% at 37,000 BVs
Gemfibrozil	Cholesterol medication	9%	94%	99%	96%, >99.8%	>95% at 40,000 BVs
Ibuprofen	Analgesics	98%	97%	97%	ND, >99.4%	>95% at 28,000 BVs
Naproxen	Analgesics	98%	>99%	97%	ND, >99%	>95% at 42,000 BVs
Primidone	Psychoactive medication	94%	ND	95%	97%, >99%	ND
Sulfamethoxazole	Antimicrobial	99%	>99%	90%	99%, >99.9%	>95% at 28,000 BVs
TCEP	Flame retardant	20%	8%	80%	93%, >97.8%	>95% at 37,000 BVs
Triclosan	Antimicrobial	97%	>99%	ND	91%, >97.8%	>95% at 93,000 BVs
Trimethoprim	Antimicrobial	97%	95%	ND	98%, >97.9%	>95% at 65,000 BVs

Source: CH2M, *Sustainable Water Recycling Initiative: Advanced Water Purification Process Feasibility Evaluation*, Report No. 3, CH2M, Newport News, VA, 2016.

Note: TOC, total organic carbon; UVAOP, ultraviolet advanced oxidation process; H$_2$O$_2$, hydrogen peroxide; NF, nanofiltration; RO, reverse osmosis; GAC, granular activated carbon; BVs, bed volumes; ND, no data.

[a] Snyder et al. (2014).

[b] Snyder et al. (2007).

[c] Drewes et al. (2013).

[d] Knoell (2014).

where we assume a balance exists between the material entering and leaving the RO system: "What comes in must equal what goes out." The concept is very helpful for evaluating biological systems and developing sampling and testing procedures, as well as many other unit processes within any treatment or processing system.

All desalination processes have two outgoing process streams: (1) the product water, which is lower in salt than the feed water, and (2) a concentrated stream that contains the salts removed from the product water. Even distillation has a "bottoms" solution that contains salt from the vaporized water. The nature of the concentrate stream depends on the salinity of the feed water, the amount of product water recovered, and the purity of the product water. To determine the volume and concentration of the two outgoing streams, a mass balance is constructed. It is necessary to know the recovery rate of water, the rejection rate of salt, and the input flow and concentration to solve equations for the flow and concentration of the product and concentrate.

REVERSE OSMOSIS CONCENTRATE DISPOSAL PRACTICES

Reverse osmosis concentrate is disposed of by several methods, including surface water discharge, sewer discharge, deep well injection, evaporation ponds, spray irrigation, and zero liquid discharge.

SURFACE WATER AND SEWER DISPOSAL

Disposal of concentrate to surface water and sewers are the two most widely used disposal options for both desalting membrane processes. Post-1992 data provide the following statistics:

Disposal Option	Percent of Desalting Plants
Surface water disposal	45%
Disposal to sewer	42%
Total	87%

This disposal option, although not always available, is the simplest option in terms of equipment involved and is frequently the lowest cost option. As will be seen, however, the design of an outfall structure for surface water disposal can be complex.

Disposal to surface water involves conveyance of the concentrate or backwash to the site of disposal and an outfall structure that typically involves a diffuser and outlet ports or valve mounted on the diffuser pipe. Factors involved in the outfall design are discussed in this section, and cost factors are presented. However, due to the large number of cost factors and the large variability in design conditions associated with surface water disposal, a relatively simple cost model cannot be developed. Disposal to surface waters requires a National Pollutant Discharge Elimination System (NPDES) permit.

Disposal to the sewer involves conveyance to the sewer site and typically a negotiated fee to be paid to the WWTP. Because the negotiated fees can range from zero to substantial, there is no model that can be presented. No disposal permits are required for this disposal option. Disposal of concentrate or backwash to the sewer, however, affects WWTP effluent that requires an NPDES permit. With regard to design considerations for disposal to surface water, a brief discussion of ambient conditions, discharge conditions, regulations, and the outfall structure are discussed below.

Because receiving waters can include rivers, lakes, estuaries, canals, oceans, and other bodies of water, the range of ambient conditions can vary greatly. *Ambient conditions* include the geometry of the receiving water bottom and the receiving water salinity, density, and velocity. Receiving water salinity, density, and velocity may vary with water depth, distance from the discharge point, and time of day and year.

Discharge conditions include the discharge geometry and the discharge flow conditions. The discharge geometry can vary from the end of the pipe to a lengthy multiple-port diffuser. The discharge can be at the water surface or submerged. The submerged outfall can be buried (except for ports) or not. Much of the historical outfall design work deals with discharges from WWTPs. These discharges can be very large—up to several hundred million gallons per day in flow. In ocean outfalls and in many inland outfalls, these discharges are of lower salinity than the receiving water, and the discharge has positive buoyancy. The less dense effluent rises in the more dense receiving water after it is discharged.

The volume of flow of membrane concentrates is on the lower side of the range of WWTP effluent volumes, extending up to perhaps 15 MGD at present. Membrane concentrate, as opposed to WWTP effluent, tends to be of higher salinity than most receiving waters, resulting in a condition of negative buoyancy where the effluent sinks after it is discharged. This raises concerns about the potential impact of the concentrate on the benthic community at the receiving water bottom. Any possible effect on the benthic community is a function of the local ecosystem, the composition of the discharge, and the degree of dilution present at the point of contact. The chance of an adverse impact is reduced by increasing the amount of dilution at the point of bottom contact through diffuser design.

With regard to concentrate discharge regulations it is important to note that receiving waters can differ substantially in their volume, flow, depth, temperature, composition, and degree of variability in these parameters. The effect of discharge of a concentrate or backwash to a receiving water can vary widely depending on these factors. The regulation of effluent disposal to receiving water involves several considerations, including the end-of-pipe characteristics of the concentrate or backwash. Comparison is made between receiving water quality standards (dependent on the classification of the receiving water) and the water quality of the effluent to determine disposal feasibility. In addition, in states such as Florida, the effluent must also pass tests where test species, chosen based on the receiving water characteristics, are exposed to various dilutions of the effluent. Because the nature of the concentrate or backwash is different than that of the receiving water, there is a region near the discharge area where mixing and subsequent dilution of the concentrate or backwash occurs.

Where conditions cannot be met at the end of the discharge pipe, a mixing zone may be granted by the regulatory agency. The mixing zone is an administrative construct that defines a limited area or volume of the receiving water where this initial dilution of the discharge is allowed to occur. The definition of an allowable mixing zone is based on receiving water modeling. The regulations require that certain conditions be met at the edge of the mixing zone in terms of concentration and toxicity.

When the mixing zone conditions have been met, the outfall structure can be properly designed and installed. Actually, the purpose of the outfall structure is to ensure that mixing conditions can be met and that discharge of the effluent, in general, will not produce any damaging effect on the receiving water, its lifeforms, wildlife, and the surrounding area.

In a highly turbulent and moving receiving water with large volume relative to the effluent discharge, simple discharge from the end of a pipe may be sufficient to ensure rapid dilution and mixing of the effluent. For most situations, however, the mixing can be improved substantially through the use of a carefully designed outfall structure. Such a design may be necessary to meet regulatory constraints. The most typical outfall structure for this purpose consists of a pipe of limited length mounted perpendicular to the end of the delivery pipe. This pipe, called a *diffuser*, has one or more discharge ports along its length.

Disposal to the Sewer

Where possible, this means of disposal is simple and usually cost effective. Disposal to a sewer does not require a permit but does require permission from the wastewater treatment plant. The impact of both the flow volume and composition of the concentrate will be considered by the WWTP, as it will affect their capacity buffer and their NPDES permit. The high volume of some concentrates prohibits their discharge to the local WWTP. In other cases, concerns are focused on the increased TDS level of the WWTP effluent that results from the concentrate discharge. The possibility of disposal to a sewer is highly site dependent. In addition to the factors mentioned, the possibility is influenced by the distance between the two facilities, by whether the two facilities are owned by the same entry, and by future capacity increases anticipated. Where disposal to a sewer is allowed, the WWTP may be required to pay fees based on volume or composition.

DEEP WELL DISPOSAL

Injection wells are a disposal option in which liquid wastes are injected into porous subsurface rock formations. Depths of the wells typically range from 1000 to 8000 feet. The rock formation receiving the waste must possess the natural ability to contain and isolate it. Paramount in the design and operation of an injection well is the ability to prevent movement of wastes into or between underground sources of drinking water. Historically, this disposal option has been referred to as *deep well injection* or *disposal to waste disposal wells*. Because of the very slow fluid movement in the injection zone, injection wells may be considered a storage method rather than a disposal method; the wastes remain there indefinitely if the injection program has been properly planned and carried out.

Because of their ability to isolate hazardous wastes from the environment, injection wells have evolved as the predominant form of hazardous waste disposal in the United States. According to one study (Gordon, 1984), almost 60% of all hazardous waste disposed of in 1981, or approximately 10 billion gallons, was injected into deep wells. By contrast, only 35% of this waste was disposed of in surface impoundments and less than 5% in landfills. The study also found that a still smaller volume of hazardous waste, under 500 million gallons, was incinerated in 1981. Although RO concentrate is not classified as hazardous, injection wells are widely used for concentrate disposal in the state of Florida.

A study prepared for the Underground Injection Practices Council showed that relatively few injection well malfunctions have resulted in contamination of water supplies (Strycker and Collins, 1987). However, other studies have documented instances of injection well failure resulting in contamination of drinking water supplies and groundwater resources (Gordon, 1984).

Injection of hazardous waste can be considered safe if the waste never migrates out of the injection zone; however, there are at least five ways a water may migrate and contaminate potable groundwater (Strycker and Collins, 1987):

- Wastes may escape through the wellbore into an underground source of drinking water because of insufficient casing or failure of the injection well casing due to corrosion or excessive injection pressure.
- Wastes may escape vertically outside of the well casing from the injection zone into an underground source of drinking water (USDW) aquifer.
- Wastes may escape vertically from the injection zone through confining beds that are inadequate because of high primary permeability, solution channels, joints, faults, or induced fractures.
- Wastes may escape vertically from the injection zone through nearby wells that are improperly cemented or plugged or that have inadequate or leaky casing.
- Wastes may contaminate groundwater directly by lateral travel of the injected wastewater from a region of saline water to a region of freshwater in the same aquifer.

EVAPORATION POND DISPOSAL

Solar evaporation, a well-established method for removing water from a concentrate solution, has been used for centuries to recover salt (sodium chloride) from seawater. There are also installations that are used for the recovery of sodium chloride and other chemicals from strong brines, such as the Great Salt Lake and the Dead Sea, and for the disposal of brines resulting from oil well operation (Office of Saline Water, 1970).

Evaporation ponds for membrane concentrate disposal are most appropriate for smaller volume flows and for regions having a relatively warm, dry climate with high evaporation rates, level terrain, and low land costs. These criteria apply predominantly in the western half of the United States—in particular, the southwestern portion. Advantages associated with evaporation ponds include the following:

- They are relatively easy and straightforward to construct.
- Properly constructed evaporation ponds are low maintenance and require little operator attention compared to mechanical equipment.
- Except for pumps to convey the wastewater to the pond, no mechanical equipment is required.

- For smaller volume flows, evaporation ponds are frequently the least costly means of disposal, especially in areas with high evaporation rates and low land costs.

Despite the inherent advances of evaporation ponds, they are not without some disadvantages that can limit their application:

- They can require large tracts of land if they are located where the evaporation rate is low or the disposal rate is high.
- Most states require impervious liners of clay or synthetic membranes such as polyvinylchloride (PVC) or Hypalon®. This requirement substantially increases the costs of evaporation ponds.
- Seepage from poorly constructed evaporation ponds can contaminate underlying potable water aquifers.
- There is little economy of scale (i.e., no cost reduction resulting from increased production) for this land-intensive disposal option. Consequently, disposal costs can be large for all but small-sized membrane plants.

In addition to the potential for contamination of groundwater, evaporation ponds have been criticized because they do not recover the water evaporated from the pond. However, the water evaporated is not "lost"; rather, it remains in the atmosphere for about 10 days and then returns to the surface of the Earth as rain or snow. This hydrologic cycle of evaporation and condensation is essential to life on land and is largely responsible for weather and climate.

For evaporation pond design, sizing of the ponds, the evaporation rate, and pond depth are important parameters. Evaporation ponds function by transferring liquid water in the pond to water vapor in the atmosphere above the pond. The rate at which an evaporation pond can transfer this water governs the size of the pond. Selection of pond size requires determination of both the surface area and the depth needed. The surface area required is dependent primarily on the evaporation rate. The pond must have adequate depth for surge capacity and water storage, storage capacity for precipitated salts, and freeboard for precipitation (rainfall) and wave action.

Proper sizing of an evaporation pond depends on accurate calculation of the annual evaporation rate. Evaporation from a freshwater body, such as a lake, is dependent on local climatological conditions, which are very site specific. To develop accurate evaporation data throughout the United States, meteorological stations have been established at which special pans simulate evaporation from large bodies of water such as lakes, reservoirs, and evaporation ponds. The pans are fabricated to standard dimensions and are situated to be as representative of a natural body of water as possible. A standard evaporation pan is referred to as a Class A pan. The standardized dimensions of the pans and the consistent methods for collecting the evaporation data allow comparatively and reasonably accurate data to be developed for the United States. The data collection must cover several years to be reasonably accurate and representative of site-specific variations in climatic conditions. Published evaporation rate databases typically cover a 10-year or greater period and are expressed in inches per year.

The pan evaporation data from each site can be compiled into a map of pan evaporation rates. Because of the small heat capacity of evaporation ponds, they tend to heat and cool more rapidly than adjacent lakes and to evaporate at a higher rate than an adjacent natural pond of water. In general, experience has shown the evaporation rate from large bodies of water to be approximately 70% of that measured in a Class A pan (Bureau of Reclamation, 1969). This percentage is referred to as the *Class A pan coefficient* and must be applied to measured pan evaporation to arrive at actual lake evaporation. Over the years, site-specific Class A pan coefficients have been developed for the entire United States. Multiplying the pan evaporation rate by the pan coefficient results in a mean annual lake evaporation rate for a specific area.

Maps depicting annual average precipitation across the United States also are available. Subtracting the mean annual evaporation from the mean annual precipitation gives the net lake surface evaporation in inches per year. This is the amount of water that will evaporate from a freshwater pond (or the amount the surface level will drop) over a year if no water other than natural precipitation enters the pond. All of these maps assume an impervious pond that allows no seepage. Note that, in some parts of the country, the results of this calculation give a negative number, and in other parts of the country it is a positive number. A negative number indicates a net loss of water from a pond over a year, or a drop in the pond surface level. A positive number indicates more precipitation than evaporation at a particular site. A freshwater pond at one of these sites would actually gain water over a year, even if no water other than natural precipitation were added. Thus, such a site would not be a candidate for an evaporation pond. It is important to realize that data of this type are representative only of the particular sites of the individual meteorological stations, which may be separated by many miles. Climatic data specific to the exact site should be obtained if at all possible before actual construction of an evaporation pond.

The evaporation data described above are for freshwater pond evaporation; however, brine density has a marked effect on the rate of solar evaporation. Most procedures for calculating evaporation rate indicate that evaporation is directly proportional to vapor pressure. Salinity reduces evaporation primarily because the vapor pressure of the saline water is lower than that of freshwater and because dissolved salts lower the free energy of the water molecules. Cohesive forces acting between the dissolved ions and the water molecules may also be responsible for inhibiting evaporation, making it more difficult for the water to escape as vapor (Miller, 1989).

The lower vapor pressure and lower evaporation rate of saline water result in a lower energy loss and, thus, a higher equilibrium temperature than that of freshwater under the same exposure conditions. The increase in temperature of the saline

DID YOU KNOW?

Reverse osmosis concentrate streams are not easily disposed of in inland areas, as surface water and sanitary sewer discharges would not be allowed, and deep well injection may not be feasible depending on geologic features.

DID YOU KNOW?

To gain an understanding of what is meant by incident solar absorption, the following definitions are provided:

- *Incident ray*—A ray of light that strikes (impinges upon) a surface. The angle between this ray and the perpendicular or normal to the surface is the angle of incidence.
- *Reflected ray*—A ray that has rebounded from a surface.
- *Angle of incidence*—The angle between the incident ray and a normal line.
- *Angle of reflection*—The angle between the reflected ray and the normal line.
- *Angle of refraction*—The angle between the refracted ray and the normal line.
- *Index of refraction*—The ratio speed of light (c) in a vacuum to its speed (v) in a given material; it is always greater than 1.

water would tend to increase evaporation, but the water is less efficient in converting radiant energy into latent heat due to the exchange of sensible heat and long-wave radiation with the atmosphere. The net result is that, with the same input of energy, the evaporation rate of saline water is lower than that of freshwater.

For water saturated with sodium chloride salt (26.4%), the solar evaporation rate is generally about 70% of the rate for freshwater (Office of Saline Water, 1970). Studies have shown that the evaporation rate from the Great Salt Lake, which has a TDS level of between 240,000 and 280,000 mg/L, is about 80 to 82% of the rate for freshwater. Other studies indicate that evaporation rates of 2%, 5%, 10%, and 20% sodium chloride solutions are 97%, 98%, 93%, and 78%, respectively, of the rates of freshwater (Bureau of Reclamation, 1969). These ratios have been determined from both experiment and theory. However, there is no simple relationship between salinity and evaporation, for there are always complex interactions among site-specific variables such as air temperature, wind velocity, relative humidity, barometric pressure, water surface temperature, heat exchange rate with the atmosphere, incident solar absorption and reflection, thermal currents in the pond, and depth of the pond. As a result, these ratios should be used only as guidelines and with discretion. It is important to recognize that salinity can significantly reduce the evaporation rate and to allow for this effect when sizing the surface area of the evaporation pond. In lieu of site-specific data, an evaporation ratio of 0.70 is a reasonable allowance for long-term evaporation reduction. This ratio is also considered to be an appropriate factor for evaporation ponds that are expected to reach salt saturation over their anticipated service life.

Pond depth is an important parameter in determining the pond evaporation rate. Studies indicate that pond depths ranging from 1 to 18 inches are optimal for maximizing evaporation rate; however, similar studies have found only a 4% reduction in the evaporation rate as pond depth is increased from 1 to 40 inches (Bureau of

> **DID YOU KNOW?**
>
> Current concentrate disposal of membrane concentrate using evaporation ponds accounts for 5% of total disposal practices.

Reclamation, 1969). Very shallow evaporation ponds are subject to drying and cracking of the liners and are not functional in long-term service for concentrate disposal. From a practical operating standpoint, an evaporation pond not only must evaporate wastewater but must also provide

- Surge capacity or contingency water storage
- Storage capacity for precipitated salts
- Freeboard for precipitation and wave action

For an evaporation pond to be a viable disposal alternative for membrane concentrate, it must be able to accept concentrate at all times and under all conditions so as not to restrict operation of the desalination plant. The pond must be able to accommodate variations in the weather and upsets in the desalination plant, which cannot be shut down because the evaporation pond level is rising faster than anticipated.

To allow for unpredictable circumstances, it is important that design contingencies be applied to the calculated pond area and depth. Experience from the design of industrial evaporation ponds has shown that discharges are largest during the first year of plant operation, are reduced during the second year, and are relatively constant thereafter. A long-term, 20% contingency may be applied to the surface areas of the pond or its capacity to continuously evaporate water. The additional contingencies above the 20% (up to 50%) during the first and second years of operation are applied to the depth holding capacity of the pond.

Freeboard for precipitation should be estimated on the basis of precipitation intensity and duration for the specific site. There may also be local codes governing freeboard requirements. In lieu of site-specific data, an allowance of 6 inches for precipitation is generally adequate where evaporation ponds are most likely to be located in the United States. Freeboard for wave action can be estimated as follows (Office of Saline Water, 1970):

$$H_w = 0.047 \times W \times \sqrt{F} \tag{12.1}$$

where

H_w = Wave height (ft).
W = Wind velocity (mph).
F = Fetch, or straight-line distance the wind can blow without obstruction (miles).

The run-up of waves on the face of the dike approaches the velocity head of the waves and can be approximated as 1.5 times the wave height (H_w). H_w is the freeboard allowance for wave action and typically ranges from 2 to 4 feet. The minimum recommended combined freeboard (for precipitation and wave action) is 2 feet. This minimum applies primarily to small ponds. Over the life of the pond (which should be sized for the same duration as the projected life of the desalination facility), the water will likely reach saturation and precipitate salts. The type and quantity of salts are highly variable and very site specific. To provide the non-engineer or non-scientist with an idea of how (for illustrative reasons only) the evaporation pond evaporation rate is mathematically determined, the following example is provided.

Estimating Rate of Evaporation Pond Evaporation Rate

In lake, reservoir, and pond management, knowledge of evaporative processes is important for understanding how water losses through evaporation are determined. Evaporation increases the storage requirement and decreases the yield of lakes and reservoirs. Several models and empirical methods used for calculating lake and reservoir evaporative processes are described in the following text.

Water Budget Model

The water budget model for lake evaporation is used to make estimations of lake evaporation in some areas. It depends on an accurate measurement of the inflow and outflow of the lake and is expressed as

$$\Delta S = P + R + G_I - G_O - E - T - O \qquad (12.2)$$

where
ΔS = Change in lake storage (mm).
P = Precipitation (mm).
R = Surface runoff or inflow (mm).
G_I = Groundwater inflow (mm).
G_O = Groundwater outflow (mm).
E = Evaporation (mm).
T = Transpiration (mm).
O = Surface water release (mm).

If a lake has little vegetation and negligible groundwater inflow and outflow, lake evaporation can be estimated by

$$E = P + R - O \pm \Delta S \qquad (12.3)$$

Note that much of the following information is adapted from Mosner and Aulenbach (2003).

Energy Budget Model

The energy budget model is recognized as being the most accurate method for determining lake evaporation, although it is also the most costly and time-consuming method (Rosenberry et al., 1993). The evaporation rate, E_{EB}, is given by (Lee and Swancar, 1996):

$$E_{EB} \, (\text{cm/day}) = \frac{Q_s - Q_r + Q_a + Q_{ar} - Q_{bs} + Q_v - Q_x}{L(1 + BR) + T_0} \qquad (12.4)$$

where

E_{EB} = Evaporation (cm/day).
Q_s = Incident shortwave radiation (cal/cm²/day).
Q_r = Reflected shortwave radiation (cal/cm²/day).
Q_a = Incident longwave radiation from atmosphere (cal/cm²/day).
Q_{ar} = Reflected longwave radiation (cal/cm²/day).
Q_{bs} = Longwave radiation emitted by lake (cal/cm²/day).
Q_v = Net energy advected by streamflow, groundwater, and precipitation (cal/cm²/day).
Q_x = Change in heat stored in water body (cal/cm²/day).
L = Latent heat of vaporization (cal/g).
BR = Bowen ratio (dimensionless).
T_0 = Water surface temperature (°C).

Priestly-Taylor Equation

The Priestly–Taylor equation is used to calculate potential evapotranspiration (Winter et al., 1995), which is a measure of the maximum possible water loss from an area under a specified set of weather conditions or evaporation as a function of latent heat of vaporization and heat flux in a water body:

$$PET = \alpha(s/s + \gamma)[(Q_n - Q_x)/L] \tag{12.5}$$

where

PET = Potential evapotranspiration (cm/day).
$\alpha = 1.26$ (dimensionless Priestly–Taylor empirically derived constant).
$(s/s + \gamma)$ = Parameters derived from the slope of the saturated vapor pressure–temperature curve at the mean air temperature; γ is the psychrometric constant, and s is the slope of the saturated vapor pressure gradient (dimensionless).
Q_n = Net radiation (cal/cm²/day).
Q_x = Change in heat stored in water body (cal/cm²/day).
L = Latent heat of vaporization (cal/g).

Penman Equation

The Penman equation for estimating potential evapotranspiration, E_0, can be written as (Winter et al., 1995):

$$E_0 = \frac{(\Delta/\gamma)H_e + E_a}{(\Delta/\gamma) + 1} \tag{12.6}$$

where

Δ = Slope of the saturation absolute humidity curve at air temperature.
γ = Psychrometric constant.
H_e = Evaporation equivalent of the net radiation.
E_a = Aerodynamic expression for evaporation.

DeBruin–Keijman Equation

The DeBruin–Keijman equation determines evaporation rates as a function of the moisture content of the air above the water body, the heat stored in the still water body, and the psychrometric constant, which is a function of atmospheric pressure and latent heat of vaporization (Winter et al., 1995):

$$PET = [(SVP/0.95SVP) + 0.63\gamma)] \times (Q_n - Q_x) \tag{12.7}$$

where SVP is the saturated vapor pressure at mean air temperature (millibars/K), and all other terms have been defined previously.

Papadakis Equation

The Papadakis equation does not account for the heat flux that occurs in the still water body to determine evaporation (Winter et al., 1995). Instead, the equation depends on the difference in the saturated vapor pressure above the water body at maximum and minimum air temperatures, and evaporation is defined as

$$PET \text{ (cm/day)} = 0.5625[E_0 max - (E_0 min - 2)] \tag{12.8}$$

where all terms have been defined previously.

SPRAY IRRIGATION DISPOSAL

Land application methods include irrigation systems, rapid infiltration, and overland flow systems (Crites et al., 2000). These methods, and in particular irrigation, were originally used to take advantage of sewage effluent as a nutrient or fertilizer source as well as to reuse the water. Membrane concentrate has been used for land application in the spray irrigation mode. Using the concentrate in lieu of fresh irrigation water helps conserve natural resources, and in areas where water conservation is of great importance spray irrigation is especially attractive. Because of the higher TDS concentration of RO concentrate, unless it is diluted (recall that dilution is the solution to pollution), the concentrate is less likely to be used for spray irrigation purposes.

Concentrate can be applied to cropland or vegetation by sprinkling or surface techniques for water conservation by exchange when lawns, parks, or golf courses are irrigated and for preservation and enlargement of green belts and open spaces. Where the nutrient concentration of the wastewater for irrigation is of little value, hydraulic loading can be maximized to the extent possible, and system costs can be minimized. Crops such as water-tolerant grasses with low potential for economic return but with high salinity tolerance are generally chosen for this type of requirement.

DID YOU KNOW?

The removal of nutrients is one advantage spray irrigation has compared to conventional disposal methods such as instream discharge.

Fundamental considerations in land application systems include knowledge of wastewater characteristics, vegetation, and public health requirements for successful design and operation. Environmental regulations at each site must be closely examined to determine if spray irrigation is feasible. Contamination of the groundwater and runoff into surface water are key concerns. Also, the quality of the concentrate—its salinity, toxicity, and the soil permeability—must be acceptable.

The principal objective in spray irrigation systems for concentrate discharge is ultimate disposal of the applied wastewater. With this objective, the hydraulic loading is usually limited by the infiltration capacity of the soil. If the site has a relatively impermeable subsurface layer or a high groundwater table, underdrains can be installed to increase the allowable loading. Grasses are usually selected for the vegetation because of their high nutrient requirements and water tolerance.

Other conditions must be met before concentrate irrigation can be considered as a practical disposal option. First, there must be a need for irrigation water in the vicinity of the membrane plant. If the need exists, a contract between the operating plant and the irrigation user would be required. Second, a backup disposal or storage method must be available during periods of heavy rainfall. Third, monitor wells must be drilled before an operating permit is obtained (Conlon, 1989).

With regard to design factors, the following considerations are applicable to spray irrigation of concentrate for ultimate disposal:

- Salt, trace metals, and salinity
- Site selection
- Preapplication treatment
- Hydraulic loading rates
- Land requirements
- Vegetation selection
- Distribution techniques
- Surface runoff control

Salt, Trace Metals, and Salinity

Three factors that affect an irrigation source's long-term influence on soil permeability are the sodium content relative to calcium and magnesium, the carbonate and bicarbonate content, and the total salt concentration of the irrigation water. Sodium salts remain in the soil and may adversely affect its structure. High sodium concentrations in clay-bearing soils disperse soil particles and decrease soil permeability, thus reducing the rate at which water moves into the soil and reducing aeration. If the soil permeability, or infiltration rate, is greatly reduced, then the vegetation on the irrigation site cannot survive. The hardness level (calcium and magnesium) will form insoluble precipitates with carbonates when the water is concentrated. This buildup of solids can eventually block the migration of water through the soil.

The U.S. Department of Agriculture's Salinity Laboratory developed a sodium adsorption ratio (SAR) to determine the sodium limit. It is defined as follows:

$$SAR = Na/[Ca + Mg)/2]^{1/2} \tag{12.9}$$

where
Na = Sodium (milliequivalent per liter, meq/L).
Ca = Calcium (meq/L).
Mg = Magnesium (meq/L).

High SAR values (>9) may adversely affect the permeability of fine-textured soils and can sometimes be toxic to plants.

Trace elements are essential for plant growth; however, at higher levels, some become toxic to both plants and microorganisms. The retention capacity for most metals in most soils is generally high, especially for pH above 7. Under low pH conditions, some metals can leach out of soils and may adversely affect the surface waters in the area.

Salinity is the most important parameter in determining the impact of the concentrate on the soil. High concentrations of salts whose accumulation is potentially harmful will be continually added to the soil with irrigation water. The rate of salt accumulation depends on the quantity applied and the rate at which it is removed from the soil by leaching. The salt levels in many brackish reverse osmosis concentrates can be between 5000 and 10,000 parts per million, a range that normally rules out spray irrigation.

In addition to the effects of total salinity on vegetation and soil, individual ions can cause a reduction in plant growth. Toxicity occurs when a specific ion is taken up and accumulated by the vegetation, ultimately resulting in damage to it. The ions of most concern in wastewater effluent irrigation are sodium, chloride, and boron. Other heavy metals can be very harmful, even if present only in small quantities. These include copper, iron, barium, lead, and manganese. These all have strict environmental regulations in many states.

In addition to the influence on the soil, the effect of the salt concentrations on the groundwater must be considered. The possible impact on groundwater sources may be a difficult obstacle where soil saturation is high and the water table is close to the surface. The chance of increasing background TDS levels of the groundwater is high with the concentrate. Due to this consideration, spray irrigation requires a runoff control system. An underdrain or piping distribution system may have to be installed under the full areas of irrigation to collect excess seepage through the soil and protect the groundwater sources. If high-salinity concentrate is being used, scaling of the underdrain may become a problem. The piping perforations used to collect the water can be easily scaled because the openings are generally small. Vulnerability to scaling must be carefully evaluated before a project is undertaken.

DID YOU KNOW?

Soluble salts in a water solution will conduct an electric current; thus, changes in electrical conductivity (EC) can be used to measure the water's salt content in electrical resistance units (deciSiemens per meter, or dS/m).

TABLE 12.12

Site Selection Factors and Criteria

Factor	Criterion
Soil	
Type	Loamy soils are preferred, but most soils from sands to clays are acceptable.
Drainability	Well-drained soil is preferred.
Depth	Uniformly 5 to 6 feet or more throughout sites is preferred.
Groundwater	
Depth to groundwater	A minimum of 5 feet is preferred.
Groundwater control	Control may be necessary to ensure renovation if the water table is less than 10 feet from the surface.
Groundwater movement	Velocity and direction of movement must be determined.
Slopes	Slopes of up to 20% are acceptable with or without terracing.
Underground formations	Formations should be mapped and analyzed with respect to interference with groundwater or percolating water movements.
Isolation	Moderate isolation from public is preferred; the degree of isolation depends on wastewater characteristics, method of application, and crop.
Distance from source of wastewater	An appropriate distance is a matter of economics.

Site Selection

Site selection factors and criteria for effluent irrigation are presented in Table 12.12. A moderately permeable soil capable of infiltration up to 2 inches per day on an intermittent basis is preferable. The total amount of land required for land application is highly variable but primarily depends on application rates.

Preapplication Treatment

Factors that should be considered when assessing the need for preapplication treatment include whether the concentrate is mixed with additional wastewaters before application, the type of vegetation grown, the degree of contact with the wastewater by the public, and the method of application. At four Florida sites, concentrate is aerated before discharge, because each plant discharges to a retention pond or ponds before irrigation. Aeration by increasing dissolved oxygen prevents stagnation and algae growth in the ponds and also supports fish populations. The ponds are required for flow equalization and mixing. Typically, concentrate is blended with biologically treated wastewater.

Hydraulic Loading Rates

Determining the hydraulic loading rate is the most critical step in designing a spray irrigation system. The loading rate is used to calculate the required irrigation area and is a function of precipitation, evapotranspiration, and percolation. The following equation represents the general water balance for hydraulic loading based on a monthly time period and assuming zero runoff:

$$HLR = ET + PER - PPT \tag{12.10}$$

where

    HLR = Hydraulic loading rate.
    ET = Evapotranspiration.
    PER = Percolation.
    PPT = Precipitation.

In most cases, surface runoff from fields irrigated with wastewater is not allowed without a permit or, at least, must be controlled; it is usually controlled just so that a permit does not have to be obtained.

Seasonal variations in each of these values would be taken into account by evaluating the water balance for each month as well as the annual balance. For precipitation, the wettest year in 10 is suggested as reasonable in most cases. Evapotranspiration will also vary from month to month, but the total for the year should be relatively constant. Percolation includes that portion of the water that, after infiltration into the soil, flows through the root zone and eventually becomes part of the groundwater. The percolation rate used in the calculation should be determined on the basis of a number of factors, including soil characteristics, underlying geologic conditions, groundwater conditions, and the length of the drying period required for satisfactory vegetation growth. The principal factor is the permeability or hydraulic conductivity of the least permeable layer in the soil profile.

Resting periods, standard in most irrigation techniques, allow the water to drain from the top few inches of soil. Aerobic conditions are thus restored, and air penetrates the soil. Resting periods may range from a portion of each day to 14 days and depend on the vegetation, the number of individual plots in the rotation cycle, and the availability of backup storage capacity.

To properly calculate an annual hydraulic loading rate, it is necessary to obtain monthly evapotranspiration, precipitation, and percolation rates. The annual hydraulic loading rate represents the sum of the monthly loading rates. Recommended loading rates range from 2 to 20 feet per year (Goigel, 1991).

Land Requirements

When a hydraulic loading rate has been determined, the required irrigation area can be calculated using the following equation:

$$A = Q \times K_1/ALR \tag{12.11}$$

where

    A = Irrigation area (acre).
    Q = Concentrate flow (gpd).
    K_1 = 0.00112 d $\times$ ft^3 $\times$ acres/(hr $\times$ gal $\times$ ft^2).
    ALR = Annual hydraulic loading rate (ft/yr).

The total land area required for spray irrigation includes allowances for buffer zones and storage and, if necessary, land for emergencies or future expansion.

For loadings of constituents such as nitrogen, which may be of interest to golf course managers who need fertilizer for the grasses, the field area requirement is calculated as follows:

$$\text{Field area (acres)} = 3040 \times C \times Q/L_c \qquad (12.12)$$

where
C = Concentration of constituent (mg/L).
Q = Flow rate (mgd).
L_c = Loading rate of constituent (lb/acre-yr).

Vegetation Selection

The important aspects of vegetation for irrigation systems are water needs and tolerances, sensitivity to wastewater constituents, public health regulations, and vegetation management considerations. The vegetation selection depends highly on the location of the irrigation site and natural conditions such as temperature, precipitation, and topsoil condition. Automated watering alone cannot always ensure vegetation propagation. Vegetation selection is the responsibility of the property owners. Woodland irrigation for growing trees is being conducted in some areas. The principal limitations on this use of wastewater include low water tolerances of certain trees and the necessity to use fixed sprinklers, which are expensive. Membrane concentrate disposal will generally be to landscape vegetation. Such application (e.g., to highway median and border strips, airport strips, golf courses, parks and recreational areas, wildlife areas) has several advantages. Problems associated with crops for consumption are avoided, and the irrigated land is already owned, so land acquisition costs are saved.

Distribution Techniques

Many different distribution techniques are available for engineered wastewater effluent applications. For irrigation, two main types, sprinkling and surface application, are used. Sprinkling systems used for spray irrigation are of two types—fixed and moving. Fixed systems, often called solid set systems, may be either on the ground surface or buried. Both types usually consist of impact sprinklers mounted on risers that are spaced along lateral pipelines, which are, in turn, connected to main pipelines. These systems are adaptable to a wide variety of terrains and may be used for irrigation of either cultivated land or woodlands. Portable aluminum pipe is normally used for aboveground systems. This pipe has the advantage of relatively low capital cost but is easily damaged, has a short expected life because of corrosion, and must be removed during cultivation and harvesting operations. Pipe used for buried systems may be buried as deep as 1.5 feet below the ground surface. Buried systems usually have the greatest capital cost; however, they are probably the most dependable and are well suited to automated control. There are a number of different moving sprinkle systems, including center-pivot, side-roll, wheel-move, rotating-boom, and winch-propelled systems.

Surface Runoff Control

Surface runoff control depends mainly on the proximity of surface water. If run-off drains to a surface water, an NPDES permit may be required. This situation should be avoided if possible due to the complication of quantifying overland runoff. Berms can be built around the irrigation field to prevent runoff. Another alternative, although expensive, is a surrounding collection system. It is best to use precautions and backup systems to ensure that overwatering and subsequent runoff do not occur in the first place.

ZERO LIQUID DISCHARGE DISPOSAL

In this approach, evaporation is used to further concentrate the membrane concentrate. For the extreme limit of processing concentrate to dry salts, the method becomes a zero discharge option. Evaporation requires major capital investment, and the high energy consumption together with the final salt or brine disposal can result in significant disposal costs. Because of this, disposal of municipal membrane concentrate by mechanical evaporation would typically be considered to be a last resort—that is, when no other disposal option is feasible. Cost aside, however, zero liquid discharge does offer some advantages:

- It may avoid a lengthy and tedious permitting process.
- It may gain quick community acceptance.
- It can be located virtually anywhere.
- It represents a positive extreme in recycling by efficiently using the water source.

When this thermal process is used following an RO system, for example, it produces additional product water by recovering high-purity distillate from the concentrate wastewater stream. The distillate can be used to help meet the system product water volume requirement. This reduces the size of the membrane system and, thus, the size of the membrane concentrate to be treated by the thermal process. In addition, because the product purity of the thermal process is so high (TDS in the range of 10 mg/L), some of the product water volume reduction of the system may be met by blending the thermal product with untreated source water. The usual concerns and considerations of using untreated water for blending must be addressed. The end result may be a system where the system product requirement is met by three streams: (1) membrane product, (2) thermal process product, and (3) bypass water.

Single- and Multiple-Effect Evaporators

Using steam as the energy source, it takes about 1000 British thermal units (Btu) to evaporate a pound of water. In a single-effect evaporator, heat released by the condensing steam is transferred across a heat exchange surface to an aqueous solution boiling at a temperature lower than that of the condensing stream. The solution absorbs heat, and part of the solution water vaporizes, causing the remaining solution to become richer in solution. The water vapor flows to a barometric or surface condenser, where it condenses as its latent heat is released to cooling water at a lower temperature. The finite temperature differences among the steam, the boiling liquid, and the condenser are the driving forces required for the heat transfer surface area to be less than infinite. Practically all of the heat removed from the condensing stream (which had been generated initially by burning fuel) is rejected to cooling water and is often dissipated to the environment without being of further use.

The water vapor that flows to the condenser in a single-effect evaporator is at a lower temperature and pressure than the heating stream but has almost as much enthalpy. Instead of releasing the latent heat to cooling water, the water vapor may be used as heating steam in another evaporator effect operating at a lower temperature and pressure than the first effect. Additional effects may be added in a similar manner, each generating additional vapor, which may be used to heat a lower temperature effect. The vapor generated in the lowest temperature effect finally is condensed by releasing its latent heat to cooling water in a condenser. The economy of a single- or multiple-effect evaporator may be expressed as the ratio of kilograms of total evaporation to kilograms of heating steam. As effects are added, the economy increases, representing more efficient energy utilization. Eventually, added effects result in marginal added benefits, and the number of effects is thus limited by both practical and economic considerations. Multiple effect evaporators increase the efficiency (economy) but add capital costs in additional evaporator bodies.

More specifically, the number of effects, and thus the economy achieved, is limited by the total temperature difference between the saturation temperature of the heating steam (or other heat source) and the temperature of the cooling water (or other heat sink). The available temperature difference may also be constrained by the temperature sensitivity of the solution to be evaporated. The total temperature difference, less any losses, becomes allocated among effects in proportion to their resistance to heat transfer, the effects being thermal resistances in series.

The heat transfer surface area for each effect is inversely proportional to the net temperature difference available for that effect. Increasing the number of effects reduces the temperature difference and evaporation duty per effect, which increases the total area of the evaporator in rough proportion to the number of effects. The temperature difference available to each effect is reduced by boiling point elevation and by the decrease in vapor saturation temperature due to pressure drop. The boiling point elevation of a solution is the increase in boiling point of the solution compared to the boiling point of pure water at the same pressure; it depends on the nature of the solute and increases with increasing solute concentration. In a multiple-effect evaporator, the boiling point elevation and vapor pressure drop losses for all the effects must be summed and subtracted from the overall temperature difference between the heat source and sink to determine the net driving force available for heat transfer.

Vapor Compression Evaporator Systems (Brine Concentrators)

A vapor compression evaporator system, or brine concentrator, is similar to a conventional single-effect evaporator, except that the vapor released from the boiling solution is compressed in a compressor. Compression raises the pressure and saturation temperature of the vapor so that it may be returned to the evaporator steam

DID YOU KNOW?

In the British system of units, the unit of heat is the British thermal unit, or Btu. One Btu is the amount of heat required to raise 1 pound of water 1°F at normal atmospheric pressure (1 atm).

chest to be used as heating steam. The latent heat of the vapor is used to evaporate more water instead of being rejected to cooling water. The compressor adds energy to the vapor to raise its saturation temperature above the boiling temperature of the solution by whatever net temperature difference is desired. The compressor is not completely efficient, as it is subject to small losses due to mechanical friction and larger losses due to non-isentropic compression. However, the additional energy required because of non-isentropic compression is not lost from the evaporator system; instead, it serves to superheat the compressed vapor. The compression energy added to the vapor is of the same magnitude as energy required to raise feed to the boiling point and make up for radiation and venting losses. By exchanging heat between the condensed vapors (distillate) and the product with the feed, it is usually possible to operate with little or no makeup heat in addition to the energy necessary to drive the compressor. The compressor power is proportional to the increase in saturation temperature produced by the compressor. The evaporator design must trade off between compressor power consumption and heat transfer surface area. Using the vapor compression approach to evaporate water requires only about 100 Btu to evaporate a pound of water. Thus, one evaporator body driven by mechanical vapor compression is equivalent to 10 effects, or a 10-body system driven by steam.

Although most brine concentrators have been used to process cooling water, concentrators have also been used to concentrate reject from RO plants. Approximately 90% of these concentrators operate with a seeded slurry process that allows the reject to be concentrated as much as 40 to 1 without scaling problems developing in the evaporator. Brine concentrators also produce a distilled product water that can be used for high-purity purposes or for blending with other water supplies. Because of the ability to achieve such high levels of concentration, brine concentrators can reduce or eliminate the need for alternative disposal methods such as deep well injection or solar evaporation ponds. When operated in conjunction with crystallizers or spray dryers, brine concentrators can achieve zero liquid discharge of RO concentrate under all climatic conditions.

Individual brine concentrator units range in capacity from approximately 10 to 700 gpm of feedwater flow. Units below 150 gpm of capacity are usually skid mounted, and larger units are field fabricated. A majority of operating brine concentrators are single-effect, vertical-tube, falling-film evaporators that use a calcium sulfate-seeded slurry process. Energy input to the brine concentrator can be provided by an electric-driven vapor compressor or by process steam from a host industrial facility. Steam-driven systems can be configured with multiple effects to minimize energy consumption. Product water quality is normally less than 10 mg/L TDS. Brine reject from the concentrator typically ranges between 2 and 10% of the feedwater flow, with TDS concentrations as high as 250,000 mg/L.

DID YOU KNOW?

Solids in water occur either in solution or in suspension and are distinguished by passing the water sample through a glass-fiber filter. The suspended solids are retained on top of the filter, and the dissolved solids pass through the filter with the water. When the filtered portion of the water sample is placed in a small dish and then evaporated, the solids in the water that remain as residue in the evaporating dish are the total dissolved solids (TDS). Dissolved solids may be organic or inorganic. Water may come into contact with these substances within the soil, on surfaces, and in the atmosphere. The organic dissolved constituents of water are from the decay products of vegetation, from organic chemicals, and from organic gases. Removing these dissolved minerals, gases, and organic constituents is desirable, because they may cause physiological effects and produce aesthetically displeasing color, taste, and odors.

Because of the corrosive nature of many wastewater brines, brine concentrators are usually constructed of high-quality materials, including titanium evaporator tubes and stainless-steel vessels suitable for 30-year evaporator life. For conditions of high chloride concentrations or other more corrosive environments, brine concentrators can be constructed of materials such as AL-6XN®, Incoloy® 825, or other exotic metals to meet performance and reliability requirements.

Crystallizers

Crystallizer technology has been used for many years to concentrate feed streams in industrial processes. More recently, as the need to concentrate wastewaters has increased, this technology has been applied to reject from desalination processes, such as brine concentrate evaporators, to reduce wastewater to a transportable solid. Crystallizer technology is especially applicable in areas where solar evaporation pond construction costs are high, solar evaporation rates are negative, or deep well disposal is costly, geologically not feasible, or not permitted.

Crystallizers used for wastewater disposal range in capacity from about 2 to 50 gpm. These units have vertical cylindrical vessels with heat input from vapor compressors or an available steam supply. For small systems ranging from 2 to 6 gpm, steam-driven crystallizers are more economical. Steam can be supplied by a package boiler or a process source, if one is available. For larger systems, electrically driven vapor compressors are normally used to supply heat for evaporation.

Typically, the crystallizer requires a purge stream of about 2% of the feed to the crystallizer. This is necessary to prevent extremely soluble species (such as calcium chloride) from building up in the vapor body and to prevent production of dry

DID YOU KNOW?

The method of evaporation selected is based on the characteristics of the RO membrane concentrate and the type of energy source to be used.

cake solids. The suggested disposal of this stream is to a small evaporation pond. The crystallizer produces considerable solids that can be disposed of to commercial landfill. The first crystallizers applied to power plant wastewater disposal experienced problems related to materials selection and process stability, but subsequent design changes and operating experience have produced reliable technology. For RO concentrate disposal, crystallizers would normally be operated with a brine concentrator evaporator to reduce brine concentrator blowdown to a transportable solid. Crystallizers can be used to concentrate RO reject directly, but their capital costs and energy usage are much higher than for a brine concentrator of equivalent capacity.

Spray Dryers

Spray dryers provide an alternative to crystallizers for concentration of wastewater brines to dryness. Spray dryers are generally more cost effective for smaller feed flows of less than 10 gpm.

PATHOGEN REMOVAL

Various states have developed different approaches to regulating pathogen removal by indirect potable reuse plants. For example, Virginia permitted the UOSA indirect potable reuse plant and the Broad Run Water Reclamation Facility based on achieving a nondetect concentration of *Escherichia coli* (less than 2 cfu/100 mL). Other states have taken a different approach. For example, California requires 12-log reduction of viruses and 10-log reduction of *Cryptosporidium* and *Giardia* from the raw wastewater to the advanced water treatment plant finished water.

Pathogen removal by each of the three proposed treatment trains is significant and would result in nondetect concentrations for all indicator organisms typically used in wastewater treatment (e.g., *E. coli*, total coliform, fecal coliform) assuming proper operation. Therefore, compliance with a UOSA-type permit or Florida's pathogen-related indirect potable reuse regulations would be met by all three proposed treatment trains. Compliance with the California regulations may be more challenging, especially for GAC-based treatment, because of the high log reduction requirements. Tables 12.13 and 12.14 show the calculated log reduction credits for each of the proposed treatment trains. Note that log reduction credits associated with 6 months of subsurface travel time as water moves through the aquifer (soil aquifer treatment) have been assumed in the calculations. Discussion with the Virginia regulators is necessary to determine how the proposed HRSD groundwater recharge project would be regulated with respect to pathogen removal.

DISINFECTION BYPRODUCTS

Excessive formation of trihalomethanes (THMs) at WWTPs is fairly common, especially for plants that provide good nitrification. Low effluent NH_3 concentrations at these plants lead to the formation of free chlorine (rather than chloramines) during chlorine disinfection, which, when reacted with bulk organics, has a propensity to form high levels of THMs. NDMA, another disinfection byproduct, can also form in significant concentrations during the disinfection process at WWTPs depending

TABLE 12.13

Estimated Log Reduction Credits for the MF–RO/NF–UVAOP–Cl$_2$–SAT Process

Parameter	Microfiltration (MF)	Reverse Osmosis (RO)/ Nanofiltration (NF)	Log Reduction Credits Ultraviolet Advanced Oxidation Process (UVAOP)	Chlorine (Cl$_2$)	Soil Aquifer Treatment (SAT)	Total
Enteric viruses	0	2	6	4	6	18
Cryptosporidium	4	2	6	4	6	18
Giardia	4	2	6	0	6	18

Source: CH2M, *Sustainable Water Recycling Initiative: Advanced Water Purification Process Feasibility Evaluation*, Report No. 3, CH2M, Newport News, VA, 2016.

Note: (1) Four-log removal of *Cryptosporidium* and *Giardia* has been assumed for microfiltration based on credit commonly granted by states for membranes passing daily membrane integrity tests. (2) Two-log removal of viruses, *Cryptosporidium*, and *Giardia* has been assumed for reverse osmosis based on credit commonly granted by states for online monitoring of conductivity. (3) Two-log removal of viruses, *Cryptosporidium*, and *Giardia* has been assumed for reverse osmosis based on credit commonly granted by states for only monitoring of conductivity or TOC. (4) Six-log removal of viruses, *Cryptosporidium*, and *Giardia* has been assumed for UVAOP based on an ultraviolet dose in excess of 500 mJ/cm^2. (5) A travel time of 6 months in the aquifer has been assumed for SAT processes, and a 1-log virus removal credit for each month of travel, in accordance with California's indirect potable reuse regulations, has been assumed. One-log removal of *Cryptosporidium* and *Giardia* for each month of travel has also been assumed because the size of these organisms is significantly larger than that of viruses.

TABLE 12.14

Estimated Log Reduction Credits for the Coagulation/Sedimentation–Ozone–BAC–GAC–UV–Cl$_2$–SAT Process

			Log Reduction Credits					
Parameter	Coagulation/ Sedimentation (+ BAC)	Ozone	Biological Activated Carbon (BAC)	Granular Activated Carbon (GAC)	Ultraviolet (UV)	Chlorine (Cl$_2$)	Soil Aquifer Treatment (SAT)	Total
Enteric viruses	2	4	0	0	0.5	4	6	16.5
Cryptosporidium	2	0	0	0	3.0	0	6	11.0
Giardia	2.5	3	0	0	3.0	0	6	14.5

Source: CH2M, *Sustainable Water Recycling Initiative: Advanced Water Purification Process Feasibility Evaluation*, Report No. 3, CH2M, Newport News, VA, 2016.

Note: (1) Coagulation/sedimentation/biological activated carbon (BAC) provides 2-log virus removal, 2-log *Cryptosporidium* removal, and 2.5-log *Giardia* removal per the Safe Water Treatment Rule (SWTR) for a well-operated plant. (2) Ozone provides 4-log virus removal and 3-log *Giardia* removal for a CT of 1.4 mg/L-min at 10°C per SWTR. (3) UV provides 0.5-log virus removal, 3-log *Cryptosporidium* removal, and 3-log *Giardia* removal at a dose of 40 mJ/cm². (4) Free chlorine provides 4-log virus removal for CT of 6 mg/L-min at a temperature of 10°C. (5) See Table 12.13 for a description of log reduction credits.

on the precursors in the water and the type of chlorination practiced. Little NDMA forms in the presence of free chlorine, but significant concentrations typically form in the presence of chloramines, with dichloramine resulting in more rapid formation kinetics than monochloramine. Both THMs and NDMA can be removed by AWT processes through specialized design, but a more cost-efficient approach is to prevent their formation by withdrawing the water from the WWTP prior to disinfection (upstream of the chlorine contact basin). Specific withdrawal points at each WWTP and the potential treatment required for TTHMs and NDMA removal should be considered in the next stage of this project.

ANTICIPATED IMPROVEMENTS TO HRSD'S EXISTING WWTPS

Some operational and capital improvements to the existing WWTPs may be required depending on the AWT train selected for implementation and the final effluent water quality produced at each WWTP. Table 12.15 shows the improvements that will likely be required. Analysis regarding WWTP improvements should be ongoing.

THE BOTTOM LINE

From the data and evaluations presented to this point, it is obvious that the bottom line is that any of the three advanced water treatment trains—RO-based, NF-based, or GAC-based—is likely to be viable for groundwater recharge of effluent generated from HRSD's WWTPs. Finished water quality produced by each train will be excellent with respect to pathogen and organics removal, but use of the RO-based or NF-based treatment train is necessary if TDS reduction is required. Partial RO or NF treatment could be used depending on the degree of TDS reduction required. BNR improvements will be required at some of the WWTPs to reduce the TN concentration and the propensity for organic fouling membranes in the RO- and NF-based trains.

Selection of the advanced water treatment train to be implemented at each WWTP should be based on numerous factors, such as finished water quality, wastewater discharge requirements, operability, sustainability, site-specific characteristics (e.g., space, existing infrastructure, hydraulics), and capital and operating costs (discussed later). Ultimate selection of the advanced water treatment train will also be dictated by regulatory requirements related to treatment, finished water quality, and wastewater discharge requirements that have not yet been established; therefore, engaging the appropriate regulatory agencies is important during the next phase of this project. Treatment selection may also be influenced by public perception.

Because HRSD's SWIFT project is a work in progress, with time for adjustment here, there, and almost anywhere, other action items that will influence advanced water treatment train selection should be considered, including the following:

- Regularly sample at each WWTP for COD, sCOD, TOC, DOC, all contaminants regulated by primary MCLs, selected CECs, and parameters specific to the design of RO and NF treatment (i.e., barium, strontium, fluoride, silica, alkalinity, pH).

TABLE 12.15

Required Wastewater Treatment Plant (WWTP) Improvements to Address Advanced Water Treatment (AWT) Operational Impacts and/or Finished Water Quality Deficiency

WWTP	MF–RO–UVAOP	MF–NF–UVAOP	FS–O3–BAC–GAC–UV
Army Base	None	None	*AWT finished water deficiency:* TDS > 750 mg/L; *Required WWTP improvements:* If possible, reduce influent TDS to WWTP
Boat Harbor	*AWT operational impact:* Greater membrane fouling requiring more conservative design due to lack of BNR; *AWT finished water deficiency:* None; *Required WWTP improvements:* Add nitrification/denitrification at WWTP	*AWT operational impact:* Greater membrane fouling requiring more conservative design due to lack of BNR; *AWT finished water deficiency:* TN > 10 mg/L; *Required WWTP improvements:* Add nitrification/denitrification at WWTP	*AWT finished water deficiency:* TN > 10 mg/L; total hardness > 150 mg/L; *Required WWTP improvements:* Add nitrification/denitrification at WWTP; if determined necessary, add softening to AWT
James River	None	None	None
Nansemond	None	None	None
Virginia Initiative Plant (VIP) and Williamsburg	None	*AWT finished water deficiency:* TN approaching 10 mg/L; *Required WWTP improvements:* Improve nitrification/denitrification at WWTP	*AWT finished water deficiency:* TN approaching 10 mg/L; total hardness > 150 mg/L (VIP only); TDS > 750 mg/L; *Required WWTP improvements:* Improve nitrification/denitrification at WWTP and reduce influent TDS to WWTP (VIP only), if possible; if determined necessary, add softening to AWT (VIP only)
York River	None	None	*AWT finished water deficiency:* Hardness > 150 mg/L; *Required WWTP improvements:* If determined necessary, add softening to AWT

Source: CH2M, *Sustainable Water Recycling Initiative: Advanced Water Purification Process Feasibility Evaluation.* Report No. 3, CH2M, Newport News, VA, 2016.

Note: MF, microfiltration; RO, reverse osmosis; UVAOP, ultraviolet advanced oxidation process; NF, nanofiltration; FS, ferric sulfate; BAC, biological activated carbon; GAC, granular activated carbon; SAT, soil aquifer treatment; TDS, total dissolved solids; BNR, biological nutrient removal.

- Regularly measure water quality (e.g., pH, alkalinity, TDS, hardness) in numerous Potomac Aquifer product wells.
- Evaluate site-specific conditions at each WWTP that may influence AWT train selection, including site space, hydraulics, geotechnical conditions, electrical service, and use of existing infrastructure for AWT.
- Conduct an industrial and commercial water quality discharge study to characterize risk and to identify chemicals of concern that may be discharged to the collection system.
- Determine potential causes for high TDS concentrations at WWTPs where effluent TDS is greater than 500 mg/L.

REFERENCES AND RECOMMENDED READING

Anderman, E.R. and Hill, M.C. (2000). *MODFLOW-2000, the U.S. Geological Survey Modular Ground-Water Model—Documentation of the Hydrogeologic-Unit Flow (HUF) Package*, Open-File Report 2000-342. Reston, VA: U.S. Geological Survey (https://pubs.er.usgs.gov/publication/ofr00342).

Bahremand, A. and De Smedt, F. (2008). Distributed hydrological modeling and sensitivity analysis in Torysa watershed, Slovakia. *Water Resources Management*, 22(3): 293–408.

Bair, E.S., Springer, A.E., and Roadcap, G.S. (1992). *CAPZONE*. Columbus: Ohio State University.

Brown, D.L. and Silvey, W.D. (1977). *Artificial Recharge to a Freshwater-Sensitive Brackish-Water Sand Aquifer*, U.S. Geological Survey Professional Paper 939. Reston, VA: U.S. Geological Survey.

Bureau of Reclamation. (1969). *Disposal of Brine Effluents from Desalting Plants*. Washington, DC: U.S. Department of the Interior.

Carlson, E., Giwercam, A., Keiding, N., and Skakkebaek, N.E. (1992). Evidence for decreasing quality of semen during past 50 years. *British Medical Journal*, 305: 609–612.

CH2M. (2016a). *Sustainable Water Recycling Initiative: Groundwater Injection Hydraulic Feasibility Evaluation*, Report No. 1. Newport News, VA: CH2M.

CH2M. (2016b). *Sustainable Water Recycling Initiative: Groundwater Injection Geochemical Compatibility Feasibility Evaluation*, Report No. 2. Newport News, VA: CH2M.

CH2M. (2016c). *Sustainable Water Recycling Initiative: Advanced Water Purification Process Feasibility Evaluation*, Report No. 3. Newport News, VA: CH2M.

Civan, F. (2000). *Reservoir Formation Damage: Fundamentals, Modeling, Assessment, and Mitigation*. Houston, TX: Gulf Publishing.

Colborn, T., vom Saal, F.S., and Soto, A.M. (1993). Developmental effects of endocrine-disrupting chemicals in wildlife and humans. *Environmental Health Perspectives*, 101(5): 378–384.

Conlon, W.J. (1989). Disposal of concentrate from membrane process plants. *AWWA Journal*, 80(4): 67.

Crites, R.W., Reed, S.C., and Bastian, R.K. (2000). *Land Treatment Systems for Municipal and Industrial Wastes*. McGraw-Hill, New York.

Davies, C.W. (1962). *Ion Association*. Washington, DC: Butterworths.

Debye, P. and Huckel, E. (1923). On the theory of electrolytes. I. Freezing point depression and related phenomena. *Physikalische Zeitschrift*, 24(9): 185–206.

Drever, J.M. (1982). *The Geochemistry of Natural Waters*. Upper Saddle River, NJ: Prentice Hall.

Drever, J.M. (1988) *The Geochemistry of Natural Waters*, 2nd ed. Upper Saddle River, NJ: Prentice Hall.

Drewes, J. (2013). *Predictive Models to Aid in Design of Membrane Systems for Organic Micropollutant Removal*. Alexandria, VA: WateReuse Research Foundation.

Dzombak, D.D. and Morel, F.M.M. (1990). *Surface Complex Modeling*. New York: John Wiley & Sons.

Evangelou, V.P. (1995). *Pyrite Oxidation and Its Control*. New York: CRC Press.

Eykhoff, P. (1974). *System Identification Parameter and State Estimation*. New York: John Wiley & Sons.

Focazio, M.J., Speiran, G.K., and Rowan, M.E. (1993). *Quality of Groundwater in the Coastal Plain Physiographic Province of Virginia*, Water-Resources Investigations Report 92-4175. Reston, VA: U.S. Geological Survey.

Goigel, J.F. (1991). Regulatory Investigation and Cost Analysis of Concentrate Disposal from Membrane Plants, master's thesis, University of Central Florida, Orlando.

Gordon, W. (1984). *A Citizen's Handbook on Ground Water Protection*. New York: Natural Resources Defense Council.

Gray, D.H. and Rex, R.W. (1966). *Formation Damage in Sandstones Caused by Clay Dispersion and Migration*. La Habra, CA: Chevron Research Company.

Hamilton, P.A. and Larson, J.D. (1988). *Hydrogeology and Analysis of the Ground-Water-Flow System in the Coastal Plain of Southeastern Virginia*, Water Resources Investigations Report 87-4240. Reston, VA: U.S. Geological Survey.

Hantash, J.E. and Jacob, C.E. (1955). Nonsteady radial flow in an infinite leaky aquifer. *Transactions of the American Geophysical Union*, 36(1): 95–100.

Hewitt, E.J. (1963). Mineral nutrition of plants in culture media. In: *Plant Physiology* (Steward, F.C., Ed.), Vol. III, pp. 97–133. New York: Academic Press.

Heywood, C.E. and Pope, J.P. (2009). *Simulation of Groundwater Flow in the Coastal Plain Aquifer System of Virginia*, Scientific Investigations Report 2009-5039. Reston, VA: U.S. Geological Survey (http://pubs.usgs.gov/sir/2009/5039/).

Hill, M. and Tiedeman, C. (2007). *Effective Groundwater Model Calibration, with Analysis of Data, Sensitivities, Prediction, and Uncertainty*. New York: John Wiley & Sons.

Hill, M., Kavetski, D., Clark, M., Ye, M., Arabic, M., Lu, D., Foglia, L., and Mehl, S. (2015). Practical use of computationally frugal model analysis methods. *Groundwater*, 54(2): 159–170.

Kavlock, R.J. et al. (1996). Research needs for the risk assessment of health and environmental effects of endocrine disruptors: a report of the U.S. EPA-sponsored workshop. *Environmental Health Perspectives*, 104(Suppl. 4): 715–740.

Knoell, T. (2014). RO element replacement strategy for a large-scale southern California water reuse facility. *Industrial Water Treatment*, 31(5): 23–29.

Laczniak, R.J. and Meng III, A.A. (1988). *Ground-Water Resources of the York–James Peninsula of Virginia*. Water Resources Investigations Report 88-4059, Reston, VA: U.S. Geological Survey.

Langevin, C.D., Thorne, Jr., D.T., Dausman, A.M. et al. (2008). *SEWAT Version 4: A Computer Program for Simulation of Multi-Species Solute and Heat Transport*. Reston, VA: U.S. Geological Survey (https://pubs.usgs.gov/tm/tm6a22/).

Langmuir, D. (1997). *Aqueous Environmental Geochemistry*. Upper Saddle River, NJ: Prentice Hall.

Leamer, E. (1978). *Specification Searches: Ad Hoc Inferences with Nonexperimental Data*. New York: John Wiley & Sons.

Lee, T.M. and Swancar, A. (1996). *Influence of Evaporation, Ground Water, and Uncertainty in the Hydrologic Budget of Lake Lucerne, a Seepage Lake in Polk County, Florida*, Water-Supply Paper 2439. Washington, DC: U.S. Geologic Survey.

McFarland, E.R. (2013). *Sediment Distribution and Hydrologic Conditions of the Potomac Aquifer in Virginia and Parts of Maryland and North Carolina*. Scientific Investigations Report 2013-5116. Reston, VA: U.S. Geological Survey.

McFarland, E.R. and Bruce, T.S. (2006). *The Virginian Coastal Plain Hydrogeologic Framework*, U.S. Geological Survey Professional Paper 1731. Reston, VA: U.S. Geologic Survey.

McGill, K. and Lucas, M.C. (2009). Mitigating Specific Capacity Losses in Aquifer Storage and Recovery Wells in the New Jersey Coastal Plain, paper presented at New Jersey American Water Works Association Annual Conference, Atlantic City, NJ, March 31.

Meade, R.H. (1964). *Removal of Water and Rearrangement of Particles During the Compaction of Clayey Sediments*, U.S. Geological Survey Professional Paper 497-B. Reston, VA: U.S. Geological Survey.

Meng, A.A. and Harsh, J.F. (1988). *Hydrogeologic Framework of the Virginia Coastal Plain*. U.S. Geological Survey Professional Paper 1404-C. Reston, VA: U.S. Geological Survey.

Miller, W. (1989). Estimating evaporation from Utah's Great Salt Lake using thermal infrared satellite imagery. *Water Research Bulletin*, 25: 541–542.

Mosner, M.S. and Aulenbach, B.T. (2003). *Comparison of Methods Used to Estimate Lake Evaporation for a Water Budget of Lake Seminole, Southwestern Georgia and Northwestern Florida*. Atlanta, GA: U.S. Geological Survey.

Office of Saline Water. (1970). *Disposal of Brine by Solar Evaporation: Field Experiments*. Washington, DC: U.S. Department of the Interior.

Pannell, D.J. (1997). Sensitivity analysis of normative economic models: theoretical framework and practical strategies. *Agricultural Economics*, 16: 139–152.

Parkhurst, D.D. (1995). *User's Guide to PHREEQC—A Computer Program for Speciation, Reaction-Path, Advective-Transport, and Inverse Geochemical Calculations*, Water-Resources Investigations Report 95-422. Reston, VA: U.S. Geological Survey.

Postma, D. (1982). Pyrite and siderite formation in brackish and freshwater swamp sediments. *American Journal of Science*, 282: 1154–1183.

Pyne, D.G. (1995). *Groundwater Recharge and Wells*. Ann Arbor, MI: Lewis Publishers.

Pyne, D.G. (2005). *Aquifer Storage and Recovery: A Guide to Groundwater Recharge through Wells*. Gainesville, FL: ASR Press.

Reed, M.G. (1972). Stabilization of formation clays with hydroxy-aluminum solutions. *Journal of Petroleum Technology*, 24(7): 860–864 (https://www.onepetro.org/journal-paper/SPE-3694-PA).

Rosenberry, D.O., Sturrock, A.M., and Winter, T.C. (1993). Evaluation of the energy budget method of determining evaporation at Williams Lake, Minnesota, using alternative instrumentation and study approaches. *Water Resources Research*, 29(8): 2473–2248.

Smith, B.S. (1999). *The Potential for Saltwater Intrusion in the Potomac Aquifers of the York–James Peninsula*, Water-Resources Investigations Report 98-4187. Reston, VA: U.S. Geological Survey.

Snyder, S.A., Wert, E.C., Hongxia, L. et al. (2007). *Removal of EDCs and Pharmaceuticals in Drinking and Reuse Treatment Processes*. Denver, CO: AWWA Research Foundation.

Snyder, S.A., von Gunten, U., Amy, G. et al. (2014). *Use of Ozone in Water Reclamation for Contaminant Oxidation*. Alexandra, VA: WateReuse Resource Foundation.

Strycker, A. and Collins, A.G. (1987). *State-of-the-Art Report: Injection of Hazardous Wastes into Deep Wells*. Ada, OK: Robert S. Kerr Environmental Research Laboratory.

Stuyfzand, P.J. (1993). Hydrochemistry and Hydrology of the Coastal Dune Area of the Western Netherlands, PhD thesis, Vrije Universiteit, Amsterdam, Netherlands.

Teifke, R.H. (1973). Stratigraphic units of the Lower Cretaceous through Miocene series. In: *Geologic Studies, Coastal Plain of Virginia*, Bulletin 83, pp. 1–78. Charlottesville: Virginia Division of Mineral Resources.

Theis, C.V. (1935). The relation between the lowering of the piezometric surface and the rate and duration of discharge of a well using ground-water storage. *Eos*, 16(2): 519–524.

Truesdell, A.H. and Jones, B.F. (1974). WATEQ, a computer program for calculating chemical equilibria of natural waters. *U.S. Geological Survey Journal of Research*, 2: 233–274.

USEPA. (2006). *Land Treatment of Municipal Wastewater Treatment.* Cincinnati, OH: U.S. Environmental Protection Agency.

USEPA. (2016). *Aquifer Recharge and Aquifer Storage and Recovery.* Washington, DC: U.S. Environmental Protection Agency (https://www.epa.gov/uic/aquifer-recharge-and -aquifer-storage-and-recovery).

Virginia DEQ. (2006a). *Status of Virginia's Water Resources.* Richmond: Virginia Department of Environmental Quality, Office of Water Supply.

Virginia DEQ. (2006b). *Virginia Coastal Plain Model 2005 Withdrawals Simulation.* Richmond: Virginia Department of Environmental Quality, Office of Water Supply.

Warner, D.L. and Lehr, J. (1981). *Subsurface Wastewater Injection: The Technology of Injecting Wastewater into Deep Wells for Disposal.* Berkeley, CA: Premier Press.

Winter, T.C., Rosenberry, D.O., and Sturrock, A.M. (1995). Evaluation of eleven equations for determining evaporation for a small lake in the north central United States. *Water Resources Research*, 31(4): 983–993.

13 Membrane Concentrate Management

Reverse osmosis is not a proper filtration method. In reverse osmosis, an applied pressure is used to overcome osmotic pressure, a colligative property that is driven by chemical potential.

INTRODUCTION[*]

Reverse osmosis (RO) and nanofiltration (NF) treatment processes associated with a proposed advanced water treatment plant (AWTP) generate a wastewater stream (concentrate) having elevated concentrations of many water quality parameters that can make discharge costly and environmentally challenging. This chapter identifies potential regulatory requirements for the surface water discharge of such concentrates. Hampton Roads Sanitation District's Sustainable Water Initiative for Tomorrow (SWIFT) project used the following approach to identify potential regulatory requirements for surface water discharge of concentrates and to assess the feasibility of selected processes to treat the concentrate prior to discharge:

1. Review and evaluate Department of Environmental Quality (DEQ) fact sheets and current discharge permits for existing Hampton Roads Sanitation District (HRSD) wastewater treatment plants (WWTPs) relative to Chesapeake Bay watershed nutrient loading requirements.
2. Use the RO concentrate discharge permitting strategy for the City of Chesapeake Northwest River Water Treatment Plant as applicable and appropriate.
3. Estimate probable discharge permit requirements for the disposal of RO concentrate to the James River and York River.
4. Identify water quality goals and regulatory requirements for surface water discharge of wastewater streams, such as RO concentrate.

EVALUATION RESULTS

REGULATORY SETTING FOR SURFACE WATER DISCHARGES

HRSD operates nine larger WWTPs that are permitted to discharge more than 1 million gallons per day (mgd) of treated effluent in the Hampton Roads area of Virginia. The majority of these facilities discharge directly or indirectly to the James River

---

[*] Much of the material in this chapter is adapted from CH2M, *Sustainable Water Recycling Initiative: Reverse Osmosis and Nanofiltration Treatment Membrane Concentrate Management Feasibility Evaluation*, Report No. 4, CH2M, Newport News, VA, 2016.

HRSD Service Area

A Political Subdivision of the Common wealth of Virginia

Major facilities include the following:

Serving the Cities of
Chesapeake, Hampton,
Newport News, Norfolk,
Poquoson, Portsmouth,
Suffolk, Virginia Beach,
Williamsburg, and the
Counties of Gloucester,
Isle of Wight, James City,
King and Queen,
King William, Mathews,
Middlesex, Surry* and York

*Excluding the Town of Claremont

1. Atlantic, Virginia Beach
2. Chesapeake-Elizabeth, Va.Beach
3. Army Base, Norfolk
4. Virginia Initiative, Norfolk
5. Nansemond, Suffolk
6. Boat Harbor, Newport News
7. James River, Newport News
8. Williamsburg, James City County
9. York River, York County
10. West Point, King William County
11. Central Middlesex, Middlesex County
12. Urbanna, Middlesex County
13. King William, King William County

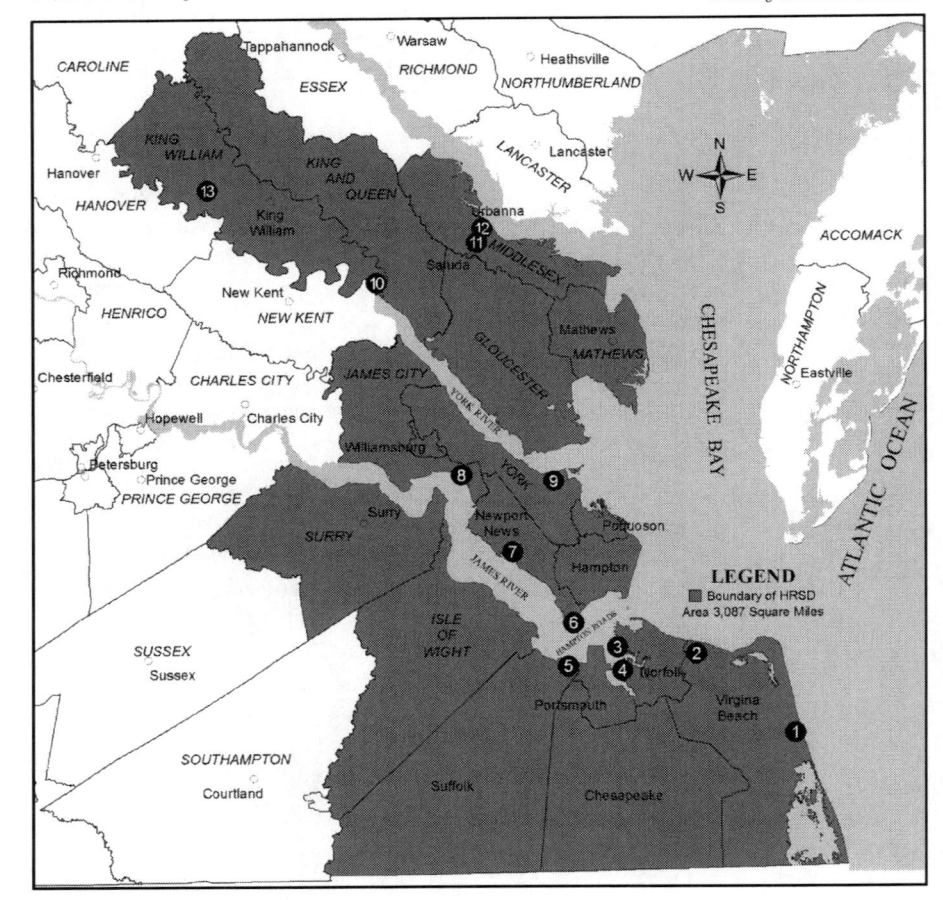

FIGURE 13.1 Location of HRSD wastewater treatment facilities. (From CH2M, *Sustainable Water Recycling Initiative: Reverse Osmosis and Nanofiltration Treatment Membrane Concentrate Management Feasibility Evaluation*, Report No. 4, CH2M, Newport News, VA, 2016.)

or York River with the exception of one facility in Virginia Beach that discharges directly to the Atlantic Ocean. HRSD also operates four smaller WWTPs located in the Middle Peninsula area. Figure 13.1 shows the location of these facilities.

Each of these treatment plants has a Virginia Pollutant Discharge Elimination System (VPDES) permit that contains specific limitations for pollutant discharge. Three of the VPDES permits—James River WWTP (VA0091272), Nansemond WWTP (VA0081299), and York River WWTP (VA0081311)—and the associated DEQ fact sheets were reviewed to evaluate the potential regulatory requirements for eliminating the current discharges and replacing these with concentrate discharges resulting from the application of either RO or NF membrane treatment process for the purpose of treating the effluents from the WWTPs for groundwater recharge.

As a general rule, each of the permits reviewed included secondary treatment requirements: minimum technology-based requirements for publicly owned treatment works (POTWs), nutrient concentration technology-based limits as a result of the discharges being to "nutrient-enriched waters" pursuant to 9 VAC 25-40, and requirements to protect recreational uses of receiving waters based on bacteriological indicators. Each of these permits also includes a total residual chlorine limitation to avoid toxic impacts from the discharge and requires HRSD to conduct annual acute effluent toxicity testing. Table 13.1 summarizes these limits for three of the HRSD facilities that are considered representative of the range of requirements for the other WWTPs.

Furthermore, the facilities in the James River and York River basins are subject to total nitrogen (TN) and total phosphorus (TP) waste load allocations for implementing the total maximum daily load requirements in the Chesapeake Bay watershed. These limitations are based on average annual loads and are independent from the TN and TP concentration limitations that are included in the individual VPDES permits. Table 13.2 summarizes the TN and TP load limitations based on the Registration Lists under the General VPDES Watershed Permit Regulation for the Chesapeake Bay watershed in Virginia (9VAC 25-820-70). Because the nutrient management strategy for Chesapeake Bay is an adaptive approach, the load allocations listed in Table 13.2 decrease in the future. The load allocations shown are the requirements that were effective January 1, 2012. More stringent limits for TP became effective for York River facilities on January 1, 2016, and more stringent limits for TN became effective for James River facilities on January 1, 2017, as noted in Table 13.2.

POTENTIAL FUTURE MEMBRANE CONCENTRATE PERMIT REQUIREMENTS

Future permit requirements for concentrate discharges to surface waters have been developed based on the following two different premises:

- Discharge in accordance with permits allowing similar pollutant discharge as allowed by the current permits
- Discharges based on additional concentrate treatment to achieve significant reductions in key pollutants—primarily TN and TP—to receiving waters and Chesapeake Bay

Each of these permitting scenarios is discussed below.

TABLE 13.1
Summary of Current VPDES

Effluent	Example Current Permit Requirements			Comment
	Nansemond: Hampton Roads Harbor (Lower James River)	James River: Warwick River (Lower James River)	York River	
Design flow (mgd)	30	20	15	—
BOD$_5$ monthly average				Secondary treatment requirements
(mg/L)	30	30	30	for POTW including minimum
(kg/d)	3406	2271	1703	of 85% removal
BOD$_5$ weekly average				
(mg/L)	45	45	45	
(kg/d)	5110	3406	2555	
Total suspended solids monthly average				Secondary treatment requirements
(mg/L)	30	30	30	for POTW including minimum
(kg/d)	3406	2271	1703	of 85% removal
Total suspended solids weekly average				
(mg/L)	45	45	45	
(kg/d)	5110	3406	2555	

(continued)

Total phosphorus, calendar year				—
(mg/L)	2.0	2.0	0.7	
(lb/yr)[a]	63,887	42,591	—	
Total nitrogen, calendar year				—
(mg/L)	8.0	NL	8.0	
(lb/yr)[a]	750,000	1,250,000	—	
Fecal coliform monthly average (geo mean) (N/MCL)	NL	NL	200	—
Enterococci monthly average (geo mean) (N/MCL)	35	35	35	—
Total residual chlorine monthly average (mg/L)	0.2	0.2	0.2	—
Total residual chlorine weekly average (mg/L)	2.4	0.60	1.3	—

Source: CH2M, *Sustainable Water Recycling Initiative: Reverse Osmosis and Nanofiltration Treatment Membrane Concentrate Management Feasibility Evaluation,* Report No. 4, CH2M, Newport News, VA, 2016.

Notes: BOD_5, 5-day biochemical oxygen demand; NL, no limit.

[a] Annual mass loads based on general VPDES permit requirements for Chesapeake Bay watershed.

TABLE 13.2

Nutrient Load Allocations for General VPDES Watershed Permit Regulation for the Chesapeake Bay Watershed in Virginia

Facility	Flow (mgd)	Watershed-Based Loads[a]		VPDES Concentration Limits	
		TN (lb/yr)	TP (lb/yr)	TN (mg/L)	TP (mg/L)
		James River Basin			
Boat Harbor	25	**740,000**	**53,299**	9.7	0.7
James River	20	**1,250,000**	**42,591**	—	2.0
Williamsburg	22.5	**800,000**	**47,915**	11.7	0.7
Nansemond	30	**750,000**	**63,887**	**8.0**	2.0
Army Base	18	**610,000**	**38,332**	11.1	0.7
Virginia Initiative Plant	40	**750,000**	**85,183**	6.2	0.7
Chesapeake–Elizabeth	24	**1,100,000**	**51,110**	15.0	0.7
Total	—	**6,000,000**	373,247	—	—
		York River Basin			
York River	15	**275,927**	**32,191**	**8.0**	0.7
West Point	0.6	**10,964**	**1279**	6.0	0.7
King William	0.025	**1424**	**190**	18.7	2.5
Total	—	**288,315**	**33,660**	—	—

Source: CH2M, *Sustainable Water Recycling Initiative: Reverse Osmosis and Nanofiltration Treatment Membrane Concentrate Management Feasibility Evaluation*, Report No. 4, CH2M, Newport News, VA, 2016.

Note: Bold values are from VPDES permits; others are calculated based on loads and permitted flow.

[a] Watershed load limits effective as of 1/1/2012. More stringent limits for TP came into effect on 1/1/2016 for the York River facilities (aggregate TP load of 19,315 lb/yr) and for TN on 1/1/2017 for the James River facilities (aggregate TN load of 4,400,000 lb/yr).

Discharges in Accordance with Permits Allowing Similar Pollutant Discharge

The following summarizes key considerations for this permitting scenario:

- The concentrate discharges would be replacing the current WWTP discharge and thus would not be considered a new discharge for regulatory purposes. Therefore, it can be assumed that the permits would require that the concentrate discharge meet the mass loading constraints of the current permits. This would not be the case for a concentrate discharge from other RO treatment facilities that was not replacing an existing discharge, as the new concentrate discharge would then need to meet antidegradation requirements, as was the case for the City of Chesapeake Northwest River Water Treatment Plant RO concentrate surface water discharge situation that was reviewed.

- Total dissolved solids in the discharge should not be an issue because these discharges are in tidal areas with varying total dissolved solids between close to freshwater and seawater (35,000 mg/L) depending on flow in the James River and York River.
- The current discharges have secondary limits, and the assumed technology basis for the concentrate discharges will probably be best professional judgment (BPJ) based on the equivalent mass loading for the secondary facilities for biochemical oxygen demand (BOD) and total suspended solids (TSS).
- The currently applicable mass loading basis for TN and TP is acceptable, and the current nutrient concentration limits would be determined to be no longer appropriate because they appear to be BPJ based on the current level of treatment at the WWTP and generally preceded the watershed-based mass loading summarized in Table 13.2.
- Because none of the facility's VPDES permits currently includes metals limits, the metals present in the concentrate discharge should be able to be addressed through a mixing zone analysis, or, depending on the outfall and diffuser hydraulics, the existing diffuser arrangements may require some changes to achieve mixing zone requirements. The ability of the current diffusers to meet dilution requirements under the concentrate waste characteristics and discharge flows must be addressed in a subsequent phase of this evaluation. As shown in Table 13.3, predicted metals concentrations for the RO concentrate are either below or close to applicable water quality criteria for saltwater systems and, thus, should be able to be addressed through an appropriately designed diffuser. Because NF will have lower rejection of metals than RO, the metals concentrations in the NF concentrate will be less than that shown here for RO. Therefore, the assumption that water quality criteria can be addressed through a mixing zone analysis or diffuser is also valid for NF concentrate.
- The concentrate would require disinfection and total residual chlorine requirements similar to the current permits.

The permitting approach described above is feasible and can provide the potential collateral benefit of reducing the discharge of pollutants to the Chesapeake Bay watershed.

Discharges Based on Additional Concentrate Treatment to Reduce Key Pollutants

The production of reclaimed water for the purpose of arresting subsidence and for indirect potable reuse is the primary goal for this project, but a significant benefit is also realized through the reduction of pollutants discharged to the Chesapeake Bay. This is especially important for TN and TP discharges because the Chesapeake Bay is considered impaired because of excessive current and historical nutrient loading, and DEQ has developed wasteload allocations for point source discharges and load allocations for nonpoint sources to reduce loading to the tributary rivers and estuaries and, ultimately, the Chesapeake Bay. The following are aspects for this scenario:

TABLE 13.3

Comparison of Projected Reverse Osmosis (RO) Concentrate Metals vs. Virginia Saltwater Quality Criteria and Any Required Dilution to Meet Criteria

| Metal | Unit | Projected RO Concentrate Quality[a] | | Virginia Saltwater Water Quality Criteria[b,c] | | Nansemond Dilution Required (% Effluent) | | James River Dilution Required (% Effluent) | |
		Nansemond	James River	Acute	Chronic	Acute	Chronic	Acute	Chronic
Flow	mgd	2.4	1.9	—	—	—	—	—	—
Aluminum	mg/L	0.64	ND	—	—	—	—	—	—
Iron	mg/L	19.4	13.2	—	—	—	—	—	—
Antimony	μg/L	ND	ND	—	—	—	—	—	—
Arsenic	μg/L	6.4	4.5	69	36	—	—	—	—
Beryllium	μg/L	ND	ND	—	—	—	—	—	—
Cadmium	μg/L	ND	ND	40	8.8	—	—	—	—
Chromium	μg/L	6.19	ND	1100	50	—	—	—	—
Copper	μg/L	7.5	2.9	9.3	6	—	80	—	—
Lead	μg/L	0.83	1.6	240	9.3	—	—	—	—

(continued)

	Units								
Mercury	µg/L	0.09	ND	1.8	0.94	—	—	—	—
Molybdenum	µg/L	11.7	3.7	—	—	—	—	—	—
Nickel	µg/L	35.8	19.2	74	8.2	23	—	—	43%
Selenium	µg/L	ND	0.64	290	71	—	—	—	—
Silver	µg/L	ND	ND	1.9	—	—	—	—	—
Thallium	µg/L	ND	ND	—	—	—	—	—	—
Zinc	µg/L	137	141	90	81	59	66	64%	57%

Source: CH2M, *Sustainable Water Recycling Initiative: Reverse Osmosis and Nanofiltration Treatment Membrane Concentrate Management Feasibility Evaluation*, Report No. 4, CH2M, Newport News, VA, 2016.

Note: ND, not detected.

a RO concentrate discharge concentrations from Nansemond and James River wastewater treatment plants (WWTPs) are based on three total data points taken from 2011 though 2013. Estimated concentrations shown here were estimated by using a mass balance approach based on best practice for the treatment technologies selected. Additional WWTP effluent sampling is recommended to provide more accurate estimated concentrations and heavy metal discharges.

b All criteria are defined as the value/water effect ratio.

c Metals criteria are based on dissolved fraction.

- Although all pollutant reductions from the current discharge situation are considered beneficial, TN and TP reductions are of primary importance.
- The degree of pollutant reduction is dependent on the AWTP process and the concentrate treatment processes selected.
- The reductions as a result of concentration treatment must be incorporated into VPDES permits to be tradable (and have monetary value).

TREATMENT OF REVERSE OSMOSIS AND NANOFILTRATION CONCENTRATE

The reverse osmosis and nanofiltration treatment processes generate a wastestream that is concentrated with inorganic and organic compounds originating in the WWTP final effluent. As discussed earlier, discharge of these wastestreams to the York River or James River will likely be subjected to permits that either (1) require compliance with current pollutant mass loading or (2) require a reduced pollutant loading with respect to TN and TP.

TREATMENT REQUIREMENTS BASED ON DISCHARGE OBJECTIVE

The treatment requirements will depend on the overall discharge objective as previously noted and discussed below.

Discharges in Accordance with Current Permits

The envisioned reverse osmosis and nanofiltration concentrate streams are projected to produce mass loadings to the receiving streams that would be slightly lower than current values (most of the pollutants are removed by reverse osmosis or nanofiltration, but some pass through into the permeate). Therefore, the projected discharge characteristics would achieve compliance with the current permit requirements and would not require further treatment. As mentioned previously, the existing WWTP diffusers would need to be reevaluated for the flow and waste concentration of the concentrate, and adjustments could be required.

Discharges Based on Additional Concentrate Treatment

A further reduction in pollutant discharge can be achieved through treatment of the reverse osmosis and nanofiltration concentrate streams. Treatment of the concentrate streams can be provided by the implementation of chemical and, in the case of reverse osmosis, biological treatment of the concentrate. Figure 13.2 shows the proposed unit processes for treating the reverse osmosis and nanofiltration concentrated streams. Treatment of the reverse osmosis concentrate would be accomplished with a sequencing batch reactor that would provide nitrification, denitrification, and both biologically and chemically enhanced phosphorus removal.

- *Nitrification* is the biological oxidation of ammonium ion to nitrate through a two-step process by two species of bacteria: *Nitrosomonas* and *Nitrobacter*. In the first step, ammonium ions are converted to nitrite by *Nitrosomonas* sp. The second stop involves the conversion of nitrite to

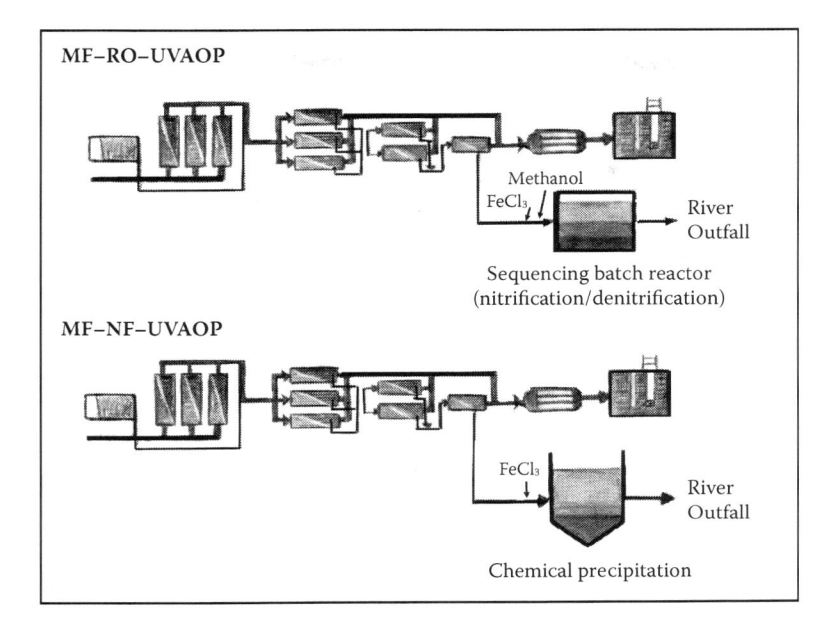

FIGURE 13.2   Proposed concentrate treatment process for RO- and NF-based treatment trains. (Illustration by F.R. Spellman and Kathern Welsh, adapted from CH2M, *Sustainable Water Recycling Initiative: Reverse Osmosis and Nanofiltration Treatment Membrane Concentrate Management Feasibility Evaluation*, Report No. 4, CH2M, Newport News, VA, 2016.)

nitrate by *Nitrobacter* sp. Both of these species are considered autotrophic bacteria because they use carbon dioxide (CO_2) as the source of carbon for building cell tissue. The two-step reaction is usually very rapid. Because of this, it is rare to find nitrite levels higher than 1.0 mg/L in water. The nitrate formed by nitrification is, in the nitrogen cycle, used by pollutants as a nitrogen source (synthesis) or reduced to N_2 gas through the process of denitrification. Nitrate can, however contaminate groundwater if it is not used for synthesis or reduced through denitrification.

- In *denitrification,* nitrate can be transformed to nitrogen gas under conditions where dissolved oxygen is absent (i.e., anoxic conditions) by heterotrophic bacteria (those that use organic carbon for building cell tissue). In order for denitrification to occur, there must be no dissolved oxygen present. If dissolved oxygen is present, the organisms will use it rather than the nitrate-bound oxygen in their metabolism. In this case, nitrogen in the form of nitrates would remain and pass into and through the soil, eventually ending up in groundwater.

Treatment of the nanofiltration concentrate would include only a chemical precipitation clarifier for phosphorus removal, as nitrogen removal is not required after nanofiltration. Nanofiltration results in much lower levels in the concentrate compared to reverse osmosis concentrate.

TABLE 13.4

River Outfall Mass Nitrogen and Phosphorus Loadings: Current WWTP Effluent and Treated Concentrate from MF–RO–UVAOP Treatment Train

Parameter	Unit	AB	BH	JR	NA	VIP	WB	YR
				Total Nitrogen (TN)				
Existing effluent	lb/yr	945,715	993,165	230,680	321,930	776,355	219,363	206,225
Treated concentrate	lb/yr	112,055	232,140	68,255	64,240	129,210	39,785	39,785
Reduction[a]	%	88.1	76.6	70.3	80.1	83.4	81.9	80.9
				Total Phosphorus (TP)				
Existing effluent	lb/yr	33,215	29,930	32,485	64,970	52,925	29,930	25,550
Treated concentrate	lb/yr	2446	3139	2993	3687	6716	2008	2701
Reduction[a]	%	92.6	89.5	90.8	94.3	87.3	93.3	89.3

Source: CH2M, *Sustainable Water Recycling Initiative: Reverse Osmosis and Nanofiltration Treatment Membrane Concentrate Management Feasibility Evaluation*, Report No. 4, CH2M, Newport News, VA, 2016.

Notes: AB, Army Base; BH, Boat Harbor; JR, James River; NA, Nansemond; VIP, Virginia Initiative Plant; WB, Williamsburg; YR, York River; MF, microfiltration; RO, reverse osmosis; UVAOP, ultraviolet advanced oxidation process.

[a] The nutrient reductions shown are based on the assumption that all flow is captured at each plant and that the AWTP operates 100% of the time.

Nutrient Discharge Comparisons for Current and Projected Treatment Options

Tables 13.4 and 13.5 show the estimated reduction in nitrogen and phosphorus being discharged to the York River and James River based on discharge of the reverse osmosis- and nanofiltration-treated concentrates compared to existing effluent discharges from the WWTPs (based on the analysis of data for 2011 through 2014 presented earlier in this text). A significant reduction in both nutrients is achieved at all WWTP locations using the proposed concentrate treatment approaches. Note that the granular activated carbon (GAC)-based treatment approach reduces the nutrient loading to the rivers by 100% (assuming 100% capture of wastewater effluent by the advanced treatment plant) because no wastestreams from this process need to be discharged to the rivers.

PROJECTED POLLUTANT LOADING FROM TREATMENT OPTIONS

The mass balances for the reverse osmosis, nanofiltration, and granular activated carbon treatment trains were discussed earlier for eight of HRSD WWTPs evaluated. Table 13.4 and 13.5 illustrate reductions from current loading conditions (using

TABLE 13.5

River Outfall Nitrogen and Phosphorus Mass Loadings: Treated Concentrate from MF–NF–UVAOP Treatment Train

Parameter	Unit	AB	BH	JR	NA	VIP	WB	YR
				Total Nitrogen (TN)				
Existing effluent	lb/yr	945,715	993,165	230,680	321,930	776,355	219,363	206,225
Treated concentrate	lb/yr	399,350	415,735	64,240	85,045	162,425	53,655	55,845
Reduction*	%	57.8	58.1	72.2	73.5	79.1	75.5	73.0
				Total Phosphorus (TP)				
Existing effluent	lb/yr	33,215	29,930	32,485	64,970	52,925	29,930	25,550
Treated concentrate	lb/yr	2,446	3,139	2,993	3,687	6,716	2,008	2,701
Reduction*	%	92.6	89.5	90.8	94.3	87.3	93.3	89.3

Source: CH2M, *Sustainable Water Recycling Initiative: Reverse Osmosis and Nanofiltration Treatment Membrane Concentrate Management Feasibility Evaluation*, Report No. 4, CH2M, Newport News, VA, 2016.

Notes: AB, Army Base; BH, Boat Harbor; JR, James River; NA, Nansemond; VIP, Virginia Initiative Plant; WB, Williamsburg; YR, York River; MF, microfiltration; RO, reverse osmosis; UVAOP, ultraviolet advanced oxidation process.

[a] The nutrient reductions shown are based on the assumption that all flow is captured at each plant and that the AWTP operates 100% of the time.

data from 2013 and 2014) for TN and TP in concentrates from the reverse osmosis- and nanofiltration-based AWTPs. For the GAC-based AWTP, the load reduction to the estuaries would be 100% because the TN and TP present in the AWTP feedwater either passes through the treatment to the groundwater (at acceptable levels) or is removed as a solid residual waste and disposed of in a landfill. To put the projected loads in perspective, Table 13.6 shows the estimated loads from the concentrate treatment options in comparison with current loading and wasteload allocations. This comparison shows a very significant reduction in nutrients and values well below the current and future wasteload allocations.

THE BOTTOM LINE

HRSD is counting on the using existing VPDES permits as a basis for moving forward with permitting of concentrate discharges for its SWIFT project. If the permitting approach for the concentrate discharges were perceived by Virginia Department of Environmental Quality as a "new discharge," this would complicate VPDES permitting and bring state antidegradation considerations into play for permitting. The bottom line on the preliminary findings in this chapter is as follows:

TABLE 13.6

Comparison of Concentrate Treatment Alternative Estimated Discharge with Current and Allocated Loadings

Location	Units	Wasteload Allocation[a]	Current[b]	Concentrate Treatment RO Estimated Discharge	NF Estimated Discharge	GAC Estimated Discharge
Total Nitrogen (TN)						
James River Basin	lb/yr	4,900,000 (3,811,699)	3,693,435	645,685	1,180,410	0
	% Current	—	—	17%	32%	0%
York River Basin	lb/yr	275,927	206,225	39,785	39,785	0
	% Current	—	—	19%	19%	0%
Total Phosphorus (TP)						
James River Basin	lb/yr	322,127	243,455	20,988	20,988	0
	% Current	—	—	9%	9%	0%
York River Basin	lb/yr	32,191 (18,295)	25,550	2701	2701	0
	% Current	—	—	11%	11%	0%

Source: CH2M, *Sustainable Water Recycling Initiative: Reverse Osmosis and Nanofiltration Treatment Membrane Concentrate Management Feasibility Evaluation*, Report No. 4, CH2M, Newport News, VA, 2016.

[a] Wasteload allocation based on the facilities evaluated in Tables 13.14 and 13.5. Values in parentheses are limits effective 1/1/2017 for TN and 1/1/2016 for TP for the facilities evaluated.

[b] Based on data evaluation of eight HRSD facilities presented earlier for 2011 through 2014.

- Replacing the current effluent discharge at one or more of HRSD's WWTPs with a discharge of concentrate produced from the application of an RO- or NF-based advanced treatment train appears feasible. The evaluation indicates that an approach based on current pollutant loading could be used to create a VPDES permit for the concentrate loading.
- Treating the reverse osmosis or nanofiltration streams to reduce both TN and TP or implementing a GAC-based AWTP approach could achieve significant reductions in both nutrients discharged to the Chesapeake Bay.

REFERENCES AND RECOMMENDED READING

CH2M. (2016a). *Sustainable Water Recycling Initiative: Groundwater Injection Hydraulic Feasibility Evaluation*, Report No. 1. Newport News, VA: CH2M.

CH2M. (2016b). *Sustainable Water Recycling Initiative: Groundwater Injection Geochemical Compatibility Feasibility Evaluation*, Report No. 2. Newport News, VA: CH2M.

CH2M. (2016c). *Sustainable Water Recycling Initiative: Advanced Water Purification Process Feasibility Evaluation*, Report No. 3. Newport News, VA: CH2M.

CH2M. (2016d). *Sustainable Water Recycling Initiative: Reverse Osmosis and Nanofiltration Treatment Membrane Concentrate Management Feasibility Evaluation*, Report No. 4. Newport News, VA: CH2M.

14 Cost Estimates

> A cost estimate is the summation of individual cost elements, using established methods and valid data, to estimate the future costs of a program, based on what is known today.
>
> **GAO (2009)**

INTRODUCTION[*]

Costs are an important factor in assessing the ultimate approach for Hampton Roads Sanitation District's Sustainable Water Initiative for Tomorrow (SWIFT) project. Because HRSD is a political subdivision of the Commonwealth of Virginia, it must receive approval for its operational budget from and account for total operational expenses to its governor-appointed board of commissioners. Moreover, with the exception of possible grants, all costs for construction, organizational operations, and maintenance are provided for by local rate payers; therefore, an accurate accounting of total expenses must be maintained and published for both commission and rate payer review.

The purpose of this chapter is to provide an estimate of capital and operating costs associated with collecting, treating, and injecting clean water into the aquifer. During the SWIFT concept cost evaluation process (conducted by HRSD's engineering contractor, CH2M), a cost estimating approach for treatment trains and injection well facilities was developed and assumptions were made with regard to standard project allowances, capital costs, plant facility models, injection well facility costs, operations and maintenance costs, and design criteria costs.

COST ESTIMATING APPROACH

TREATMENT TRAINS

Cost estimates were prepared for the three advanced water treatment (AWT) trains—reverse osmosis, nanofiltration, and biological activated carbon—discussed earlier and shown again in the simplified process-flow diagram in Figure 14.1.

INJECTION WELL FACILITIES

As mentioned earlier, subsidence control wells are injection wells whose primary objective is to reduce or eliminate the loss of land surface elevation due to removal of groundwater providing subsurface support. Land subsidence control is achieved by injecting water to an underground formation to maintain fluid pressure and avoid

---

[*] Much of this chapter is adapted from CH2M, *Sustainable Water Recycling Initiative: Concept Cost Evaluation*, Report No. 5, CH2M, Newport News, VA, 2016.

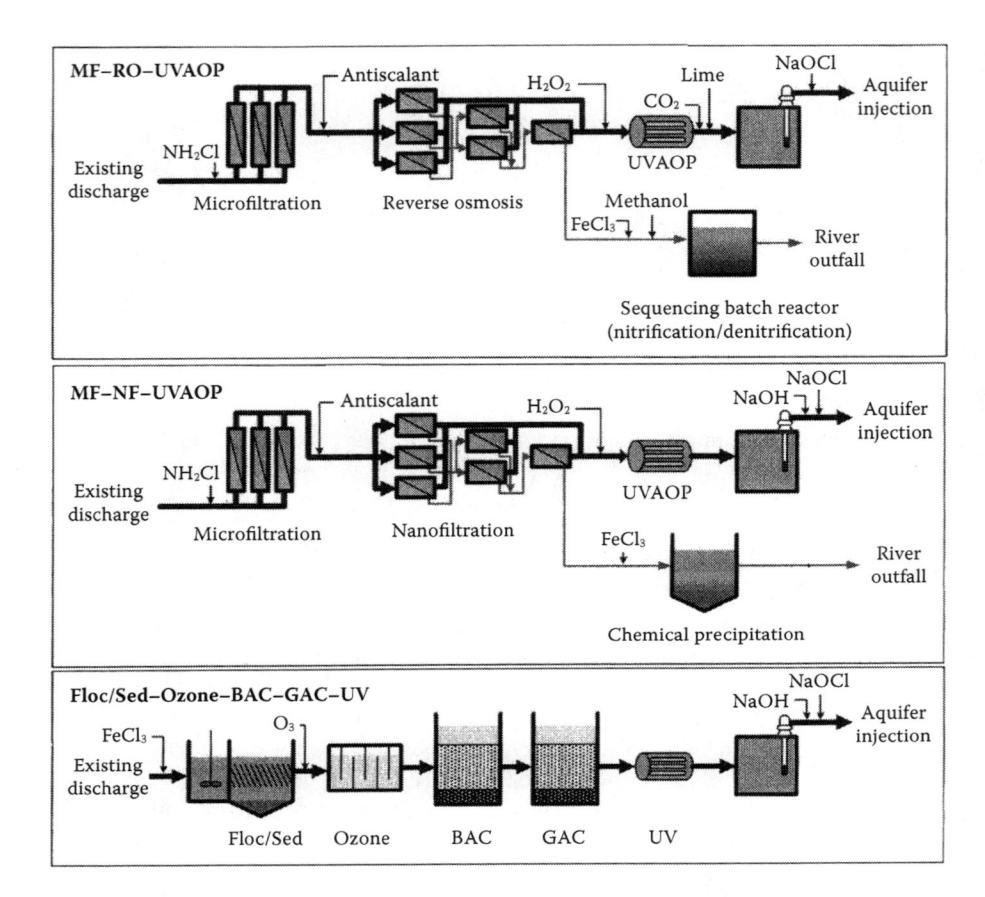

FIGURE 14.1 Process-flow diagrams for three advanced water treatment (AWT) trains. (From CH2M, *Sustainable Water Recycling Initiative: Concept Cost Evaluation*, Report No. 5, CH2M, Newport News, VA, 2016.)

compaction. HRSD's planned use of injection wells has a multifaceted objective that involves preventing, rebounding from, and mitigating land subsidence due to withdrawal of groundwater from local aquifers; the conveyance of water treated to water quality standards to replenish the potable water supply in the Potomac Aquifer; preventing the outfalling of nutrients into the lower Chesapeake Bay; and preventing saltwater intrusion into groundwater aquifers.

Cost estimates were prepared for injection and monitoring wells that fully access the three aquifers of the Potomac Aquifer System (PAS) as discussed earlier. The injection wells will convey treated clean water into the PAS, and injection and monitoring will extend to nearly 1580 feet below grade at a typical wastewater treatment plant (WWTP) in the HRSD service area. Injection well facilities will include wellhead piping, valving, instrumentation (e.g., pressure transducers, flowmeters), and a production well pump to conduct routine backflushing operations. A central building will house the motor control centers (MCCs), variable-frequency drives

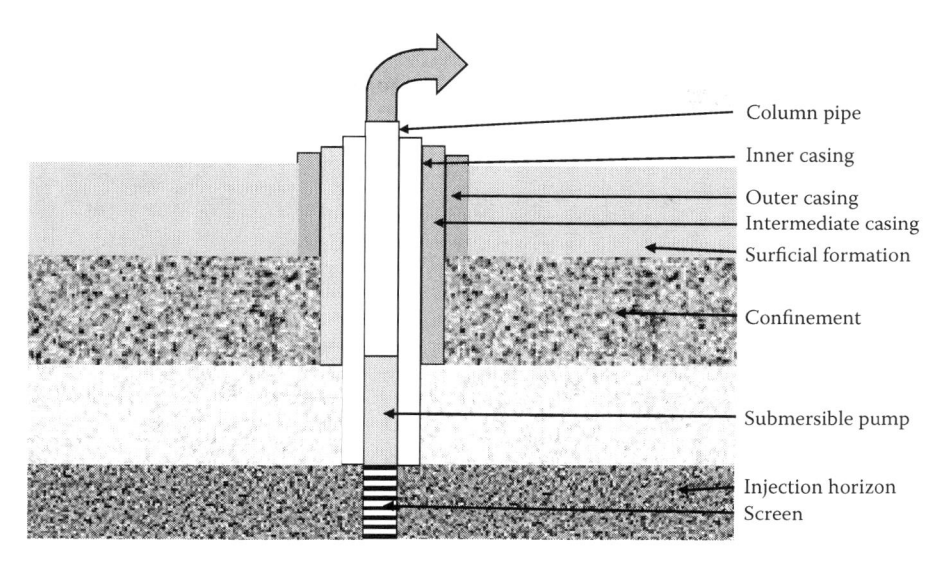

FIGURE 14.2 Schematic of a typical injection well. (From CH2M, *Sustainable Water Recycling Initiative: Concept Cost Evaluation*, Report No. 5, CH2M, Newport News, VA, 2016.)

(VFDs), a single-chemical feed system, a sampling sink, and a display panel allowing monitoring and operation of the wells. Figure 14.2 provides a schematic of a typical injection well.

Approach and Assumptions

Capital and annual operations and maintenance (O&M) cost estimates for each proposed advanced water treatment train were developed by HRSD's engineering contractor, CH2M; the contractor used its parametric cost estimating program for water and wastewater treatment plants. The parametric cost estimating program uses fundamental design criteria for treatment processes, general arrangement drawings based on actual plant designs, and an extensive water treatment cost database from constructed plants to generate detailed quantity takeoffs and reliable cost estimates. The costs are for a complete and fully operational advanced treatment plant, excluding any improvements to secondary treatment at the WWTP, with the necessary site development, electrical, computer, O&M buildings, and miscellaneous support infrastructure included in a typical plant. The overall costs have been developed using the approach discussed in the following subsections.

Standard Project Allowances

Standard project allowances (standard percentages) for items such as contractor overhead and profit, contingencies, engineering, and bonds and insurance are applied to the construction cost estimate to generated a total capital cost estimate. These percentages, as well as site allowance percentages, are shown in Table 14.1.

TABLE 14.1

Site Allowances, Contractor Markups, and Nonconstruction Costs

Item	Allowance for Granular Activated Charcoal (GAC)-Based Treatment	Allowance for Reverse Osmosis (RO)-Based Treatment
Site Allowances		
Site work (roads, fences, landscaping, etc.)	6%	5%
Plant computer (SCADA)	2%	3%
Yard electrical (primary feed, switchgear, generator)	5%	6%
Yard piping (process piping, fire loop, service water, natural gas)	15%	11%
Contractor Markups		
Overhead	7%	—
Profit	10%	—
Mobilization, bonds, and insurance	3%	—
Contingency	30%	—
Nonconstruction Costs		
Engineering	10%	—
Construction management	10%	—
Commissioning and startup	2%	—

Source: CH2M, *Sustainable Water Recycling Initiative: Concept Cost Evaluation*, Report No. 5, CH2M, Newport News, VA, 2016.

Different site allowances were used for membrane-based advanced treatment and granular activated carbon (GAC)-based advanced treatment because of the different site requirements for each.

Capital Cost Plant Facility Model

Capital costs for all treatment trains were developed for a model generic advanced water treatment plant facility with a capacity of 20 million gallons per day (mgd). Annual O&M costs were also based on an average flow of 20 mgd, which assumes that the advanced water treatment plant is operated at 100% reliability; this assumption may require adjustment as the project proceeds further into design.

Injection Well Facility Costs

Conventional estimating techniques were used to estimate capital costs for the injection well facilities. Disposing of 20 mgd of WWTP effluent will require seven wells that will inject effluent at rates approaching 2.9 mgd each. An additional two injection wells will serve as back-up systems when other wells are removed from service for repair or rehabilitation. To conduct routine backflushing operations, each injection well will contain a pump capable of pumping at 3 mgd. A small-diameter (4-inch) monitoring well extending to the same depth and containing an identical amount of screen will be matched with each injection well for monitoring water levels and collecting samples.

Operations and Maintenance Costs

Annual O&M costs were developed for both the advanced water treatment plant and injection well facilities per the discussion described herein.

Advanced Water Treatment Plant Operations and Maintenance Costs

Annual O&M costs for the advanced water treatment plant include labor, power, consumables, and regular replacement for items with an expected life of less than 30 years (e.g., membranes). Labor costs were developed assuming the addition of one operator, two maintenance staff, and one instrument technician to the existing WWTP staff. Further consideration should be given to the appropriate level of staff required at each WWTP where advanced treatment is implemented. Power costs were estimated by calculating the equipment and building electrical power draw and applying a unit power cost of $0.06 per kilowatt hour (kWh). The cost for consumables (e.g., chemicals) was calculated using the annual average usage times a unit cost for each consumable. Unit costs for chemicals were obtained from HRSD, where available; the remaining unit costs for chemicals were obtained from CH2M's treatment plant database. A plant life of 30 years and a discount rate of 5% were used for the net present value (NPV) analysis.

Injection Well Facilities Operations and Maintenance Costs

Annual O&M costs for the injection well facilities include labor, consumables, sampling, and invasive rehabilitation for at least one well per year over the 30-year service life of the treatment facility. Labor costs were estimated for one operator working 20 hours per week to conduct performance tracking, monitoring of water level, sampling, backflushing, overseeing rehabilitation events, and other essential activities. Power costs were estimated under the injection pump station. Similarly, the cost of running a single-chemical feed system (chemicals) was covered under other facilities. The cost estimates presented here are considered to fall between the Class 5 and Class 4 cost estimate definitions for process industries as defined by American Association of Code Enforcement (AACE). Class 5 estimates are based on a design that is 0% to 2% complete and have a stated accuracy of +100%/−50%; Class 4 estimates are based on a design that is 1% to 15% complete and have a stated accuracy of +50%/−30%. Costs that have not been estimated but may be significant include land acquisition, right of way and easement procurement, environmental mitigation, permitting, and program management. The cost estimates presented in this report should be refined when more information is gathered regarding the specific WWTPs at which advanced treatment will be implemented. In addition, some WWTPs may require improvements to the secondary processes to allow implementation of advanced treatment; these costs should be added to the cost estimates presented in this report.

TABLE 14.2
Major Design Criteria for Reverse Osmosis (RO)- and Nanofiltration (NF)-Based Treatment Trains

Parameter	Value
Secondary effluent equalization basin volume	1 million gallons
Overall plant recovery	83%
MF flux	35 gpm/sf per day
Number of MF trains	10 duty, 1 standby
RO (or NF) recovery	85%
RO (or NF) flux	12 gpm/sf per day
Average RO and NF feed pressure	RO, 190 psi; NF, 95 psi
Number of RO (or NF) trains	5 duty, 0 standby
UVAOP	>500 mJ/cm^2
Number of UVAOP trains	4 duty, 0 standby
Finished water equalization and pumping	180,000-gallon equalization tank with 2 duty and 1 standby 200-hp pumps
Chemical doses (averages)	Monochloramine, 3 mg/L; sulfuric acid, 20 mg/L; scale inhibitor, 2.5 mg/L; hydrogen peroxide, 3 mg/L; lime, 30 mg/L (RO only); carbon dioxide, 30 mg/L (RO only); sodium hydroxide (20 mg/L (NF only); sodium hypochlorite, 3 mg/L

Source: CH2M, *Sustainable Water Recycling Initiative: Concept Cost Evaluation*, Report No. 5, CH2M, Newport News, VA, 2016.

Notes: MF, microfiltration; RO, reverse osmosis; NF, nanofiltration; UVAOP, ultraviolet advanced oxidation process; gpm/sf, gallons per minute per square foot; psi, pounds per square inch.

DESIGN CRITERIA

Detailed design criteria for each advanced water treatment train were selected to conceptually design each process facility, which then allowed development of material quantity take-offs and capital and O&M cost estimates. Analysis of aquifer characteristics local to the HRSD WWTPs and groundwater flow modeling were used to determine capacities for injection wells installed in the Potomac Aquifer System. The number of injection wells remained identical for each advanced water treatment train. Major design criteria for each advanced water treatment train and injection well facilities are shown in Tables 14.2. through 14.4.

COST ESTIMATES

CAPITAL AND ANNUAL OPERATING COSTS

The capital costs for the three proposed advanced water treatment trains are shown in Tables 14.5, 14.6, and 14.7. Project contingency, overhead, and site allowance costs are included in the facility costs.

TABLE 14.3
Major Design Criteria for Granular Activated Carbon (GAC)-Based Treatment Train

Parameter	Value
Secondary effluent equalization basin volume	1 million gallons
Overall plant recovery	96%
Flocculation time	20 minutes via three stages
Sedimentation type and loading rate	Inclined plate settlers; 0.32 gpm/sf (projected area)
Number of flocculation/sedimentation trains	2 duty, 0 standby
Ozone contact time	8 minutes
Ozone dose	6 mg/L
BAC filtration loading rate	9 gpm/sf
Number of BAC filters	3 duty, 1 standby
GAC contactor loading rate	6 gpm/sf
Number of GAC contactors	3 duty, 1 standby
UV disinfection dose	22 mJ/cm^2
Finished water equalization and pumping	420,000-gallon equalization tank with 2 duty and 1 standby 200-hp pumps
Chemical doses (averages)	Ferric chloride, 30 mg/L; sodium hydroxide, 20 mg/L; sodium hypochlorite, 3 mg/L

Source: CH2M, *Sustainable Water Recycling Initiative: Concept Cost Evaluation*, Report No. 5, CH2M, Newport News, VA, 2016.

Note: BAC, biological activated carbon; GAC, granular activated carbon; UV, ultraviolet; gpm/sf, gallons per minute per square foot.

TABLE 14.4
Major Design Criteria for Injection Well Facilities

Parameter	Value
Injection wells	9 wells: 7 duty, 2 standby
Injection well capacity	3 mgd
Backflushing pumps with VFD	9 × 3-mgd, 200-hp motor
Monitoring well	9, each 1580 feet deep; 310 feet stainless steel screen
Chemical feed system	1 × pH adjustment, or disinfection
Dual direction injection wellhead	9 × flowmeters (2 each), bag or sand filtration system, 1 FCV, 1 PCV
Monitoring well sampling pumps	9 × 3 hp
Control building	1 x chemical feed system, 9 × MCC and VFDs, control panel, sampling sink
Invasive rehabilitation	1 well per year for 30 years

Source: CH2M, *Sustainable Water Recycling Initiative: Concept Cost Evaluation*, Report No. 5, CH2M, Newport News, VA, 2016.

Note: VFD, variable-frequency drive; FCV, flow control valve; PCV, pressure control valve; MCC, motor control center.

TABLE 14.5
Capital and Annual Operating and Maintenance Costs for Reverse Osmosis-Based Treatment Train

Facility	Capital Cost	Annual O&M Cost
Secondary equalization and pumping	$4,210,000	$97,000
Microfiltration	$31,466,000	$512,000
Reverse osmosis break tank	$739,000	$0
Reverse osmosis	$38,031,000	$2,323,000
UVAOP	$22,484,000	$2,232,000
Finished water equalization and injection pump station	$3,047,000	$196,000
Injection well facilities	$21,960,000	$381,000
Chemical storage and feed facilities	$7,806,000	$695,000
Microfiltration backwash equalization and pumping	$1,421,000	$4000
RO concentration treatment	$8,008,000	$408,000
Subtotal	$139,172,000	$6,848,000
Engineering	$13,918,000	$0
Engineering services during construction	$13,918,000	$0
Commissioning and startup	$2,784,000	$0
Plant labor	$0	$324,480
Total	$170,000,000	$7,200,000

Source: CH2M, *Sustainable Water Recycling Initiative: Concept Cost Evaluation*, Report No. 5, CH2M, Newport News, VA, 2016.

Note: Total capital costs have been rounded up to the millions digit: total annual O&M costs have been rounded up to the nearest hundred thousands digit. UVAOP, ultraviolet advanced oxidation process.

NET PRESENT VALUE[*]

To better understand the net present value (NPV) of HRSD's SWIFT project for three advanced water treatment trains, a brief primer on NPV is provided here. Keep in mind that managers involved with construction costs that arise at the beginning of a project (e.g., capital costs) and future costs that are necessary to implement and maintain the project after the initial construction period (e.g., annual O&M costs, periodic costs) must be knowledgeable about these costs. Present value analysis is a method to evaluate expenditures, either capital or O&M, that occur over different time periods. This standard methodology allows for cost comparisons of different project alternatives on the basis of a single cost figure for each alternative. This single number, referred to as the *present value*, is the amount that must be set aside at the initial point in time (base year) to ensure that necessary funds will be available in the future, assuming certain economic conditions. A present value analysis involves four basic steps:

[*] Information in this section is based on USEPA, *A Guide to Developing and Documenting Cost Estimates During the Feasibility Study*, U.S. Environmental Protection Agency, Washington, DC, 2000.

TABLE 14.6
Capital and Annual Operating and Maintenance Costs for Nanofiltration-Based Treatment Train

Facility	Capital Cost	Annual O&M Cost
Secondary equalization and pumping	$4,210,000	$97,000
Microfiltration	$31,466,000	$512,000
Nanofiltration break tank	$739,000	$0
Nanofiltration	$36,622,000	$1,776,000
UVAOP	$22,484,000	$2,232,000
Finished water equalization and injection pump station	$3,047,000	$196,000
Injection well facilities	$21,960,000	$381,000
Chemical storage and feed facilities	$3,184,000	$706,000
Microfiltration backwash equalization and pumping	$1,421,000	$4,000
Nanofiltration concentrate treatment	$3,348,000	$123,000
Subtotal	$128,481,000	$6,027,000
Engineering	$12,849,000	$0
Engineering services during construction	$12,849,000	$0
Commissioning and startup	$2,570,000	$0
Plant labor	$0	$324,480
Total	$157,000,000	$6,400,000

Source: CH2M, *Sustainable Water Recycling Initiative: Concept Cost Evaluation*, Report No. 5, CH2M, Newport News, VA, 2016.

Note: Total capital costs have been rounded up to the millions digit; total annual O&M costs have been rounded up to the nearest hundred thousands digit. UVAOP, ultraviolet advanced oxidation process.

1. Define the period of analysis.
2. Calculate the cash outflows (payments) for each year of the project.
3. Select a discount rate to use in the present value calculation.
4. Calculate the present value.

Define Period of Analysis

The period of analysis is the period of time over which present value is calculated. In general, the period of analysis should be equivalent to the project duration, resulting in a complete life-cycle cost estimate for implementing the construction project. The project duration generally begins with the planning, design, and construction of the project; continues through short- and long-term O&M; and ends with project completion and closeout. Each step involved may have a different project duration. For example, one alternative may have a two-year construction period and no future O&M. Another alterative may have no construction period and many years of O&M.

TABLE 14.7

Capital and Annual Operating and Maintenance Costs for Granular Activated Carbon-Based Treatment Train

Facility	Capital Cost	Annual O&M Cost
Secondary equalization and pumping	$4,324,000	$99,000
Rapid mix for coagulant addition	$1,892,000	$7,000
Flocculation	$5,505,000	$37,000
Sedimentation	$6,270,000	$33,000
Ozone oxidation	$16,911,000	$523,000
Biological activated carbon filtration	$12,432,000	$41,000
Granular activated carbon influent pump station	$2,904,000	$97,000
Granular activated carbon contactors	$17,726,000	$586,000
Ultraviolet disinfection	$2,499,000	$54,000
Finished waste equalization and injection pump station	$5,768,000	$251,000
Injection well facilities	$21,960,000	$381,000
Chemical storage and feed facilities	$2,521,000	$1,050,000
Filter backwash equalization and return pumping station	$3,542,000	$10,000
Subtotal	$104,253,000	$3,169,000
Engineering	$10,426,000	$0
Engineering services during construction	$10,426,000	$0
Commissioning and startup	$2,086,000	$0
Plant labor	$0	$325,000
Total	$128,000,000	$3,500,000

Source: CH2M, *Sustainable Water Recycling Initiative: Concept Cost Evaluation*, Report No. 5, CH2M, Newport News, VA, 2016.

Notes: Total capital costs have been rounded up to the millions digit: total annual O&M costs have been rounded up to the nearest hundred thousands digit.

Calculate Annual Cash Outflows

The second step of the present value analysis is to add up the capital and O&M cash outflows for each year of the project (i.e., annual cash outflow). These include capital costs to construct the project, annual O&M costs to operate and maintain the remedial alternative over its planned life, and periodic costs for those capital or O&M costs that occur only once every few years. Usually, most or all of the capital costs are expended during the construction and startup of the project, before annual O&M begins. Although the present value of period costs is small for those that occur near the end of the project duration (e.g., closeout costs), these costs should be included in the present value analysis. Most feasibility study cost analyses begin with a simplifying assumption that the duration of initial construction and startup will be less than one year (i.e., construction work will occur in "year zero" of the project). This "year zero" assumption can be modified if a preliminary project schedule has been developed and it is known that capital construction costs will be expended beyond one year.

Select a Discount Rate

The third step in the present value analysis is to select a discount rate. A discount rate, which is similar to an interest rate, is used to account for the *time value of money*, which is the idea that a dollar today is worth more than the same dollar in the future due to its potential earning capacity (e.g., interest). Thus, discounting reflects the productivity of capital. If the capital were not employed in a specific use, it would have productive value in alternative uses. The choice of a discount rate is important because the selected rate directly impacts the present value of a cost estimate, which is then used when making a remedy selection decision. The higher the discount rate, the lower the present value of the cost estimate. For feasibility study cost analyses, the same discount rate should be used to evaluate all alternatives for a site, even if the period of analysis differs from one to another.

Calculate the Present Value

The last step is to calculate the present value. The value of a remedial alternative represents the sum of the present values of all future payments associated with the project. For example, if the project will entail capital and O&M costs each year for 12 years, the present value is the sum of the present values of each of the 12 payments, or expenditures. The present value of a future payment is the actual value that will be disbursed, discounted at an appropriate rate of interest. Present value for payment x_t in year t at a discount rate of i is calculated using Equation 14.1:

$$PV = [1/(1 + i)^t] \times x_t \qquad (14.1)$$

The first operand in Equation (14.1), $1/(1 + i)^t$, can be referred as a *discount factor*. Note that, for present value analyses during the feasibility study, distinction is generally not made of what time of the year the total cost for each year is incurred (i.e., beginning, middle, or end). This simplifying assumption would not necessarily be used for budgeting purposes but is appropriate for feasibility study cost estimating purposes.

Net Present Value for Each AWT Train

Table 14.8 summarizes the total capital cost, total annual operating cost, and total net present value (NPV) for each advanced water treatment train. The NPV is based on a period of 30 years and a discount rate of 5%.

TABLE 14.8
Net Present Value for Three Advanced Water Treatment Trains

Treatment Train	Total Capital Cost	Total Annual O&M	Total Net Present Value
Reverse osmosis	$170,000,000	$7,200,000	$281,000,000
Nanofiltration	$157,000,000	$6,400,000	$256,000,000
Granular activated carbon	$128,000,000	$3,500,000	$182,000,000

Source: CH2M, *Sustainable Water Recycling Initiative: Concept Cost Evaluation*, Report No. 5, CH2M, Newport News, VA, 2016.

THE BOTTOM LINE

The candidate treatment process cost estimates indicate a significant difference between membrane-based (RO and NF) treatment approaches vs. non-membrane (BAC and GAC) processes, in terms of both capital and annual costs. Costs are an important factor in assessing the ultimate approach for SWIFT.

REFERENCES AND RECOMMENDED READING

CH2M. (2016a). *Sustainable Water Recycling Initiative: Groundwater Injection Hydraulic Feasibility Evaluation*, Report No. 1. Newport News, VA: CH2M.

CH2M. (2016b). *Sustainable Water Recycling Initiative: Groundwater Injection Geochemical Compatibility Feasibility Evaluation*, Report No. 2. Newport News, VA: CH2M.

CH2M. (2016c). *Sustainable Water Recycling Initiative: Advanced Water Purification Process Feasibility Evaluation*, Report No. 3. Newport News, VA: CH2M.

CH2M. (2016d). *Sustainable Water Recycling Initiative: Reverse Osmosis and Nanofiltration Treatment Membrane Concentrate Management Feasibility Evaluation*, Report No. 4. Newport News, VA: CH2M.

CH2M. (2016e). *Sustainable Water Recycling Initiative: Concept Cost Evaluation*, Report No. 5. Newport News, VA: CH2M.

GAO. (2009). *GAO Cost Estimating and Assessment Guide: Best Practices for Developing and Managing Capital Program Costs*, GAO-09-3SP. Washington, DC: U.S. Government Accountability Office.

Afterword or Beforehand

Date: Today

Time: Right now

Place: First Landing State Park, Virginia Beach, Virginia, at the very spot of land settled by the English in 1607

Surroundings: On the edge of Chesapeake Bay in a maritime forest filled with lagoons, bald cypress swamps, and rare flora and fauna all of which have withstood strong winds, periodic flooding, and salt spray, although today it is bathed in full sunshine

Protagonists: Mr. Rabbit and Mr. Grasshopper

Today, during a Mr. Grasshopper and Mr. Rabbit conversation, Mr. Rabbit said to Mr. Grasshopper, "Well, what do you think, Mr. Grasshopper? What do you think about the author's account of land subsidence, groundwater, and that wonderful organization, HRSD? How about HRSD's far-thinking proposal, its ongoing project to save the area from further subsidence, to restock groundwater supplies, to prevent saltwater intrusion, and to prevent outfalling of nutrients into our wonderful Bay?"

Before responding to Mr. Rabbit's query, Mr. Grasshopper made some noise by rubbing a row of pegs on his hind legs against the edges of his forewings (for you humans out there, this maneuver is called *stridulating*). Anyway, Mr. Grasshopper finally replied, "You know, I really appreciated the author talking about those dinosaurs, and how it was not the Chesapeake Bay bolide that did them in. No, sir, they were long gone before then. Just like those human-types, I like dinosaurs, too. My ancestors used to ride on their heads, and backs, but never on their tails, ha, ha, way back in what human-kind calls the Triassic, about 250 million years ago."

Mr. Grasshopper paused and then continued, "Yep, it was quite a tale about that bolide and what happened. And many of its effects are still around to this very day. I guess my only question at this point in time is, do you think, Mr. Rabbit, do you think that HRSD's SWIFT program will really work?"

After a thoughtful moment of silence, Mr. Rabbit finally replied, "I am not sure. I think their efforts to save the Bay, and the surrounding area, and the native groundwater are to be applauded or, in my case, thumped about a bit. Anyway, I noticed at the end of his narrative, right here, that the author could not decide whether to call our conversation an afterword or a beforehand."

Mr. Rabbit had Mr. Grasshopper's full attention.

Mr. Grasshopper replied, "You know, I agree with your assessment there. When you come right down to it, an afterword is sort of an ending, like a conclusion or an epilogue or a postscript. But, by throwing in the term beforehand, the author is trying to make two points quite clear: First, HRSD's SWIFT program is a work in progress, and, second, the jury is still out on whether it will work. What do you think, Mr. Rabbit?"

There was another moment of silence.

Mr. Rabbit then replied in his characteristically succinct manner: "Right on, grasshopper. Right on! I applaud these humans' efforts and hope they work."

Glossary

Absorption: Any process by which one substance penetrates the interior of another substance.

Acid: Refers to water or other liquid that has a pH less than 5.5; pH modifier used in the U.S. Fish and Wildlife Service wetland classification system; in common usage, acidic water has a pH less than 7.

Acid rain: Precipitation with higher than normal acidity, caused primarily by sulfur and nitrogen dioxide air pollution.

Acidic deposition: The transfer of acidic or acidifying substances from the atmosphere to the surface of the Earth or to objects on its surface. Transfer can be either by wet deposition (rain, snow, dew, fog, frost, hail) or by dry deposition (gases, aerosols, or fine to coarse particles).

Acre-foot (acre-ft.): The volume of water required to cover an acre of land to a depth of 1 foot; equivalent to 43,560 cubic feet or 32,851 gallons.

Activated carbon: A very porous material that after being subjected to intense heat to drive off impurities can be used to adsorb pollutants from water.

Adsorption: The process by which one substance is attracted to and adheres to the surface of another substance, without actually penetrating its internal structure.

Aeration: A physical treatment method that promotes biological degradation of organic matter. The process may be passive (when waste is exposed to air), or active (when a mixing or bubbling device introduces the air).

Aerobic bacteria: A type of bacteria that requires free oxygen to carry out metabolic function.

Algae: Chlorophyll-bearing nonvascular, primarily aquatic species that have no true roots, stems, or leaves; most algae are microscopic, but some species can be as large as vascular plants.

Algal bloom: The rapid proliferation of passively floating, simple plant life, such as blue–green algae, in and on a body of water.

Alkaline: Refers to water or other liquid that has a pH greater than 7; pH modifier in the U.S. Fish and Wildlife Service wetland classification system; in common usage, a pH of water greater than 7.4.

Alluvial aquifer: A water-bearing deposit of unconsolidated material (sand and gravel) left behind by a river or other flowing water.

Alluvium: General term for sediments of gravel, sand, silt clay, or other particulate rock material deposited by flowing water, usually in the beds of rivers and streams, on a flood plain, on a delta, or at the base of a mountain.

Alpine snow glade: Marshy clearing between slopes above the timberline in mountains.

Amalgamation: The dissolving or blending of a metal (commonly gold and silver) in mercury to separate it from its parent material.

Ammonia: A compound of nitrogen and hydrogen (NH_3) that is a common byproduct of animal waste. Ammonia readily converts to nitrate in soils and streams.

Anaerobic: Pertaining to, taking place in, or caused by the absence of oxygen.

Anomalies: As related to fish, externally visible skin or subcutaneous disorders, including deformities, eroded fins, lesions, and tumors.

Anthropogenic: Having to do with or caused by humans.

Anticline: A fold in the Earth's crust, convex upward, whose core contains stratigraphically older rocks.

Aquaculture: The science of faming organisms that live in water, such as fish, shellfish, and algae.

Aquatic: Living or growing in or on water.

Aquatic guidelines: Specific levels of water quality which, if reached, may adversely affect aquatic life. These are nonenforceable guidelines issued by a governmental agency or other institution.

Aquifer: A geologic formation, group of formations, or part of a formation that contains sufficient saturated permeable material to yield significant quantities of water to springs and wells.

Aquitard: A saturated, but poorly permeable, geologic unit that impedes groundwater movement and does not yield water freely to wells, but which may transmit appreciable water to and from adjacent aquifers and, where sufficiently thick, may constitute an important groundwater storage unit. Areally extensive aquitards may function regionally as confined units within aquifer systems.

Arroyo: A small, deep, flat-floored channel or gully of an ephemeral or intermittent stream, usually with nearly vertical banks cut into unconsolidated material.

Artesian: Refers to confined aquifers. Sometimes the term is used to denote a portion of a confined aquifer where the altitudes of the potentiometric surface are above land surface (flowing wells and artesian wells are synonymous in this usage). But more generally the term indicates that the altitudes of the potentiometric surface are above the altitude of the base or the confining unit (in which case artesian wells and flowing wells are not synonymous).

Artificial recharge: Augmentation of natural replenishment of groundwater storage by some method of construction, by spreading the water, or by pumping water directly into an aquifer.

Atmospheric deposition: The transfer of substances from the air to the surface of the Earth, either in wet form (rain, fog, snow, dew, frost, hail) or in dry form (gases, aerosols, particles).

Atmospheric pressure: The pressure exerted by the atmosphere on any surface beneath or within it; equal to 14.7 pounds per square inch at sea level.

Average discharge: As used by the U.S. Geological Survey, the arithmetic average of all complete water years of record of surface water discharge whether consecutive or not. The term "average" generally is reserved for average of record and "mean" is used for averages of shorter periods—namely, daily, monthly, or annual mean discharges.

Background concentration: Concentration of a substance in a particular environment that is indicative of minimal influence by human (anthropogenic)sources.

Backwater: A body of water in which the flow is slowed or turned back by an obstruction such as a bridge or dam, an opposing current, or the movement of the tide.

Bacteria: Single-celled microscopic organisms.

Bank: The sloping ground that borders a stream and confines the water in the natural channel when the water level, or flow, is normal.

Bank storage: The change in the amount of water stored in an aquifer adjacent to a surface water body resulting from a change in stage of the surface water body.

Barrier bar: An elongate offshore ridge, submerged at least at high tide and built up by the action of waves or currents.

Base flow: The sustained low flow of a stream, usually groundwater inflow to the stream channel.

Basic: The opposite of acidic; refers to water that has a pH greater than 7.

Basin and range physiography: A region characterized by a series of generally north-trending mountain ranges separated by alluvial valleys.

Bed material: Sediment comprising the streambed.

Bed sediment: The material that temporarily is stationary in the bottom of a stream or other watercourse.

Bedload: Sediment that moves on or near the streambed and is in almost continuous contact with the bed.

Bedrock: A general term used for solid rock that underlies soils or other unconsolidated material.

Benthic invertebrates: Insects, mollusk, crustaceans, worms, and other organisms without a backbone that live in, on, or near the bottom of lakes, streams, or oceans.

Benthic organism: A form of aquatic life that lives on or near the bottom of stream, lakes, or oceans.

Bioaccumulation: The biological sequestering of a substance at a higher concentration than that at which it occurs in the surrounding environment or medium. Also, the process whereby a substance enters organisms through gills, epithelial tissues, diet, or other means.

Bioavailability: The capacity of a chemical constituent to be taken up by living organisms either through physical contact or by ingestion.

Biochemical: Refers to chemical processes that occur inside or are mediated by living organisms.

Biochemical oxygen demand (BOD): The amount of oxygen required by bacteria to stabilize decomposable organic matter under aerobic conditions.

Biochemical process: A process characterized by, produced by, or involving chemical reactions in living organism.

Biodegradation: Transformation of a substance into new compounds through biochemical reactions or the actions of microorganisms such as bacteria.

Biological treatment: A process that uses living organisms to bring about chemical changes.

Biomass: The amount of living matter, in the form of organisms, present in a particular habitat, usually expressed as weight per unit area.

Biota: All living organisms of an area.

Blue hole: A subsurface void, usually a solution sinkhole, developed in carbonate rocks that are open to the Earth's surface and contains tidally influenced waters of fresh, marine, or mixed chemistry.

Bog: A nutrient-poor, acidic wetland dominated by a waterlogged, spongy mat of sphagnum moss that ultimately forms a thick layer of acidic peat; generally has no inflow or outflow and is fed primarily by rain water.

Brackish water: Water with a salinity intermediate between seawater and freshwater (containing from 1000 to 10,000 mg/L of dissolved solids).

Breakdown product: A compound derived by chemical, biological, or physical action upon a pesticide. The breakdown is a natural process that may result in a more toxic or a less toxic compound and a more persistent or less persistent compound.

Breakpoint chlorination: The addition of chlorine to water until the chlorine demand has been satisfied and free chlorine residual is available for disinfection.

$C \times T$ value: The product of the residual disinfectant concentration (C), in milligrams per liter, and the corresponding disinfectant contact time (T), in minutes. Minimum $C \times T$ values are specified by the Surface Water Treatment Rule as a means of ensuring adequate killing or inactivation of pathogenic microorganisms in water.

Calcareous: Refers to substance formed of calcium carbonate or magnesium carbonate by biological deposition or inorganic precipitation, or containing those minerals in sufficient quantities to effervesce when treated with cold hydrochloric acid.

Capillary fringe: The zone above the water table in which water is held by surface tension. Water in the capillary fringe is under a pressure less than atmospheric.

Carbonate rocks: Rocks (such as limestone or dolostone) that are composed primarily of minerals (such as calcite and dolomite) containing a carbonate ion.

Cenote: Steep-walled natural well that extends below the water table; generally caused by collapse of a cave roof; term reserved for features found in the Yucatan Peninsula of Mexico.

Center pivot irrigation: An automated sprinkler system with a rotating pipe or boom that supplies water to a circular area of an agricultural field through sprinkler heads or nozzles.

Channel scour: Erosion by flowing water and sediment on a stream channel; results in removal of mud, silt, and sand on the outside curve of a stream bend and the bed material of a stream channel.

Channelization: The straightening and deepening of a stream channel to permit the water to move faster or to drain a wet area for farming.

Chemical treatment: A process that results in the formation of a new substance or substances. The most common chemical water treatment processes include coagulation, disinfection, water softening, and filtration.

Chlordane: Octachlor-4,7-methanotetrahydroindane; an organochlorine insecticide no longer registered for use in the United States. Technical chlordane is a mixture in which the primary components are *cis*- and *trans*-chlordane, *cis*- and *trans*-nonachlor, and heptachlor.

Chlorinated solvent: A volatile organic compound containing chlorine; some common solvents are trichloroethylene, tetrachloroethylene, and carbon tetrachloride.

Chlorofluorocarbons: A class of volatile compounds consisting of carbon, chlorine, and fluorine; commonly called *freons*, which have been used in refrigeration mechanisms, as blowing agents in the fabrication of flexible and rigid foams, and, until banned from use several years ago, as propellants in spray cans.

Chlorination: The process of adding chlorine to water to kill disease-causing organisms or to act as an oxidizing agent.

Chlorine demand: A measure of the amount of chlorine that will combine with impurities and is therefore unavailable to act as a disinfectant.

Cienaga: A marshy area where the ground is wet due to the presence of seepage of springs.

Clean Water Act (CWA): Federal law passed in 1972 (with subsequent amendments) intended to restore and maintain the chemical, physical, and biological integrity of the nation's waters. Its long-range goal is to eliminate the discharge of pollutants into navigable waters and to make national waters fishable and swimmable.

Climate: The sum total of the meteorological elements that characterize the average and extreme conditions of the atmosphere over a long period of time at any one place or region of the Earth's surface.

Coagulants: Chemicals that cause small particles to stick together to form larger particles.

Coagulation: A chemical water treatment method that causes very small suspended particles to attract one another and form larger particles. This is accomplished by the addition of a coagulant, which neutralizes the electrostatic charges causing the particles to repel each other.

Coliform bacteria: A group of bacteria predominantly inhabiting the intestines of humans or animals but also occasionally found elsewhere. The presence of these bacteria in water is used as an indicator of fecal contamination (contamination by animal or human wastes).

Color: A physical characteristic of water. Color is most commonly tan or brown from oxidized iron, but contaminants may cause other colors, such as green or blue. Color differs from turbidity, which is the cloudiness of the water.

Combined sewer overflow: The discharge of untreated sewage and stormwater to a stream when the capacity of a combined storm/sanitary sewer system is exceeded by storm runoff.

Communicable diseases: Usually caused by *microbes*—microscopic organisms including bacteria, protozoa, and viruses. Most microbes are essential components of our environment and do not cause disease. Those that do are called *pathogenic organisms*, or simply *pathogens*.

Community: In ecology, the species that interact in a common area.

Community water system: A public water system that serves at least 15 service connections used by year-round residents or regularly serves at least 25 year-round residents.

Compaction: In its geologic sense, compaction refers to the inelastic compression of the aquifer system. Compaction of the aquifer system reflects the rearrangement of the mineral grain pore structure and largely nonrecoverable reduction of the porosity under stresses greater than the preconsolidation stress.

Compaction, as used here, is synonymous with the term virgin consolidation used by soils engineers. The term refers to both the process and the measured change in thickness. As a practical matter, a very small amount (1 to 5%) of the compaction is recoverable as a slight elastic rebound of the compacted material if stresses are reduced.

Compaction, residual: Compaction that would ultimately occur if a given increase in applied stress were maintained until steady-state pore pressures were achieved. Residual compaction may also be defined as the difference between (1) the amount of compaction that will occur ultimately for a given increase in applied stress, and (2) that which has occurred at a specified time.

Composite sample: A series of individual or grab samples taken at different times from the same sampling point and mixed together.

Compression: Refers to the decrease in thickness of sediments as a result of an increase in vertical compressive stress. Compression many be elastic (fully recoverable) or inelastic (nonrecoverable).

Concentration: The ratio of the quantity of any substance present in a sample of a given volume or a given weight compared to the volume or weight of the sample.

Cone of depression: The depression of heads around a pumping well caused by withdrawal of water.

Confined aquifer (artesian aquifer): An aquifer that is completely filled with water under pressure and that is overlain by material that restricts the movement of water.

Confining bed: A layer of rock having very low hydraulic conductivity that hampers the movement of water into and out of an aquifer.

Confining layer: A body of impermeable or distinctly less permeable material located stratigraphically adjacent to one or more aquifers; this layer restricts the movement of water into and out of the aquifers.

Confining unit: A saturated, relatively low-permeability geologic unit that is areally extensive and serves to confine an adjacent artesian aquifer or aquifers. Leaky confining units may transmit appreciable water to and from adjacent aquifers.

Confluence: The flowing together of two or more streams; the place where a tributary joins the main stream.

Conglomerate: A coarse-grained sedimentary rock composed of fragments larger than 2 millimeters in diameter.

Consolidation: In soil mechanics, the adjustment of a saturated soil in response to increased load, involving the squeezing of water from the pores and decrease in void ratio or porosity of the soul. In this book, the geologic term compaction is used in preference to consolidation.

Constituent: A chemical or biological substance in water, sediment, or biota that can be measured by an analytical method.

Consumptive use: The quantity of water that is not available for immediate rescue because it has been evaporated, transpired, or incorporated into products, plant tissue, or animal tissue.

Contact recreation: Recreational activities, such as swimming and kayaking, in which contact with water is prolonged or intimate and in which there is a likelihood of ingesting water.

Contaminant: A toxic material found as an unwanted residue in or on a substance.

Contamination: Degradation of water quality compared to original or natural conditions due to human activity.

Contributing area: The area in a drainage basin that contributes water to streamflow or recharge to an aquifer.

Core sample: A sample of rock, soil, or other material obtained by driving a hollow tube into the undisturbed medium and withdrawing it with its contained sample.

Criterion: A standard rule or test on which a judgment or decision can be based.

Cross connection: Any connection between safe drinking water and a nonpotable water or fluid.

Datum plane: A horizontal plane to which ground elevations or water surface elevations are referenced.

Deepwater habitat: Permanently flooded lands lying below the deepwater boundary of wetlands.

Degradation products: Compounds resulting from transformation of an organic substance through chemical, photochemical, and/or biochemical reactions.

Denitrification: A process by which oxidized forms of nitrogen such as nitrate are reduced to form nitrites, nitrogen oxides, ammonia, or free nitrogen; commonly brought about by the action of denitrifying bacteria and usually resulting in the escape of nitrogen to the air.

Detection limit: The concentration of a constituent or analyte below which a particular analytical method cannot determine, with a high degree of certainty, the concentration.

Diatoms: Single-celled, colonial, or filamentous algae with siliceous cell walls constructed of two overlapping parts.

Direct runoff: The runoff entering stream channels promptly after rainfall or snowmelt.

Discharge: The volume of fluid passing a point per unit of time, commonly expressed in cubic feet per second, million gallons per day, gallons per minute, or seconds per minute per day.

Discharge area (groundwater): Area where subsurface water is discharged to the land surface, to surface water, or to the atmosphere.

Disinfectants and disinfection byproducts (DBPs): A term used in connection with state and federal regulations designed to protect public health by limiting the concentration of either disinfectants or the byproducts formed by the reaction of disinfectants with other substances in the water (such as trihalomethanes, or THMs).

Disinfection: A chemical treatment method involving the addition of a substance (e.g., chlorine, ozone, hydrogen peroxide) that destroys or inactivates harmful microorganisms or inhibits their activity.

Dispersion: The extent to which a liquid substance introduced into a groundwater system spreads as it moves through the system.

Dissociate: The process of ion separation that occurs when an ionic solid is dissolved in water.

Dissolved constituent: Operationally defined as a constituent that passes through a 0.45-micrometer filter.

Dissolved oxygen (DO): The oxygen dissolved in water, usually expressed in milligrams per liter, parts per million, or percent of saturation.

Dissolved solids: Any material that can dissolve in water and be recovered by evaporating the water after filtering the suspended material.

Diversion: A turning aside or alteration of the natural course of a flow of water, normally considered to be the water physically leaving the natural channel. In some areas, this can be consumptive use direct from another stream, such as for livestock watering. In other areas, a diversion consists of such actions as taking water through a canal, pipe, or conduit.

Dolomite: A sedimentary rock consisting chiefly of magnesium carbonate.

Domestic withdrawals: Water used for normal household purposes, such as drinking, food preparation, bathing, washing clothes and dishes, flushing toilets, and watering lawns and gardens. The water may be obtained from a public supplier or may be self-supplied. Also called *residential water use.*

Drainage area (of a stream): At a specified location, the area, measured in a horizontal plane, enclosed by a drainage divide.

Drainage basin: The land area drained by a river or stream.

Drainage divide: Boundary between adjoining drainage basins.

Drawdown: The difference between the water level in a well before pumping and the water level in the well during pumping. Also, for flowing wells, the reduction of the pressure head as a result of the discharge of water.

Drinking water standards: Water quality standards addressing, for example, suspended solids, unpleasant taste, and microbes harmful to human health. Drinking water standards are included in state water quality rules.

Drinking water supply: Any raw or finished water source that is or may be used as a public water system or as drinking water by one or more individuals.

Drip irrigation: An irrigation system in which water is applied directly to the root zone of plants by means of applicators (orifices, emitters, porous tubing, or perforate pipe) operated under low pressure. The applicators can be placed on or below the surface of the ground or can be suspended from supports.

Drought: A prolonged period of less-than-normal precipitation such that the lack of water causes a serious hydrologic imbalance.

Ecoregion: An area of similar climate, landform, soil, potential natural vegetation, hydrology, or other ecologically relevant variables.

Ecosystem: Community of organisms considered together with the nonliving factors of its environment.

Effluent: Outflow from a particular source, such as a stream that flows from a lake or liquid waste that flows from a factory or sewage-treatment plant.

Effluent limitations: Standards developed by the U.S. Environmental Protection Agency to define the levels of pollutants that could be discharged into surface waters.

Electrodialysis: The process of separating substances in a solution by dialysis, using an electric field as the driving force.

Electronegativity: The tendency for atoms that do not have a complete octet of electrons in their outer shell to become negatively charged.

Ellipsoid, Earth: A mathematically determined three-dimensional surface obtained by rotating an ellipse about its semi-minor axis. In the case of the Earth, the ellipsoid is the modeled shape of its surface, which is relatively flattened in the polar axis.

Ellipsoid, height: The distance of a point above the ellipsoid measured perpendicular to the surface of the ellipsoid.

Emergent plants: Erect, rooted, herbaceous plants that may be temporarily or permanently flooded at the base but do not tolerate prolonged inundation of the entire plant.

Enhanced Surface Water Treatment Rule (ESWTR): A revision of the original Surface Water Treatment Rule that includes new technology and requirements to deal with newly identified problems.

Environment: The sum of all conditions and influences affecting the life of organisms.

Environmental sample: A water sample collected from an aquifer or stream for the purpose of chemical, physical, or biological characterization of the sampled resource.

Environmental setting: Land area characterized by a unique combination of natural and human-related factors, such as row-crop cultivation or glacial-till soils.

Ephemeral stream: A stream or part of a stream that flows only in direct response to precipitation; it receives little or no water from springs, melting snow, or other sources, and its channel is at all times above the water table.

EPT richness index: An index based on the sum of the number of taxa in three insect orders, Ephemeroptera (mayflies), Plecoptera (stoneflies), and Trichoptera (caddisflies), which are composed primarily of species considered to be relatively intolerant to environmental alterations.

Equipotential line: A line on a map or cross-section along which total heads are the same.

Erosion: The process whereby materials of the Earth's crust are loosened, dissolved, or worn away and simultaneously moved from one place to another.

Eutrophication: The process by which water becomes enriched with plant nutrients, most commonly phosphorus and nitrogen.

Evaporite minerals (deposits): Minerals or deposits of minerals formed by evaporation of water containing salts. These deposits are common in arid climates.

Evaporites: Class of sedimentary rocks composed primarily of minerals precipitated from a saline solution as a result of extensive or total evaporation of water.

Evapotranspiration: The process by which water is discharged to the atmosphere as a result of evaporation from the soil and surface-water bodies and transpiration by plants.

Exfoliation: The process by which concentric scales, plates, or shells of rock, ranging from less than a centimeter to several meters in thickness, are stripped from the bare surface of a large rock mass.

Facultative bacteria: A type of anaerobic bacteria that can metabolize its food either aerobically or anaerobically.

Fall line: Imaginary line marking the boundary between the ancient, resistant crystalline rocks of the Piedmont province of the Appalachian Mountains and the younger, softer sediments of the Atlantic Coastal Plain province in the Eastern United States. Along rivers, this line commonly is reflected by waterfalls.

Fecal bacteria: Microscopic single-celled organisms (primarily fecal coliforms and fecal streptococci) found in the wastes of warm-blooded animals. Their presence in water is used to assess the sanitary quality of water for body-contact recreation or for consumption. Their presence indicates contamination by the wastes of warm-blooded animals and the possible presence of pathogenic (disease-producing) organisms.

Federal Water Pollution Control Act (1972): The objective of the Act is "to restore and maintain the chemical, physical, and biological integrity of the nation's waters." This Act and subsequent Clean Water Act amendments represent the most far-reaching water pollution control legislation ever enacted. They provide for comprehensive programs for water pollution control, uniform laws, and interstate cooperation, as well as grants for research, investigations, and training on national programs on surveillance, the effects of pollutants, pollution control, and the identification and measurement of pollutants. Additionally, they allot grants and loans for the construction of treatment works. The Act established national discharge standards with enforcement provisions, as well as several milestone achievement dates. It required secondary treatment of domestic waste by publicly owned treatment works and the application of "best practicable" water pollution control technology by industry by 1977. Virtually all industrial sources have achieved compliance. (Because of economic difficulties and cumbersome federal requirements, certain publicly owned treatment works obtained an extension to 1988 for compliance.) The Act mandates a strong pretreatment program to control toxic pollutants discharged by industry into publicly owned treatment works and called for new levels of technology to be imposed during the 1980s and 1990s, particularly for controlling toxic pollutants. The 1987 amendments require regulation of stormwater from industrial activity.

Fertilizer: Any of a large number of natural or synthetic materials, including manure and nitrogen, phosphorus, and potassium compounds, spread on or worked into soil to increase its fertility.

Filtrate: Liquid that has been passed through a filter.

Filtration: A physical treatment method for removing solid (particulate) matter from water by passing the water through porous media such as sand or a man-made filter.

Flocculation: Water treatment process that follows coagulation; it uses gentle stirring to bring suspended particles together so they will form larger, more settleable clumps called *floc*.

Flood: Any relatively high streamflow that overflows the natural or artificial banks of a stream.

Flood attenuation: A weakening or reduction in the force or intensity of a flood.

Flood irrigation: The application of irrigation water whereby the entire surface of the soil is covered by ponded water.

Flood plain: A strip of relatively flat land bordering a stream channel that is inundated at times of high water.

Flow line: The idealized path followed by particles of water.

Flow net: The grid pattern formed by a network of flow lines and equipotential lines.

Flowpath: An underground route for groundwater movement, extending from a recharge (intake) zone to a discharge (output) zone such as a shallow stream.

Fluvial: Pertaining to a river or stream.

Freshwater: Water that contains less than 1000 mg/L dissolved solids.

Freshwater chronic criteria: The highest concentration of a contaminant that freshwater aquatic organisms can be exposed to for an extended period of time (4 days) without adverse effects.

Geodetic datum: A set of constants specifying the coordinate system used for geodetic control (e.g., for calculating the coordinates of points on the Earth).

Geoid, Earth: The sea-level equipotential surface or figure of the Earth. If the Earth were completely covered by a shallow sea, the surface of this sea would conform to the geoid shaped by the hydrodynamic equilibrium of the water subject to gravitational and rotational forces. Mountains and valleys are departures from the reference geoid.

Grab sample: A single water sample collected at one time from a single point.

Groundwater: Freshwater found under the surface of the Earth, usually in aquifers. Groundwater is a major source of drinking water and a source of growing concern in areas where leaching agricultural or industrial pollutants or substances from leaking underground storage tanks are contaminating it.

Habitat: The part of the physical environment in which a plant or animal lives.

Hardness: A characteristic of water caused primarily by the salts of calcium and magnesium. It causes deposition of scale in boilers, damage in some industrial processes, and sometimes an objectionable taste. It can also decrease the effectiveness of soap.

Head: The height above a datum plane of a column of water; in a groundwater system, it is composed of elevation head and pressure head.

Headwaters: The source and upper part of a stream.

Hydraulic conductivity: The capacity of a rock to transmit water. It is expressed as the volume of water at the existing kinematic viscosity that will move in unit time under a unit hydraulic gradient through a unit area measured at right angles to the direction of flow.

Hydraulic gradient: Change of hydraulic head per unit of distance in a given direction.

Hydrocompaction: The process of volume decrease and density increase that occurs when certain moisture-deficient deposits compact as they are wetted for the first time since burial. The vertical downward movement of the land surface that results from this process has also been termed shallow subsidence and near-surface subsidence.

Hydrogen bonding: The term used to describe the weak but effective attraction that occurs between polar covalent molecules.

Hydrograph: Graph showing variation of water elevation, velocity, streamflow, or other property of water with respect to time.

Hydrologic cycle: Literally, the water–earth cycle; the movement of water in all three physical forms through the various environmental media (air, water, biota, soil).

Hydrology: The science that deals with water as it occurs in the atmosphere, on the surface of the ground, and underground.

Hydrostatic pressure: The pressure exerted by water at any given point in a body of water at rest.

Hygroscopic: Refers to a substance that readily absorbs moisture.

Impermeability: The incapacity of a rock to transmit a fluid.

Index of Biotic Integrity (IBI): An aggregated number, or index, based on several attributes or metrics of a fish community that provides an assessment of biological conditions.

Indicator sites: Stream sampling sites located at outlets of drainage basins with relatively homogeneous land use and physiographic conditions; most indicator-site basins have drainage areas ranging form 20 to 200 square miles.

Infiltration: The downward movement of water from the atmosphere into soil or porous rock.

Influent: Water flowing into a reservoir, basin, or treatment plant.

Inorganic: Containing no carbon; matter other than plant or animal.

Inorganic chemical: A chemical substance of mineral origin not having carbon in its molecular structure.

Inorganic soil: Soil with less than 20% organic matter in the upper 16 inches.

Ionic bond: The attractive forces between oppositely charged ions (e.g., the forces between sodium and chloride ions in a sodium chloride crystal).

Instantaneous discharge: The volume of water that passes a point at a particular instant of time.

Instream use: Water use taking place within the stream channel for such purposes as hydroelectric power generation, navigation, water quality improvement, fish propagation, and recreation. Sometimes called *nonwithdrawal use* or *in-channel use*.

Intermittent stream: A stream that flows only when it receives water from rainfall runoff or springs, or from some surface source such as melting snow.

Internal drainage: Surface drainage whereby the water does not reach the ocean, such as drainage toward the lowermost or central part of an interior basin or closed depression.

Intertidal: Alternately flooded and exposed by tides.

Intolerant organisms: Organisms that are not adaptable to human alterations to the environment and thus decline in numbers where alterations occur.

Invertebrate: An animal having no backbone or spinal column.

Ion: A positively or negatively charged atom or group of atoms.

Ionic bond: The attractive forces between oppositely charged ions, such as the forces between the sodium and chloride ions in a sodium chloride crystal.

Irrigation: Controlled application of water to arable land to supply requirements of crops not satisfied by rainfall.

Irrigation return flow: The part of irrigation applied to the surface that is not consumed by evapotranspiration or taken up by plants and that migrates to an aquifer or surface water body.

Irrigation withdrawals: Withdrawals of water for application on land to assist in the growing of crops and pastures or to maintain recreational lands.

Karst: A type of topography that is formed on limestone, dolomite, gypsum, and other rocks, primarily by dissolution, and that is characterized by sinkholes, caves, and subterranean drainage.

Karst, mantled: A terrane of karst features, usually subdued, and covered by soil or a thin alluvium.

Karstification: Action by water, mainly chemical but also mechanical, that produces features of a karst topography.

Kill: Dutch term for stream or creek.

Lacustrine: Pertaining to, produced by, or formed in a lake.

Leachate: A liquid that has percolated through soil containing soluble substances and that contains certain amounts of these substances in solution.

Leaching: The removal of materials in solution from soil or rock; also refers to movement of pesticides or nutrients from land surfaces to groundwater.

Limnetic: The deepwater zone (greater than 2 meters deep).

Littoral: The shallow-water zone (less than 2 meters deep).

Load: Material that is moved or carried by streams, reported as weight of material transported during a specified time period, such as tons per year.

Main stem: The principal trunk of a river or a stream.

Marsh: A water-saturated, poorly drained area, intermittently or permanently water covered, having aquatic and grasslike vegetation.

Maturity (stream): The stage in the development of a stream at which it has reached its maximum efficiency, when velocity is just sufficient to carry the sediment delivered to it by tributaries; characterized by a broad, open, flat-floored valley having a moderate gradient and gentle slope.

Maximum contaminant level (MCL): The maximum allowable concentration of a contaminant in drinking water, as established by state and/or federal regulations. Primary MCLs are health related and mandatory. Secondary MCLs are related to the aesthetics of the water and are highly recommended but not required.

Mean discharge: The arithmetic mean of individual daily mean discharges of a stream during a specific period, usually daily, monthly, or annually.

Membrane filter method: A laboratory method used for coliform testing. The procedure uses an ultrathin filter with a uniform pore size smaller than bacteria (less than a micron). After water is forced through the filter, the filter is incubated in a medium that promotes the growth of coliform bacteria. Bacterial colonies with a green–gold sheen indicate the presence of coliform bacteria.

Method detection limit: The minimum concentration of a substance that can be accurately identified and measured with current lab technologies.

Midge: A small fly in the family Chironomidae. The larval (juvenile) life stages are aquatic.

Minimum reporting level (MRL): The smallest measured concentration of a constituent that may be reliably reported using a given analytical method. In many cases, the MRL is used when documentation for the method detection limit is not available.

Mitigation: Actions taken to avoid, reduce, or compensate for the effects of human-induced environmental damage.

Modes of transmission of disease: The ways in which diseases spread from one person to another.

Monitoring: Repeated observation, measurement, or sampling at a site on a scheduled or event basis for a particular purpose.

Monitoring well: A well designed for measuring water levels and testing groundwater quality.

Multiple-tube fermentation method: A laboratory method used for coliform testing, which uses a nutrient broth placed in a culture tubes. Gas production indicates the presence of coliform bacteria.

National Pollutant Discharge Elimination System (NPDES): Permit program authorized by the Clean Water Act that controls water pollution by regulating discharges to any water body. It sets the highest permissible effluent limits prior to making any discharge.

National Primary Drinking Water Regulations (NPDWRs): Regulations developed under the Safe Drinking Water Act that establish maximum contaminant levels, monitoring requirements, and reporting procedures for contaminants in drinking water that endanger human health.

Near Coastal Water Initiative: Initiative developed in 1985 to provide for management of specific problems in waters near coastlines that are not dealt with in other programs.

Nitrate: An ion consisting of nitrogen and oxygen (NO_3). Nitrate is a plant nutrient that is very mobile in soils.

Nonbiodegradable: Substances that do not break down easily in the environment.

Non-point source: A source (of any water-carried material) from a broad area, rather than from discrete points.

Non-point source contaminant: A substance that pollutes or degrades water that comes from lawn or cropland runoff, the atmosphere, roadways, and other diffuse sources.

Non-point source water pollution: Water contamination that cannot be traced to a discrete source (such as a discharge pipe); instead, it originates from a broad area (such as leaching of agricultural chemicals from cropland) and enters the water resource diffusely over a large area.

Nonpolar covalently bonded: Refers to a molecule composed of atoms that share their electrons equally, resulting in a molecule that does not have polarity.

Nutrient: Any inorganic or organic compound needed to sustain plant life.

Organic: Containing carbon, but possibly also containing hydrogen, oxygen, chlorine, nitrogen, and other elements.

Organic chemical: A chemical substance of animal or vegetable origin having carbon in its molecular structure.

Organic detritus: Any loose organic material in streams (such as leaves, bark, or twigs) removed and transported by mechanical means, such as disintegration or abrasion.

Organic soil: Soil that contains more than 20% organic matter in the upper 16 inches.

Organochlorine compound: Synthetic organic compounds containing chlorine. As generally used, term refers to compounds containing mostly or exclusively carbon, hydrogen, and chlorine.

Outwash: Soil material washed down a hillside by rainwater and deposited upon more gently sloping land.

Overdraft: Any withdrawal of groundwater in excess of the safe yield.

Overland flow: The flow of rainwater or snowmelt over the land surface toward stream channels.

Oxidation: A chemical treatment method where a substance either gains oxygen or loses hydrogen or electrons through a chemical reaction.

Oxidizer: A substance that oxidizes another substance.

Paleokarst: A karstified area that has been buried by later deposition of sediments.

Parts per million (ppm): The number of weight or volume units of a constituent present in each 1 million units of a solution or mixture. Formerly used to express the results of most water and wastewater analyses, ppm is being replaced by milligrams per liter (mg/L). For drinking water analyses, concentrations expressed in parts per million and milligrams per liter are equivalent. A single ppm can be compared to a shot glass full of water inside a swimming pool.

Pathogens: Types of microorganisms that can cause disease.

Perched groundwater: Unconfined groundwater separated from an underlying main body of groundwater by an unsaturated zone.

Percolation: The movement, under hydrostatic pressure, of water through interstices of a rock or soil (except the movement through large openings such as caves).

Perennial stream: A stream that normally has water in its channel at all times.

Periphyton: Microorganisms that coat rocks, plants, and other surfaces on lake bottoms.

Permeability: The capacity of a rock for transmitting a fluid; a measure of the relative ease with which a porous medium can transmit a liquid; the quality of the soil that enables water to move downward through the soil profile. Permeability is measured as the number of inches per hour that water moves downward through the saturated soil. Terms describing permeability are as follows:

Very slow	Less than 0.06 in./hr
Slow	0.06–0.2 in./hr
Moderately slow	0.2–0.6 in./hr
Moderate	0.6–2.0 in./hr
Moderately rapid	2.0–6.0 in./hr
Rapid	6.0–20 in.//hr
Very rapid	More than 20 in./hr

pH: A measure of the acidity (less than 7) or alkalinity (greater than 7) of a solution; a pH of 7 is considered neutral.

Phosphorus: A nutrient essential for growth that can play a key role in stimulating aquatic growth in lakes and streams.

Photosynthesis: The synthesis of compounds with the aid of light.

Physical treatment: Any water treatment process that does not produce a new substance (e.g., screening, adsorption, aeration, sedimentation, filtration).

Plutonic: A loosely defined term with a number of current usages. Here, it is used to describe igneous rock bodies that crystallized at great depth or, more generally, any intrusive igneous rock.

Point source: A discrete source of water pollutants.

Polar covalent bond: Occurs when the shared pair of electrons between two atoms are not equally held; thus, one of the atoms becomes slightly positively charged and the other atom becomes slightly negatively charged.

Polar covalent molecule: Molecule containing one or more polar covalent bonds; exhibits partial positive and negative poles, causing them to behave like tiny magnets. Water is the most common polar covalent substance.

Pollutant: Any substance introduced into the environment that adversely affects the usefulness of the resource.

Pollution: The presence of matter or energy whose nature, location, or quantity produces undesired environmental effects. Under the Clean Water Act, the term is defined as a manmade or human-induced alteration of the physical, biological, or radiological integrity of water.

Polychlorinated biphenyls (PCBs): A mixture of chlorinated derivatives of biphenyl, marketed under the trade name Aroclor, with a number designating the chlorine content (e.g., Aroclor 1260). PCBs were used in transformers and capacitors for insulating purposes and in gas pipeline systems as a lubricant. Further sales or new uses were banned by law in 1979.

Polycyclic aromatic hydrocarbons (PAHs): A class of organic compounds with a fused-ring aromatic structure. PAHs result from incomplete combustion of organic carbon (including wood), municipal solid waste, and fossil fuels, as well as from natural or anthropogenic introduction of uncombusted coal and oil. PAHs included benzo(*a*)pyrene, fluoranthene, and pyrene.

Population: A collection of individuals of one species or mixed species making up the residents of a prescribed area.

Porosity: (1) The ratio of the volume of voids in a rock or soil to the total volume, usually stated as a percent. (2) A measure of the water-bearing capacity of subsurface rock. With respect to water movement, it is not just the total magnitude of porosity that is important but also the size of the voids and the extent to which they are interconnected, as the pores in a formation may be open, or interconnected, or closed and isolated. For example, clay may have a very high porosity with respect to potential water content, but it constitutes a poor medium as an aquifer because the pores are usually so small.

Potable water: Water that is safe and palatable for human consumption.

Potentiometric surface: A surface that represents the total head in an aquifer; that is, it represents the height above a datum plane at which the water level stands in tightly cased wells that penetrate the aquifer.

Precipitation: Any or all forms of water particles that fall from the atmosphere, such as rain, snow, hail, and sleet. The act or process of producing a solid phase within a liquid medium.

Pretreatment: Any physical, chemical, or mechanical process used before the main water treatment processes. It can include screening, presedimentation, and chemical addition.

Primary Drinking Water Standards: Regulations on drinking water quality (under the Safe Drinking Water Act) considered essential for preservation of public health.

Primary treatment: The first step of treatment at a municipal wastewater treatment plant; typically involves screening and sedimentation to remove materials that float or settle.

Public supply withdrawals: Water withdrawn by public and private water suppliers for use within a general community. The water is used for a variety of purposes, such as domestic, commercial, industrial, and public water use.

Public water system: As defined by the Safe Drinking Water Act, any system, publicly or privately owned, that serves at least 15 service connections 60 days out of the year, or an average of 25 people at least 60 days out of the year.

Publicly owned treatment works (POTW): A waste treatment works owned by a state, local government unit, or Indian tribe, usually designed to treat domestic wastewaters.

Rain shadow: A dry region on the lee side of a topographic obstacle, usually a mountain range, where rainfall is noticeably less than on the windward side.

Reach: A continuous part of a stream between two specified points.

Reaeration: The replenishment of oxygen in water from which oxygen has been removed.

Receiving waters: A river, lake, ocean, stream, or other water source into which wastewater or treated effluent is discharged.

Recharge: The process by which water is added to a zone of saturation, usually by percolation from the soil surface.

Recharge area (groundwater): An area within which water infiltrates the ground and reaches the zone of saturation.

Reference dose (RfD): An estimate of the amount of a chemical that a person can be exposed to on a daily basis that is not anticipated to cause adverse systemic health effects over the person's lifetime.

Representative sample: A sample containing all of the constituents present in the water from which it was taken.

Return flow: That part of irrigation water that is not consumed by evapotranspiration and that returns to its source or another body of water.

Reverse osmosis (RO): Solutions of differing ion concentration are separated by a semipermeable membrane. Typically, water flows from the chamber with lesser ion concentration into the chamber with the greater ion concentration, resulting in hydrostatic or osmotic pressure. In RO, enough external pressure is applied to overcome this hydrostatic pressure, thus reversing the flow of water. This results in the water on the other side of the membrane becoming depleted in ions and demineralized.

Riffle: A shallow part of the stream where water flows swiftly over completely or partially submerged obstructions to produce surface agitation.

Riparian: Pertaining to or situated on the bank of a natural body of flowing water.

Riparian rights: A concept of water law under which authorization to use water in a stream is based on ownership of the land adjacent to the stream.

Riparian zone: Pertaining to or located on the bank of a body of water, especially a stream.

Rock: Any naturally formed, consolidated, or unconsolidated material (but not soil) consisting of two or more minerals.

Runoff: That part of precipitation or snowmelt that appears in streams or surface water bodies.

Rural withdrawals: Water used in suburban or farm areas for domestic and livestock needs. The water generally is self-supplied and includes domestic use, drinking water for livestock, and other uses such as dairy sanitation, evaporation from stock-watering ponds, and cleaning and waste disposal.

Safe Drinking Water Act (SDWA): A federal law passed in 1974 with the goal of establishing federal standards for drinking water quality, protecting underground sources of water, and setting up a system of state and federal cooperation to ensure compliance with the law.

Saline water: Water that is considered unsuitable for human consumption or for irrigation because of its high content of dissolved solids (generally expressed as mg/L of dissolved solids); seawater is generally considered to contain more than 35,000 mg/L of dissolved solids. A general salinity scale is

	Concentration of dissolved solids (mg/L)
Slightly saline	1000–3,000
Moderately saline	3000–10,000
Very saline	10,000–35,000
Brine	More than 35,000

Saturated zone: A subsurface zone in which all the interstices or voids are filled with water under pressure greater than that of the atmosphere.

Screening: A pretreatment method that uses coarse screens to remove large debris from the water to prevent clogging of pipes or channels to the treatment plant.

Secondary Drinking Water Standards: Regulations developed under the Safe Drinking Water Act that established maximum levels of substances affecting the aesthetic characteristics (taste, color, or odor) of drinking water.

Secondary maximum contaminant level (SMCL): The maximum level of a contaminant or undesirable constituent in public water systems that, in the judgment of the U.S. Environmental Protection Agency, is required to protect the public welfare. SMCLs are secondary (nonenforceable) drinking water regulations established by the USEPA for contaminants that may adversely affect the odor or appearance of such water.

Secondary treatment: The second step of treatment at a municipal wastewater treatment plant. This step uses growing numbers of microorganisms to digest organic matter and reduce the amount of organic waste. Water leaving this process is chlorinated to destroy any disease-causing microorganisms before its release.

Sedimentation: A physical treatment method that involves reducing the velocity of water in basins so the suspended material can settle out by gravity.

Seep: A small area where water percolates slowly to the land surface.

Seiche: A sudden oscillation, caused by the wind, of a moderate-size body of water.

Sinkhole: A depression in a karst area. At land surface, its shape is general circular and its size measured in meters to tens of meters; underground, it is commonly funnel shaped and associated with subterranean drainage.

Sinuosity: The ratio of the channel length between two points on a channel to the straight-line distance between the same two points; a measure of meandering.

Soil: The layer of material at the land surface that supports plant growth.

Soil horizon: A layer of soil that is distinguishable from adjacent layers by characteristic physical and chemical properties.

Soil moisture: Water occurring in the pore spaces between the soil particles in the unsaturated zone from which water is discharged by the transpiration of plants or by evaporation from the soil.

Solution: Formed when a solid, gas, or another liquid in contact with a liquid becomes dispersed homogeneously throughout the liquid. The substance that dissolves is called a *solute*, and the liquid in which it dissolves is called the *solvent*.

Solvated: Refers to when either a positive or negative ion becomes completely surrounded by polar solvent molecules.

Sorb: To take up and hold, by either absorption or adsorption.

Sorption: General term for the interaction (binding or association) of a solute ion or molecule with a solid.

Spall: A chip or fragment removed from a rock surface by weathering, especially by the process of exfoliation.

Specific capacity: Yield of a well per unit of drawdown.

Specific retention: Ratio of the volume of water retained in a rock after gravity drainage to the volume of the work.

Specific storage: Volume of water that an aquifer system releases or takes into storage per unit volume per unit change in head. The specific storage is equivalent to the storage coefficient divided by the thickness of the aquifer system.

Specific yield: Ratio of the volume of water that will drain under the influence of gravity to the volume of saturated rock.

Spring: Place where a concentrated discharge of groundwater flows at the ground surface.

Storage: Capacity of an aquifer, aquitard, or aquifer system to release or accept water into groundwater storage, per unit change in hydraulic head.

Storage coefficient: Volume of water released from storage in a unit prism of an aquifer when the head is lowered a unit distance.

Strain: Deformation that results from a stress; expressed in terms of the amount of deformation per inch.

Stratification: Layered structure of sedimentary rocks.

Stress and strain: In materials, stress is a measure of the deforming force applied to a body. Strain (which is often erroneously used as a synonym for stress) is really the resulting change in its shape (deformation). For perfectly elastic material, stress is proportional to strain. This relationship is explained by

Hooke's law, which states that the deformation of a body is proportional to the magnitude of the deforming force, provided that the body's elastic limit is not exceeded. If the elastic limit is not reached, the body will return to its original size once the force is removed. For example, if a spring is stretched 2 cm by a weight of 1 N, it will be stretched 4 cm by a weight of 2 N, and so on; however, once the load exceeds the elastic limit for the spring, Hooke's law will no longer be obeyed, and each successive increase in weight will result in a greater extension until the spring finally breaks. Stress forces are categorized in three ways:

1. Tension (or tensile stress), in which equal and opposite forces that act away from each other are applied to a body; tends to elongate a body
2. Compression stress, in which equal and opposite forces that act toward each other are applied to a body; tends to shorten a body
3. Shear stress, in which equal and opposite forces that do not act along the same line of action or plane are applied to a body; tends to change the shape of a body without changing its volume

Stress, geostatic (lithostatic): Total weight (per unit area) of sediments and water above some plane of reference. Geostatic stress normal to any horizontal plane of reference in a saturated deposit may also be defined as the sum of the effective stress and the fluid pressure at that depth.

Stress, preconsolidation: Maximum antecedent effective stress to which a deposit has been subjected and which it can withstand without undergoing additional permanent deformation. Stress changes in the range less than the preconsolidation stress produce elastic deformations of small magnitude. In fine-grained materials, stress increases beyond the preconsolidation stress produce much larger deformations that are principally inelastic (nonrecoverable). Synonymous with virgin stress.

Stress, seepage: Force (per unit area) transferred from the water to the medium by viscous friction when water flows through a porous medium. The forced transferred to the medium is equal to the loss of hydraulic head and is termed the seepage force exerted in the direction of flow.

Subsidence: A dropping of the land surface as a result of groundwater being pumped; cracks and fissures can appear in the land. Some consider subsidence to be a virtually irreversible process. Others, like the author of this book, think that the jury is still out on its irreversibility. HRSD's SWIFT project may prove that land subsidence can be reversed.

Surface runoff: Runoff that travels over the land surface to the nearest stream channel.

Surface tension: The molecules at the surface of water hold onto each other tightly because there are no molecules pulling on them from the air above. As the molecules on the surface stick together, they form an invisible "skin."

Surface water: All water naturally open to the atmosphere, and all springs, wells, or other collectors that are directly influenced by surface water.

Surface Water Treatment Rule (SWTR): A federal regulation established by the U.S. Environmental Protection Agency under the Safe Drinking Water Act that imposes specific monitoring and treatment requirements on all public drinking water systems that draw water from a surface water source.

Suspended sediment: Sediment that is transported in suspension by a stream.

Suspended solids: Different from suspended sediment only in the way in which the sample is collected and analyzed.

Synthetic organic chemicals (SOCs): Generally applies to manufactured chemicals (e.g., herbicides, pesticides, chemicals widely used in industry) that are not as volatile as volatile organic compounds.

Total head: The height above a datum plane of a column of water. In a groundwater system, it is composed of elevation head and pressure head.

Total suspended solids (TSS): Solids present in wastewater.

Transmissivity (groundwater): Capacity of a rock to transmit water under pressure. The coefficient of transmissibility is the rate of flow of water, at the prevailing water temperature, in gallons per day, through a vertical strip of the aquifer 1 foot wide and extending the full saturated height of the aquifer under a hydraulic gradient of 100%. A hydraulic gradient of 100% means a 1-foot drop in head in 1 foot of flow distance.

Transpiration: The process by which water passes through living organisms, primarily plants, into the atmosphere.

Trihalomethanes (THMs): A group of compounds formed when natural organic compounds from decaying vegetation and soil (such as humic and fulvic acids) react with chlorine.

Turbidity: A measure of the cloudiness of water caused by the presence of suspended matter, which shelters harmful microorganisms and reduces the effectiveness of disinfecting compounds.

Unconfined aquifer: An aquifer whose upper surface is a water table free to fluctuate under atmospheric pressure.

Unsaturated zone: A subsurface zone above the water table in which the pore spaces may contain a combination of air and water.

Vehicle of disease transmission: Any nonliving object or substance contaminated with pathogens.

Vernal pool: A small lake or pond that is filled with water for only a short time during the spring.

Vug: A small cavity or chamber in rock that may be lined with crystals.

Wastewater: The spent or used water from individual homes, a community, a farm, or an industry that contains dissolved or suspended matter.

Water budget: An accounting of the inflow to, outflow from, and storage changes of water in a hydrologic unit.

Water column: An imaginary column extending through a water body from its floor to its surface.

Water demand: Water requirements for a particular purpose, such as irrigation, power, municipal supply, plant transpiration, or storage.

Water softening: A chemical treatment method that uses either chemicals to precipitate or a zeolite to remove those metal ions (typically Ca^{2+}, Mg^{2+}, Fe^{3+}) responsible for hard water.

Water table: The top water surface of an unconfined aquifer at atmospheric pressure.

Waterborne disease: Water is a potential vehicle of disease transmission, and waterborne disease is possibly one of the most preventable types of communicable illness. The application of basic sanitary principles and technology has virtually eliminated serious outbreaks of waterborne diseases in developed countries. The most prevalent waterborne diseases include typhoid fever, dysentery, cholera, infectious hepatitis, and gastroenteritis.

Watershed: The land area that drains into a river, river system, or other body of water.

Wellhead protection: The protection of the surface and subsurface areas surrounding a water well or well field supplying a public water system from contamination by human activity.

Yield: The mass of material or constituent transported by a river in a specified period of time divided by the drainage area of the river basin.

Yield, optimal: An optimal amount of groundwater, by virtue of its use, that should be withdrawn from an aquifer system or groundwater basin each year. It is a dynamic quantity that must be determined from a set of alternative groundwater management decisions subject to goals, objectives, and constraints of the management plan.

Yield, perennial: The amount of usable water from an aquifer than can be economically consumed each year for an indefinite period of time. It is a specified amount that is commonly specified equal to the mean annual recharge to the aquifer system, which thereby limits the amount of groundwater that can be pumped for beneficial use.

Yield, safe: The amount of groundwater that can be safely withdrawn from a groundwater basin annually, without producing an undesirable result. Undesirable results include but are not limited to depletion of groundwater storage, intrusion of water of undesirable quality, contraventions of existing water rights, deterioration of the economic advantages of pumping (such as excessively lower water levels and the attendant increased pumping lifts and associated energy costs), excessive depletion of stream flow by induced infiltration, and land subsidence.

Zone of aeration: The zone above the water table; water in the zone of aeration does not flow into a well.

Zone of capillarity: The area above a water table where some or all of the interstices (pores) are filled with water that is held by capillarity.

Index